UNITEXT

La Matematica per il 3+2

Volume 165

The **UNITEXT - La Matematica per il 3+2** series is designed for undergraduate and graduate academic courses, and also includes books addressed to PhD students in mathematics, presented at a sufficiently general and advanced level so that the student or scholar interested in a more specific theme would get the necessary background to explore it.

Originally released in Italian, the series now publishes textbooks in English addressed to students in mathematics worldwide.

Some of the most successful books in the series have evolved through several editions, adapting to the evolution of teaching curricula.

Submissions must include at least 3 sample chapters, a table of contents, and a preface outlining the aims and scope of the book, how the book fits in with the current literature, and which courses the book is suitable for.

For any further information, please contact the Editor at Springer: francesca.bonadei@springer.com

THE SERIES IS INDEXED IN SCOPUS

***

UNITEXT is glad to announce a new series of free webinars and interviews handled by the Board members, who rotate in order to interview top experts in their field.

Access this link to subscribe to the events:

https://cassyni.com/s/springer-unitext

Andrea Pascucci

Probability Theory I

Random Variables and Distributions

 Springer

Andrea Pascucci (iD)
Dipartimento di Matematica
Alma Mater Studiorum – Università di
Bologna
Bologna, Italy

ISSN 2038-5714   ISSN 2532-3318   (electronic)
UNITEXT
ISSN 2038-5722   ISSN 2038-5757   (electronic)
La Matematica per il 3+2
ISBN 978-3-031-63189-4   ISBN 978-3-031-63190-0   (eBook)
https://doi.org/10.1007/978-3-031-63190-0

This book is a translation of the original Italian edition "Teoria della Probabilità" by Andrea Pascucci, published by Springer-Verlag Italia S.r.l. in 2020. The translation was done with the help of an artificial intelligence machine translation tool. A subsequent human revision was done primarily in terms of content, so that the book will read stylistically differently from a conventional translation. Springer Nature works continuously to further the development of tools for the production of books and on the related technologies to support the authors.

Translation from the Italian language edition: "Teoria della Probabilità. Variabili aleatorie e distribuzioni" by Andrea Pascucci, © Springer-Verlag Italia S.r.l., part of Springer Nature 2020. Published by Springer Milano. All Rights Reserved.

Cover illustration: Cino Valentini, Archeologia 1, 2021, affresco acrilico, private collection

This Springer imprint is published by the registered company Springer Nature Switzerland AG
The registered company address is: Gewerbestrasse 11, 6330 Cham, Switzerland

If disposing of this product, please recycle the paper.

To my family

Preface

"For over two millennia, Aristotle's logic has ruled over the thinking of western intellectuals. All precise theories, all scientific models, even models of the process of thinking itself, have in principle conformed to the straight-jacket of logic. But from its shady beginnings devising gambling strategies and counting corpses in medieval London, probability theory and statistical inference now emerge as better foundations for scientific models, especially those of the process of thinking and as essential ingredients of theoretical mathematics, even the foundations of mathematics itself. We propose that this sea change in our perspective will affect virtually all of mathematics in the next century."

D. Mumford, The Dawning of the Age of Stochasticity [30]

"In conclusion, what have Tversky and Kahneman[1] shown us with their convincing series of experiments? That the human being, even the intelligent, educated, and even with some notions of statistics, is not a probabilistic animal. Probability theory has developed very late in the history of scientific thought, it is not taught in schools, sometimes it is not understood very well even by those who should apply it."

V. D'Urso, F. Giusberti, Esperimenti di psicologia [13]

A (R)evolution in Mathematics

In "classical" mathematics (which still constitutes the majority of the content taught in high schools and universities) mathematical concepts represent and describe *deterministic* quantities: when talking, for example, about a real variable or a geometric object, one thinks respectively of a number that can be well determined and a figure that can be defined analytically and represented exactly. Mathematics has always been considered the language and the most powerful tool to describe physical and natural phenomena in order to interpret and acquire knowledge on

---

[1] Nobel Prize for economics in 2002.

multiple aspects of reality. But the models that mathematics can provide are always simplifications and almost never provide a complete description of the phenomenon to be studied.

Consider the following trivial example: if I go to the supermarket and buy 1 Kg of flour, I can be satisfied with the fact that the package weighs 1 Kg because it is written on the packaging; if I don't trust it, I can weigh it with my scale and find out that maybe it is not exactly 1 Kg but a few grams more or less; then I could also wonder if my scale is really reliable and accurate to the gram and then resign myself to the fact that maybe I will never know the *true* weight of the flour package. In this case, of course, it doesn't matter much... However, the example helps to understand that many phenomena (or perhaps all of reality) can be interpreted as the sum or combination of several factors classified as *deterministic factors* (in the sense of observable at the macroscopic level) and *stochastic factors* (in the sense of random, aleatory, unobservable, or unpredictable).

The term "stochastic" comes from the Greek στόχος which means target (of shooting) or, figuratively, conjecture. Sometimes, as in the example of flour, the deterministic factor is *prevalent* in the sense that, for various reasons, it is not worth considering other factors and we prefer to neglect them or do not have the tools to include them in our analysis: in this perhaps simplistic way, by analogy, one could describe the approach of classical physics and all theories formulated before the twentieth century that aim to give a description at the macroscopic and observable level. On the other hand, there are many phenomena in which the stochastic factor is not only not negligible but is even *dominant*: a striking example is provided by the main theories of modern physics, in particular quantum mechanics. Staying close to everyday reality, there is now no application area of mathematics in which the *stochastic factor* can be neglected: from economics to medicine, from engineering to meteorology, mathematical models must necessarily include uncertainty; in fact, the phenomenon in question may be intrinsically random like the price of a stock or the signal in a voice recognition or autonomous driving systems, or it may not be observable with precision or difficult to interpret like a disturbed radio signal, a tomographic image or the position of a subatomic particle.

There is also a more general level at which the role of probability in the development of today's society cannot be ignored: it is now considered an educational emergency, the increasingly pressing need for the dissemination and strengthening of probabilistic knowledge. A real literacy campaign in this field can prevent trivial misconceptions, such as the one of the "late" numbers in the lottery game, from having the devastating social and economic effects that we observe today: just think that, based on official data, the money spent by Italians on gambling (and we are only talking about legal games) in 2017 exceeded the ceiling of 100 billion euros, four times more than in 2004.

A positive signal is given by the evolution of probability teaching in high schools: until a few years ago, probability was absent from school curricula and is now rapidly increasing its presence in textbooks and exams, also causing some confusion among teachers due to such rapid updating of content. It is important to emphasize that stochastic mathematics (probability) does not want to dethrone classical mathematics but has its foundations in the latter and enhances it by deepening the links with other scientific disciplines. Paradoxically, the world of higher education and universities seems to have greater inertia, which tends to slow down the transition from deterministic to stochastic thinking. In part, this is understandable: the defense of the status quo is what normally happens in the face of every profound scientific revolution and, in all respects, we are talking about a real silent and irreversible revolution, which involves all areas of mathematics. In this regard, the phrase placed at the beginning of this introduction, by the Anglo-American mathematician David Mumford, Fields Medalist[2] in 1974 for his studies in the field of algebraic geometry, is illuminating. In the article from which the phrase was taken, Mumford confirms the fact that probability theory developed very late in the history of scientific thought.[3]

Probability in the Past

The term *probability* comes from the Latin *probabilitas*, which describes the characteristic of a person (e.g., a witness in a trial) to be reliable, credible, honest (*probus*). This differs in part from the modern meaning of *probability* understood as the study of methods for quantifying and estimating *random events*. Although the study of phenomena in uncertain situations has aroused interest in all ages (starting with gambling), probability theory as a mathematical discipline has relatively recent origins. The first studies of probability date back to the sixteenth century: among the first to deal with it were Gerolamo Cardano (1501–1576) and Galileo Galilei (1564–1642).

Traditionally, the birth of the modern concept of probability is attributed to Blaise Pascal (1623–1662) and Pierre de Fermat (1601–1665). Actually, the debate on

---

[2] The International Medal for Outstanding Discoveries in Mathematics, or more simply Fields Medal, is an award recognized for mathematicians who have not exceeded the age of 40 at the time of the International Congress of Mathematicians of the International Mathematical Union (IMU), which is held every four years. It is often considered the highest recognition a mathematician can receive: together with the Abel Prize, it is defined by many as the "Nobel Prize for Mathematics," although the comparison is improper for various reasons, including the age limit inherent in the awarding of the Fields Medal (source: Wikipedia).

[3] The classical subdivisions of mathematics are *geometry*, *algebra*, and *analysis*. The perception of space (through senses and muscular interaction) is the primitive element of our experience on which *geometry* is based. *Analysis*, I would argue, is the outgrowth of the human experience of force and its children, acceleration and oscillation. *Algebra* seems to stem from the grammar of actions, i.e., the fact that we carry out actions in specific orders, concatenating one after the other, and making various "higher order" actions out of simpler more basic ones. I believe there is a fourth branch of human experience which creates reproducible mental objects, hence creates math: our experience of thought itself through our conscious observation of our mind at work. The division of mathematics corresponding to this realm of experience is not logic but *probability* and *statistics* (D. Mumford, [30]).

the very nature of probability has been very long and articulated; it has interested transversally the fields of knowledge from mathematics to philosophy, and has continued until today, producing different interpretations and settings. For greater clarity and precision, it is appropriate first of all to distinguish *Probability theory* (which deals with the mathematical formalization of concepts and the development of theory based on certain assumptions) from *Statistics* (which deals with the determination or estimation of the probability of random events from data, while also utilizing the results of Probability theory). In this brief introduction, we aim to provide a succinct summary of some of the primary interpretations of the concept of probability: some of them are more motivated by calculation and others by probability theory. Let us start by considering some random events, listed in increasing order of complexity:

- E_1 = "flipping a coin, you get heads."
- E_2 = "Mr. Smith will not have car accidents in the next 12 months."
- E_3 = "within 10 years there will be completely autonomous driving cars."

Let us examine these events in light of some interpretations of the concept of probability:

- *Classical definition:* the probability of an event is the ratio between the number of favorable cases and the number of total cases. For example, in the case of E_1, the probability is equal to $\frac{1}{2} = 50\%$. This is the oldest definition of probability, attributed to Pierre Simon Laplace (1749–1827). This definition is limited to considering phenomena that admit a *finite* number of possible cases and in which the cases are *equiprobable*: with this interpretation, it is not clear how to study events E_2 and E_3.
- *Frequentist (or statistical) definition:* it is assumed that the event consists of the success of an experiment that can be reproduced an indefinite number of times (e.g., if the experiment is flipping a coin, the event could be "getting heads"). If S_n denotes the number of successes in n experiments, the probability is defined (it would be better to say, *calculated*) as

$$\lim_{n \to \infty} \frac{S_n}{n}.$$

At the basis of this definition is the Empirical Law of Chance (which, in theoretical terms, corresponds to the Law of large numbers), for which, for example, in the case of flipping a coin, it is empirically observed that $\frac{S_n}{n}$ approximates the value of 50% as n tends to infinity. The frequentist definition significantly expands the field of application to all the fields (physics, economics, medicine, etc.) in which statistical data concerning past events that have occurred under similar conditions are available: for example, the probability of event E_2 can be calculated with a statistical estimate based on historical data (as insurance companies usually do). The frequentist approach does not allow for the study of the third event, which is not the outcome of a "reproducible random experiment."

- *Subjective definition (or Bayesian[4]):* probability is defined as a measure of the degree of conviction that a subject has regarding the occurrence of an event. In this approach, probability is not an intrinsic and objective property of random phenomena but depends on the subject's evaluation. Operationally,[5] the probability of an event is defined as *the price that an individual considers fair to pay to receive* 1 *if the event occurs and* 0 *if the event does not occur*: for example, the probability of an event is equal to 70% for an individual who considers it fair to bet 70 to receive 100 if the event occurs and lose everything otherwise. The definition is made meaningful by assuming a criterion of *coherence* or rationality of the individual who must assign probabilities in such a way that it is not possible to obtain a certain gain or loss (in today's financial jargon, one would speak of the absence of arbitrage opportunities); particular attention must then be paid to avoid paradoxes of the following type: in the example of tossing a coin, an individual may be willing to bet 1 euro to receive 2 in case of "heads" and 0 in case of "tails" (thus attributing a probability of 50% to the event "heads") but the same individual might not be willing to play 1 million euros on the same bet. The subjective approach was proposed and developed by Frank P. Ramsey (1903–1930), Bruno de Finetti (1906–1985), and later by Leonard J. Savage (1917–1971): it generalizes the previous ones and allows for the definition of probabilities for events like E_3.

The debate on the possible interpretations of probability has been going on for a long time and is still open. But in the first half of the last century, there was a decisive turning point, due to the work of the Russian mathematician Andrej N. Kolmogorov (1903–1987). He was the first to lay the foundations for the *mathematical formalization* of probability, fully incorporating it into the realm of mathematical disciplines. Kolmogorov put aside the difficult problems of logical foundation and the dualism between the objective and subjective views, focusing on the development of probability as a *mathematical theory*. Kolmogorov's contribution is fundamental because, bypassing the epistemological problems, it unleashed the full power of abstract reasoning and logical-deductive thinking applied to the study of probability and thus facilitated the transition from *probability calculation* to *probability theory*. Starting from Kolmogorov's work and thanks to the contribution of many great mathematicians of the last century, profound results have been achieved and research fields still completely unexplored have been opened.

Now it is important to emphasize that the mathematical formalization of probability requires a considerable degree of abstraction. Therefore, it is absolutely natural that the theory of probability appears difficult, if not incomprehensible, at first glance. Kolmogorov uses the language of *measure theory*: an event is identified

---

[4] Thomas Bayes (1701–1761).

[5] To quantify, that is, to translate into a number, the degree of conviction of a subject about an event, the idea is to examine how the subject acts in a bet concerning the considered event.

with a *set E* whose elements represent individual possible outcomes of the random phenomenon considered; the probability $P = P(E)$ is a *measure*, that is, a set function that enjoys some properties: to fix ideas, think of the Lebesgue measure. The use of the abstract language of measure theory is viewed by some (even by some mathematicians) with suspicion because it seems to weaken intuition. However, this is the inevitable price that must be paid to be able to exploit all the power of abstract reasoning and synthetic thinking, which is ultimately the true strength of the mathematical approach.

In this book, we present the first rudiments of probability theory according to Kolmogorov's axiomatic approach. We will limit ourselves to introducing and examining the concepts of *probability space, distribution*, and *random variable*. Drawing a parallel between probability and mathematical analysis, the content of this text roughly corresponds to the introduction of real numbers in an undergraduate course in calculus: this means that we will only take the very first steps in the vast field of Probability theory. The second volume [34] complements the current material by exploring more advanced classical topics in stochastic analysis and calculus.

Probability in the Present

As stated in David Mumford's sentence at the beginning of the introduction, nowadays probability theory is considered an essential ingredient for the *theoretical development of mathematics and for the foundations of mathematics itself*. As an example, the important review article [29] tells, in great detail, the incredible developments of research in stochastic process theory from the middle of the last century onwards.

From an applied perspective, probability theory is *the tool* used to model and manage risk in all areas where phenomena are studied under uncertainty. Let us give some examples:

- **Physics and Engineering** where extensive use is made of Monte Carlo stochastic numerical methods, formalized among the first by Enrico Fermi and John von Neumann.
- **Economics and Finance**, starting from the famous Black-Scholes-Merton formula for which the authors received the Nobel Prize. Financial modeling generally requires an advanced mathematical-probabilistic-numerical background: the content of this book roughly corresponds to Appendix A.1 of [33].
- **Telecommunications**: NASA uses the Kalman-Bucy method to filter signals from satellites and probes sent into space. From [32, p. 2], "*In 1960 Kalman and in 1961 Kalman and Bucy proved what is now known as the Kalman-Bucy filter. Basically the filter gives a procedure for estimating the state of a system which satisfies a 'noisy' linear differential equation, based on a series of 'noisy' observations. Almost immediately the discovery found applications in*

aerospace engineering (Ranger, Mariner, Apollo etc.) and it now has a broad range of applications. Thus the Kalman-Bucy filter is an example of a recent mathematical discovery which has already proved to be useful—it is not just 'potentially' useful. It is also a counterexample to the assertion that 'applied mathematics is bad mathematics' and to the assertion that 'the only really useful mathematics is the elementary mathematics'. For the Kalman-Bucy filter—as the whole subject of stochastic differential equations—involves advanced, interesting and first class mathematics".

- **Medicine and Botany**: the most important stochastic process, Brownian motion, is named after Robert Brown, a botanist who around 1830 observed the irregular movement of colloidal particles in suspension. Brownian motion was used by Louis Jean Baptist Bachelier in 1900 in his doctoral thesis to model stock prices and was the subject of one of Albert Einstein's most famous papers published in 1905. The first mathematically rigorous definition of Brownian motion was given by Norbert Wiener in 1923.
- **Genetics**: it is the science that studies the transmission of traits and the mechanisms by which these are inherited. Gregor Johann Mendel (1822–1884), a Czech Augustinian monk considered the precursor of modern genetics, made a fundamental methodological contribution by applying probability theory for the first time to the study of biological inheritance.
- **Computer Science**: quantum computers exploit the laws of quantum mechanics for data processing. In a current computer, the unit of information is the *bit*: while we can always determine the state of a bit and accurately determine whether it is 0 or 1, we cannot determine with equal precision the state of a *qubit*, the quantum information unit, but only the probabilities that it takes the values 0 and 1.
- **Jurisprudence**: the verdict issued by a judge in a court is based on the probability of guilt of the defendant estimated from the information provided by the investigations. In this area, the concept of conditional probability plays a fundamental role and its incorrect use is at the basis of sensational judicial errors: some of them are recounted in [35].
- **Meteorology**: for forecasting beyond the fifth day, it is essential to have probabilistic weather models; probabilistic models generally run in the main international weather centers because they require very complex and computationally expensive statistical-mathematical procedures. Starting from 2020, the Data Center of the *European Center for Medium-Range Weather Forecasts* (ECMWF) is located in Bologna, Italy.
- **Military Applications**: from [41, p. 139]: *"In 1938, Kolmogorov had published a paper that established the basic theorems for smoothing and predicting stationary stochastic processes. An interesting comment on the secrecy of war efforts comes from Norbert Wiener (1894–1964) who, at the Massachusetts Institute of Technology, worked on applications of these methods to military problems during and after the war. These results were considered so important to America's Cold War efforts that Wiener's work was declared top secret. But all of it, Wiener insisted, could have been deduced from Kolmogorov's early paper."*

Finally, probability is at the basis of the development of the most recent *Machine Learning* technologies and all related applications to artificial intelligence, self-driving cars, voice and image recognition, etc. (see, e.g., [19] and [38]). Nowadays, an advanced knowledge of Probability theory is the minimum requirement for anyone who wants to deal with applied mathematics in one of the areas mentioned above.

In conclusion, I believe we can concur that our study of mathematics primarily stems from our passion for the subject, rather than solely seeking it for its potential to secure future employment. Undoubtedly, mathematics transcends the need for practical justification. But it is also true that we do not live on the moon and sooner or later we will have to find a job. So it is important to know the real applications of mathematics: they are numerous, require advanced, absolutely non-trivial knowledge, so much so as to satisfy even the aesthetic taste of a so-called "pure mathematician." Ultimately, for those inclined toward pure research, probability theory is certainly one of the most fascinating and least explored fields, in which the contribution of the best young minds is fundamental and highly desirable.

Bibliographic Note

There are many excellent introductory texts on Probability theory: among my favorites, and which have been the main source of inspiration and ideas, are those by Bass [3], Durrett [12], Klenke [24], and Williams [46]. Below, I list in alphabetical order other important reference texts: Baldi [1], Bass [2], Bauer [4], Biagini and Campanino [5], Billingsley [6], Caravenna and Dai Pra [8], Feller [15], Jacod and Protter [21], Kallenberg [23], Letta [28], Neveu [31], Pintacuda [36], Shiryaev [42], and Sinai [43]. Of course, it is important to recognize that this list is by no means exhaustive.

This book represents yet another endeavor to systematically gather the basic concepts of probability in a structured, unified, and comprehensive manner. Its purpose is to pave the way for more advanced studies, as delineated in the subsequent volume [34] on stochastic processes and stochastic calculus.

Readers who wish to report any errors, typos, or suggestions for improvement can do so at the following address: `andrea.pascucci@unibo.it`.

The corrections received after publication will be made available on the website at: `unibo.it/sitoweb/andrea.pascucci/`.

Bologna, Italy   Andrea Pascucci
April 2024

Contents

Frequently Used Symbols and Notations

- $A := B$ means that A is, *by definition*, equal to B
- $\uplus$ indicates the *disjoint* union
- $A_n \nearrow A$ indicates that $(A_n)_{n\in\mathbb{N}}$ is an *increasing* sequence of sets such that $A = \bigcup\limits_{n\in\mathbb{N}} A_n$
- $A_n \searrow A$ indicates that $(A_n)_{n\in\mathbb{N}}$ is a *decreasing* sequence of sets such that $A = \bigcap\limits_{n\in\mathbb{N}} A_n$
- $\sharp A$ or $|A|$ indicates the cardinality of the set A. $A \leftrightarrow B$ if $|A| = |B|$
- $\mathscr{B}_d = \mathscr{B}(\mathbb{R}^d)$ is the Borel σ-algebra in $\mathbb{R}^d$; $\mathscr{B} := \mathscr{B}_1$
- $m\mathscr{F}$ (resp. $m\mathscr{F}^+$, $b\mathscr{F}$) is the class of $\mathscr{F}$-measurable functions (resp. $\mathscr{F}$-measurable and non-negative, $\mathscr{F}$-measurable and bounded)
- $\mathscr{N}$ is the family of negligible sets (cf. Definition 1.1.16)
- Numerical sets:
  - Natural numbers: $\mathbb{N} = \{1, 2, 3, \ldots\}$, $\mathbb{N}_0 = \mathbb{N} \cup \{0\}$, $I_n := \{1, \ldots, n\}$ for $n \in \mathbb{N}$
  - Real numbers $\mathbb{R}$, extended real numbers $\bar{\mathbb{R}} = \mathbb{R} \cup \{\pm\infty\}$, positive real numbers $\mathbb{R}_{>0} =]0, +\infty[$, non-negative real numbers $\mathbb{R}_{\geq 0} = [0, +\infty[$
- Leb_d indicates the d-dimensional Lebesgue measure; $\mathrm{Leb} := \mathrm{Leb}_1$
- Indicator function of a set A

$$\mathbb{1}_A(x) := \begin{cases} 1 & \text{if } x \in A \\ 0 & \text{otherwise} \end{cases}$$

- Euclidean scalar product:

$$\langle x, y \rangle = x \cdot y = \sum_{i=1}^{d} x_i y_i, \qquad x = (x_1, \ldots, x_d),\ y = (y_1, \ldots, y_d) \in \mathbb{R}^d$$

In matrix operations, the d-dimensional vector x is identified with the $d \times 1$ column matrix.

- Maximum and minimum of real numbers:

$$x \wedge y = \min\{x, y\}, \qquad x \vee y = \max\{x, y\}$$

- Positive and negative part:

$$x^+ = x \vee 0, \qquad x^- = (-x) \vee 0$$

- Argument of the maximum and minimum of $f : A \longrightarrow \mathbb{R}$:

$$\underset{x \in A}{\arg\max}\, f(x) = \{y \in A \mid f(y) \geq f(x) \text{ for every } x \in A\}$$

$$\underset{x \in A}{\arg\min}\, f(x) = \{y \in A \mid f(y) \leq f(x) \text{ for every } x \in A\}$$

Abbreviations

r.v.   = random variable, a.s. = almost surely. A certain property holds a.s. if there exists $N \in \mathcal{N}$ (negligible set) such that the property is true for every $\omega \in \Omega \setminus N$

a.e.   = almost everywhere (with respect to the Lebesgue measure)

We indicate the importance of the results with the following symbols:

[!]   means that you should pay close attention and try to understand well because an important concept, a new idea, or a new technique is being introduced

[!!]   means that the result is very important

[!!!]   means that the result is fundamental

Points flagged with the lighter "Gray Shade" warrant attention.

Chapter 1
Measures and Probability Spaces

> *The philosophy of the foundations of probability must be divorced from mathematics and statistics, exactly as the discussion of our intuitive space concept is now divorced from geometry.*
>
> *William Feller*

Probability is generally referred to uncertain phenomena, the outcome of which is not known with certainty. As Costantini [9] points out, it is not easy to give a general definition, and over the centuries many scholars have sought answers to questions such as:

(1) what is Probability?
(2) how is Probability calculated?[1]
(3) how does Probability "work"?[2]

On the other hand, only relatively recently the different nature of such questions, and the fact that they must be investigated with specific methods and tools of different and distinct disciplines, has been understood:

(1) in *Philosophy* the concept of Probability and its possible meaning is investigated, trying to give a definition and study its nature from a general perspective. The philosophical approach has led to interpretations and definitions that are very different;

---

[1] There are many cases where it is important to calculate or at least estimate the probability of an uncertain event. For example, a gambler is interested in knowing the probability of getting a certain hand in Poker; an insurance company must estimate the probability that one of its policyholders will have one or more accidents during a year; a car manufacturer wants to estimate the probability that the price of steel will not exceed a certain value; an airline can overbook based on the probability that a certain number of travelers will not show up for boarding.

[2] In other words, is it possible to formalize the principles and general rules of Probability in rigorous mathematical terms, in analogy with what is done, for example, in Euclidean geometry?

© The Editor(s) (if applicable) and The Author(s), under exclusive license
to Springer Nature Switzerland AG 2024
A. Pascucci, *Probability Theory I*, La Matematica per il 3+2 165,
https://doi.org/10.1007/978-3-031-63190-0_1

(2) *Statistics* is the discipline that studies the methods for estimating and evaluating Probability based on observations and available data on the random phenomenon considered;

(3) *Probability theory* is the purely mathematical discipline that applies abstract reasoning and logical-deductive reasoning to formalize Probability and its rules, starting from axioms and primitive definitions (as are, by analogy, the concepts of point and line in Geometry).

When approaching the study of Probability for the first time, confusion and misunderstandings may arise from not adequately distinguishing the different approaches (philosophical, statistical, and mathematical). In this text, *we exclusively adopt the mathematical perspective*: our goal is to provide an introduction to Probability theory.

1.1 Measurable Spaces and Probability Spaces

Probability theory studies phenomena whose outcome is uncertain: these are called *random phenomena* (or *random experiments*). Trivial examples of random phenomena are the tossing of a coin or the drawing of a card from a deck. The outcomes of a random phenomenon are not necessarily all "equivalent" in the sense that, for some reason, one outcome may be more "probable" (plausible, likely, expected, etc.) than another. Note that, since by definition none of the possible outcomes can be ruled out a priori, Probability theory does not aim to *predict* the outcome of a random phenomenon (which is impossible!) but to estimate, in the sense of *measuring*, the degree of reliability (the probability) of the individual possible outcomes or the combination of some of them. This is why the mathematical tools and language on which modern Probability theory is based are those of *Measure theory*, which is also the starting point of our treatise. Section 1.1.1 is devoted to recalling the first definitions and concepts of measure theory; in the following Sect. 1.1.2 we give their probabilistic interpretation.

1.1.1 Measurable Spaces

Definition 1.1.1 (Measurable Space) A *measurable space* is a pair $(\Omega, \mathscr{F})$ where:

(i) Ω is a non-empty set;

(ii) $\mathscr{F}$ is a *σ-algebra on* Ω, that is, $\mathscr{F}$ is a non-empty family of subsets of Ω that satisfies the following properties:

   (ii-a) if $A \in \mathscr{F}$ then $A^c := \Omega \setminus A \in \mathscr{F}$;
   (ii-b) the countable union of elements of $\mathscr{F}$ belongs to $\mathscr{F}$.

Property (ii-a) is expressed by saying that $\mathscr{F}$ is a family *closed under complement*; property (ii-b) is expressed by saying that $\mathscr{F}$ is a family $\sigma - \cup$-*closed(closed under countable union)*.

Remark 1.1.2 From property (ii-b) it also follows that if $A, B \in \mathscr{F}$ then $A \cup B \in \mathscr{F}$, that is, $\mathscr{F}$ is $\cup$-*closed (closed under finite union)*. In fact, given $A, B \in \mathscr{F}$, one can construct the sequence $C_1 = A, C_n = B$ for every $n \geq 2$; then

$$A \cup B = \bigcup_{n=1}^{\infty} C_n \in \mathscr{F}.$$

A σ-algebra $\mathscr{F}$ is non-empty by definition and therefore there exists $A \in \mathscr{F}$ and, by (ii-a), we have $A^c \in \mathscr{F}$: then also $\Omega = A \cup A^c \in \mathscr{F}$ and, again by ii-a), $\emptyset \in \mathscr{F}$. We observe that $\{\emptyset, \Omega\}$ is the *smallest* σ-algebra on Ω; conversely, the set of parts $\mathscr{P}(\Omega)$ is the *largest* σ-algebra on Ω.

We also note that the finite or countable intersection of elements of a σ-algebra $\mathscr{F}$ belongs to $\mathscr{F}$: in fact, if (A_n) is a finite or countable family in $\mathscr{F}$, combining the properties ii-a) and ii-b), we have that

$$\bigcap_n A_n = \left(\bigcup_n A_n^c \right)^c \in \mathscr{F}.$$

As a consequence, we say that $\mathscr{F}$ is $\cap$-closed and σ-$\cap$-closed.

Definition 1.1.3 (Measure) A *measure* on the measurable space $(\Omega, \mathscr{F})$ is a function

$$\mu : \mathscr{F} \longrightarrow [0, +\infty]$$

such that:

(iii-a) $\mu(\emptyset) = 0$;
(iii-b) μ is σ-additive on $\mathscr{F}$, that is, for every sequence $(A_n)_{n \in \mathbb{N}}$ of disjoint elements of $\mathscr{F}$ we have[3]

$$\mu \left(\biguplus_{n=1}^{\infty} A_n \right) = \sum_{n=1}^{\infty} \mu(A_n).$$

---

[3] We recall that the symbol $\biguplus$ indicates disjoint union. We observe that $\biguplus_{n \in \mathbb{N}} A_n \in \mathscr{F}$ since $\mathscr{F}$ is a σ-algebra.

Remark 1.1.4 Every measure μ is *additive* in the sense that, for every *finite* family $A_1, \ldots, A_n$ of disjoint sets in $\mathscr{F}$, we have

$$\mu\left(\biguplus_{k=1}^{n} A_k\right) = \sum_{k=1}^{n} \mu(A_k).$$

In fact, setting $A_k = \emptyset$ for $k > n$, we have

$$\mu\left(\biguplus_{k=1}^{n} A_k\right) = \mu\left(\biguplus_{k=1}^{\infty} A_k\right) =$$

(by σ-additivity)

$$= \sum_{k=1}^{\infty} \mu(A_k) =$$

(since $\mu(\emptyset) = 0$)

$$= \sum_{k=1}^{n} \mu(A_k).$$

Definition 1.1.5 A measure μ on $(\Omega, \mathscr{F})$ is said to be *finite* if $\mu(\Omega) < \infty$ and is said to be *σ-finite* if there exists a sequence (A_n) in $\mathscr{F}$ such that

$$\Omega = \bigcup_{n \in \mathbb{N}} A_n \quad \text{and} \quad \mu(A_n) < +\infty, \quad n \in \mathbb{N}.$$

Example 1.1.6 The first example of a σ-finite measure encountered in courses of mathematical analysis is the Lebesgue measure; it is defined on the Euclidean d-dimensional space, $\Omega = \mathbb{R}^d$, equipped with the σ-algebra of Lebesgue measurable sets.

1.1.2 Probability Spaces

Definition 1.1.7 (Probability Space) A measure space $(\Omega, \mathscr{F}, \mu)$ in which $\mu(\Omega) = 1$ is called a *probability space*: in this case, we usually use the letter P instead of μ and say that P is a *probability measure* (or simply a *probability*).

In a probability space $(\Omega, \mathscr{F}, P)$, each element $\omega \in \Omega$ is called an *outcome*; each $A \in \mathscr{F}$ is called an *event* and the number $P(A)$ is called the *probability of A*. Moreover, we say that Ω is the *sample space* and $\mathscr{F}$ is the *σ-algebra of events*.

When Ω is finite or countable, we always assume $\mathscr{F} = \mathscr{P}(\Omega)$ and say that $(\Omega, \mathscr{P}(\Omega), P)$ (or, more simply, (Ω, P)) is a *discrete probability space*. If instead Ω is uncountable, we speak of a *continuous (or general) probability space*.

Example 1.1.8 [!]   Consider the random phenomenon of rolling a fair six-sided die. The sample space

$$\Omega = \{1, 2, 3, 4, 5, 6\}$$

represents the possible states (outcomes) of the considered random experiment. Intuitively, *an event is a statement about the outcome of the experiment*, for example:

  (i) $A =$ "the result of the roll is an odd number";
 (ii) $B =$ "the result of the roll is the number 4";
(iii) $C =$ "the result of the roll is greater than 7".

Each statement corresponds to a subset of Ω:

  (i) $A = \{1, 3, 5\}$;
 (ii) $B = \{4\}$;
(iii) $C = \emptyset$.

This explains why mathematically we have defined an event as a subset of Ω. In particular, B is called an *elementary event* since it consists of a single outcome. Note that we distinguish the *outcome* 4 from the *elementary event* $\{4\}$.

The *logical operations* between events have a translation in terms of *set operations*, for example:

- "A or B" corresponds to $A \cup B$;
- "A and B" corresponds to $A \cap B$;
- "not A" corresponds to $A^c = \Omega \setminus A$;
- "A but not B" corresponds to $A \setminus B$.

Example 1.1.9   A runner has a 30% chance of winning the 100-meter race, a 40% chance of winning the 200-meter race, and a 50% chance of winning at least one of the two races. What is the probability that he wins both races?

Let

  (i) $A =$ "the runner wins the 100-meter race",
 (ii) $B =$ "the runner wins the 200-meter race",

the data of the problem are: $P(A) = 30\%$, $P(B) = 40\%$, and $P(A \cup B) = 50\%$. We are asked to determine $P(A \cap B)$. Using set operations (see also the following Lemma 1.1.25) we find that

$$P(A \cap B) = P(A) + P(B) - P(A \cup B) = 20\%.$$

Remark 1.1.10 The sample space Ω is, by definition, a *generic non-empty set*: it is legitimate to ask what sense it makes to assume such a degree of generality. In fact, we will see that in classic problems, Ω will simply be a *finite set* or the *Euclidean space* $\mathbb{R}^d$. However, in the most interesting applications, it may also happen that Ω is a *functional space* (such as, for example, the space of continuous functions). Often, Ω will also have a certain structure, for example, that of a *metric space*, to have some useful tools for the development of the theory.

Example 1.1.11 (Discrete Uniform Probability) Let Ω be finite. For each $A \subseteq \Omega$, let $|A|$ denote the cardinality of A and set

$$P(A) = \frac{|A|}{|\Omega|}. \tag{1.1.1}$$

Then P is a probability measure, called *uniform probability*. By definition, we have

$$P(\{\omega\}) = \frac{1}{|\Omega|}, \qquad \omega \in \Omega,$$

that is, each outcome is "equiprobable". The uniform probability corresponds to the classical concept of probability according to Laplace, as mentioned in the preface. For example, in the case of rolling a fair six-sided die, it is natural to consider the uniform probability

$$P(\{\omega\}) = \frac{1}{6}, \qquad \omega \in \Omega := \{1, 2, 3, 4, 5, 6\}.$$

Remark 1.1.12 A probability space in which each elementary event is equiprobable and has positive probability is necessarily finite. Consequently, for example, *it is not possible to define uniform probability on $\mathbb{N}$*: in fact, it should be $P(\{n\}) = 0$ for each $n \in \mathbb{N}$ and consequently, due to the σ-additivity, also $P(\mathbb{N}) = 0$ which is absurd.

Remark 1.1.13 [!] In a discrete probability space (Ω, P), consider the function

$$p : \Omega \longrightarrow [0, 1], \qquad p(\omega) = P(\{\omega\}), \qquad \omega \in \Omega.$$

It is clear that p is a non-negative function that has the property

$$\sum_{\omega \in \Omega} p(\omega) = \sum_{\omega \in \Omega} P(\{\omega\}) = P(\Omega) = 1. \tag{1.1.2}$$

Note that the sums in (1.1.2) are series with non-negative terms and therefore their value does not depend on the order of the addends. The second equality in (1.1.2) is a consequence of the σ-additivity of P.

In fact, *there is a bijective relationship between p and P* in the sense that, given any non-negative function p such that $\sum_{\omega \in \Omega} p(\omega) = 1$, and setting

$$P(A) := \sum_{\omega \in A} p(\omega), \qquad A \subseteq \Omega,$$

then P is a discrete probability on Ω.

In other words, *a discrete probability is uniquely defined by the probabilities of individual elementary events.* From an operational standpoint, it is much simpler to define the probability of individual outcomes (i.e., p) than to explicitly define P by assigning the probability of all events. Consider that, for example, if Ω has a cardinality of 100, then p is defined by the hundred values $p(\omega)$, with $\omega \in \Omega$, while P is defined on $\mathscr{P}(\Omega)$ which has a cardinality of $2^{100} \approx 10^{30}$.

Remark 1.1.14 (Probability in High School) [!] The previous observation suggests a reasonable and synthetic way to introduce the concept of probability in high school: let us consider the case of a *finite* (or, at most, countable) sample space

$$\Omega = \{\omega_1, \ldots, \omega_N\},$$

with $N \in \mathbb{N}$; we describe the concepts of outcome and event as in Example 1.1.8. Then we can explain that introducing a probability measure P on Ω means assigning the probabilities of individual outcomes: precisely, some numbers $p_1, \ldots, p_N$ are fixed such that

$$p_1, \ldots, p_N \geq 0 \quad \text{and} \quad p_1 + \cdots + p_N = 1, \tag{1.1.3}$$

where p_i indicates the probability of the i-th elementary event, i.e.,

$$p_i = P(\{\omega_i\}), \qquad i = 1, \ldots, N.$$

Finally, by definition, for each event A we set

$$P(A) = \sum_{\omega \in A} P(\{\omega\}). \tag{1.1.4}$$

This definition of probability space (Ω, P) is equivalent to the general definition (Definition 1.1.7, obviously in the case of finite Ω). The so-called *classical (or uniform) probability* is the one in which the outcomes are equiprobable, $p_1 = p_2 = \cdots = p_N$, so from (1.1.3) it follows that their common value is $\frac{1}{N}$. Therefore, classical probability is just a very particular case, although significant, among the infinite probability measures that can be chosen: in that case, clearly (1.1.4) reduces to the formula of "favorable outcomes over total outcomes".

Example 1.1.15 We give an alternative solution to the problem of Example 1.1.9. We can use as a sample space $\Omega = \{vv, vs, sv, ss\}$, where vv is the outcome in which the runner wins both races, vs is the outcome in which the runner wins the first race and loses the second, and so on: therefore $A = \{vv, vs\}$ and $B = \{vv, sv\}$. If $p = p(\omega)$ indicates the probability of individual outcomes, based on the problem data we obtain the linear system

$$\begin{cases} p(vv) + p(vs) = 30\% \\ p(vv) + p(sv) = 40\% \\ p(vv) + p(vs) + p(sv) = 50\% \end{cases}$$

from which we derive $p(vv) = P(A \cap B) = 20\%$, $p(vs) = 10\%$, $p(sv) = 20\%$ and $p(ss) = 1 - p(vv) - p(vs) - p(sv) = 50\%$.

We conclude the section with a couple of definitions that we will often use later.

Definition 1.1.16 (Negligible Sets and Almost Certain Sets) In a probability space $(\Omega, \mathscr{F}, P)$, we say that:

- a subset N of Ω is *negligible for P* if $N \subseteq A$ with $A \in \mathscr{F}$ such that $P(A) = 0$;
- a subset C of Ω is *almost certain for P* if its complement is negligible or, equivalently, if there exists $B \in \mathscr{F}$ such that $B \subseteq C$ and $P(B) = 1$.

We denote by $\mathscr{N}$ the family of negligible sets in $(\Omega, \mathscr{F}, P)$.

Negligible sets and almost certain sets *are not necessarily events* and therefore in general the probability $P(N)$ is not defined for N negligible or almost certain.

Definition 1.1.17 (Complete Space) A probability space $(\Omega, \mathscr{F}, P)$ is *complete* if $\mathscr{N} \subseteq \mathscr{F}$.

Remark 1.1.18 In a complete space, negligible sets for P (and consequently also almost certain sets for P) are events. Therefore, in a complete space, we have that

- N is negligible if and only if $P(N) = 0$;
- C is almost certain if and only if $P(C) = 1$.

Clearly, the completeness property depends on the probability measure considered. We will see later that it is always possible to "complete" a probability space (cf. Remark 1.4.3) and explain the importance of the completeness property (see, for example, Remarks 2.1.11 and 2.1.14).

1.1.3 Algebras and σ-Algebras

The suffix "σ-" (for example, in σ-algebra or σ-additivity) is used to specify that a definition or a property is valid for *countable* quantities and not just *finite*. In analogy with the concept of σ-algebra, we give the following useful

Definition 1.1.19 (Algebra) An *algebra* is a non-empty family $\mathscr{A}$ of subsets of Ω such that:

(i) $\mathscr{A}$ is closed under complement;
(ii) $\mathscr{A}$ is $\cup$-closed (i.e., closed under finite union).

Every σ-algebra is an algebra. If $A, B \in \mathscr{A}$ then $A \cap B = (A^c \cup B^c)^c \in \mathscr{A}$ and consequently $\mathscr{A}$ is $\cap$-closed.

Example 1.1.20 [!] In $\mathbb{R}$, consider the family $\mathscr{A}$ of *finite* unions of intervals (not necessarily bounded) of the type

$$]a, b], \qquad -\infty \le a \le b \le +\infty,$$

where by convention

$$]a, a] = \emptyset, \qquad]a, b] = \{x \in \mathbb{R} \mid x > a\} \quad \text{in the case } b = +\infty.$$

We note that $\mathscr{A}$ is an algebra but not a σ-algebra since, for example, $\bigcup_{n \ge 1}]0, 1 - \frac{1}{n}] =]0, 1[\notin \mathscr{A}$.

Since it will be useful to consider measures defined on algebras, we give the following extension of the concept of measure (cf. Definition 1.1.3).

Definition 1.1.21 (Measure) Let $\mathscr{A}$ be a family of subsets of Ω such that $\emptyset \in \mathscr{A}$. A *measure* on $\mathscr{A}$ is a function

$$\mu : \mathscr{A} \longrightarrow [0, +\infty]$$

such that:

(i) $\mu(\emptyset) = 0$;
(ii) μ is σ-additive on $\mathscr{A}$ in the sense that for every sequence $(A_n)_{n \in \mathbb{N}}$ of disjoint elements of $\mathscr{A}$, *such that* $A := \biguplus_{n \in \mathbb{N}} A_n \in \mathscr{A}$, we have

$$\mu(A) = \sum_{n=1}^{\infty} \mu(A_n).$$

We prove some basic properties of measures (and therefore, in particular, of probability measures).

Proposition 1.1.22 *Let μ be a measure on an algebra $\mathscr{A}$. The following properties hold true:*

(i) Monotonicity: for every $A, B \in \mathscr{A}$ such that $A \subseteq B$ we have

$$\mu(A) \leq \mu(B); \tag{1.1.5}$$

moreover, if $\mu(A) < \infty$ then

$$\mu(B \setminus A) = \mu(B) - \mu(A). \tag{1.1.6}$$

In particular, if P is a probability measure then

$$P(A^c) = 1 - P(A); \tag{1.1.7}$$

(ii) σ-subadditivity: for every $A \in \mathscr{A}$ and $(A_n)_{n \in \mathbb{N}}$ sequence in $\mathscr{A}$, we have

$$A \subseteq \bigcup_{n \in \mathbb{N}} A_n \quad \Longrightarrow \quad \mu(A) \leq \sum_{n=1}^{\infty} \mu(A_n).$$

Proof We prove (i): if $A \subseteq B$ then, by the additivity of μ and being $B \setminus A \in \mathscr{A}$, we have

$$\mu(B) = \mu(A \uplus (B \setminus A)) = \mu(A) + \mu(B \setminus A).$$

From the fact that $\mu(B \setminus A) \geq 0$ it follows (1.1.5) and, in the particular case where $\mu(A) < \infty$, it also follows (1.1.6).

To prove (ii), we set

$$\tilde{A}_1 := A_1 \cap A, \qquad \tilde{A}_{n+1} := A \cap A_{n+1} \setminus \bigcup_{k=1}^{n} A_k.$$

Observe that $\tilde{A}_n \subseteq A_n$. Moreover, the sets $\tilde{A}_n$ belong to the algebra $\mathscr{A}$ since they are obtained through finite operations from elements of $\mathscr{A}$ and, by hypothesis, we have

$$\biguplus_{n \in \mathbb{N}} \tilde{A}_n = A \in \mathscr{A}.$$

Then, by monotonicity, we have

$$\mu(A) = \mu\left(\biguplus_{n\in\mathbb{N}} \widetilde{A}_n\right) =$$

(by σ-additivity and then again by monotonicity)

$$= \sum_{n=1}^{\infty} \mu(\widetilde{A}_n) \le \sum_{n=1}^{\infty} \mu(A_n).$$

$\square$

Example 1.1.23 Formula (1.1.7) is useful for solving problems of the following type: calculate the probability of getting at least one 6 when rolling a die 8 times. We define Ω as the set of possible sequences of rolls: then $|\Omega| = 6^8$. We can determine the probability of the event that interests us (call it A) more easily by considering A^c, that is, the set of sequences that do not contain 6: in fact, we will have $|A^c| = 5^8$ and therefore by (1.1.7)

$$P(A) = 1 - P(A^c) = 1 - \frac{5^8}{6^8}.$$

Exercise 1.1.24 Let A, B be certain events, that is, such that $P(A) = P(B) = 1$. Prove that $A \cap B$ is also a certain event.

Lemma 1.1.25 *Let $\mathscr{A}$ be an algebra. A function*

$$\mu : \mathscr{A} \longrightarrow [0, +\infty]$$

such that $\mu(\emptyset) = 0$, is additive if and only if

$$\mu(A \cup B) + \mu(A \cap B) = \mu(A) + \mu(B), \qquad A, B \in \mathscr{F}. \tag{1.1.8}$$

Proof If μ is additive then

$$\mu(A \cup B) + \mu(A \cap B) = \mu(A) + \mu(B \setminus A) + \mu(A \cap B) = \mu(A) + \mu(B).$$

Conversely, from (1.1.8) with disjoint A, B we have the additivity of μ. $\square$

Remark 1.1.26 In the case of probability measures, (1.1.8) can be usefully rewritten in the form

$$P(A \cup B) = P(A) + P(B) - P(A \cap B) \tag{1.1.9}$$

Example 1.1.27 When rolling two dice, what is the probability that at least one of the two rolls has a result less than or equal to 3?

Let $I_n = \{k \in \mathbb{N} \mid k \le n\}$ and consider the sample space $\Omega = I_6 \times I_6$ of possible pairs of roll results. Let $A = I_3 \times I_6$ (and respectively $B = I_6 \times I_3$) be the event in which the result of the first die (respectively the second die) is less than or equal to 3. We are asked to calculate the probability of $A \cup B$. Note that A, B are not disjoint and in the uniform probability P, counting the elements, we have

$$P(A) = P(B) = \frac{3 \cdot 6}{6 \cdot 6} = \frac{1}{2}, \qquad P(A \cap B) = \frac{3 \cdot 3}{6 \cdot 6} = \frac{1}{4}.$$

Then, by (1.1.9), we obtain

$$P(A \cup B) = P(A) + P(B) - P(A \cap B) = \frac{3}{4}.$$

Remark 1.1.28   (1.1.8) generalizes easily to the case of three sets $A_1, A_2, A_3 \in \mathscr{F}$:

$$P(A_1 \cup A_2 \cup A_3) = P(A_1) + P(A_2 \cup A_3) - P((A_1 \cap A_2) \cup (A_1 \cap A_3))$$
$$= P(A_1) + P(A_2) + P(A_3) - P(A_1 \cap A_2) - P(A_1 \cap A_3)$$
$$- P(A_2 \cap A_3)$$
$$+ P(A_1 \cap A_2 \cap A_3).$$

In general, the following formula is proved by induction

$$P\left(\bigcup_{k=1}^{n} A_k\right) = \sum_{k=1}^{n} (-1)^{k-1} \sum_{\{i_1,\ldots,i_k\} \subseteq \{1,\ldots,n\}} P(A_{i_1} \cap \cdots \cap A_{i_k})$$

where the last sum is intended over all subsets of $\{1, \ldots, n\}$ with k elements.

Example 1.1.29 Let A, B be events in $(\Omega, \mathscr{F}, P)$. If $P(A) = 1$ then $P(A \cap B) = P(B)$. Indeed, by the finite additivity of P we have

$$P(B) = P(A \cap B) + P(A^c \cap B) = P(A \cap B)$$

since, by (1.1.5), $P(A^c \cap B) \le P(A^c) = 0$.

1.1.4   Finite Additivity and σ-Additivity

In a general probability space, σ-additivity is a stronger property than additivity. We will soon understand, in Proposition 1.1.32, the importance of requiring σ-additivity in the definition of probability measure: this is a rather delicate point as we see in the next example.

Example 1.1.30 (Continuous Uniform Probability) Suppose we want to define the concept of uniform probability on the real interval $\Omega = [0, 1]$. From an intuitive standpoint, it is natural to set

$$P([a, b]) = b - a, \qquad 0 \le a \le b \le 1. \tag{1.1.10}$$

Then obviously $P(\Omega) = 1$ and the probability of the event $[a, b]$ (which can be interpreted as the event "a point chosen at random in $[0, 1]$ belongs to $[a, b]$") depends only on the length of $[a, b]$ and is invariant under translation. Note that $P(\{x\}) = P([x, x]) = 0$ for each $x \in [0, 1]$, that is, each outcome has zero probability, and P is nothing but the Lebesgue measure. Giuseppe Vitali proved in 1905 (cf. [45]) that it is not possible to extend the Lebesgue measure to the whole set of parts $\mathscr{P}(\Omega)$ or, in other words, there does not exist P defined on the set of parts of $[0, 1]$, which is σ-additive and satisfies (1.1.10). If this holds true, then in the context of general probability spaces, it becomes *necessary* to introduce a σ-algebra of events upon which to define P: in general, such a σ-algebra will be *smaller* than the set of parts of Ω.

In our context, Vitali's result can be stated as follows: there does not exist a probability measure P on $([0, 1], \mathscr{P}([0, 1]))$ that is invariant under translations, that is, such that $P(A) = P(A_x)$ for each $A \subseteq [0, 1]$ and $x \in [0, 1]$, where

$$A_x = \{y \in [0, 1] \mid y = a + x \text{ or } y = a + x - 1 \text{ for some } a \in A\}.$$

The proof proceeds by contradiction and is based on the axiom of choice. Consider on $[0, 1]$ the equivalence relation $x \sim y$ if and only if $(x - y) \in \mathbb{Q}$: by the axiom of choice, from each equivalence class it is possible to select a representative and having done so, we denote by A the set of such representatives. Now, by hypothesis, $P(A_q) = P(A)$ for every $q \in \mathbb{Q} \cap [0, 1]$ and moreover $A_q \cap A_p = \emptyset$ for $q \neq p$ in $\mathbb{Q} \cap [0, 1]$. Hence we obtain

$$[0, 1] = \biguplus_{q \in \mathbb{Q} \cap [0,1]} A_q$$

and if P were σ-additive, we would have

$$1 = P([0, 1]) = \sum_{q \in \mathbb{Q} \cap [0,1]} P(A_q) = \sum_{q \in \mathbb{Q} \cap [0,1]} P(A).$$

However, the last sum can only assume the value 0 (when $P(A) = 0$) or diverge (when $P(A) > 0$) and this leads to a contradiction. Note that the contradiction is a consequence of the request for *countable* additivity (i.e., σ-additivity) of P.

Notation 1.1.31 We write

$$A_n \nearrow A \quad \text{and} \quad B_n \searrow B$$

to indicate that $(A_n)_{n \in \mathbb{N}}$ is an *increasing* sequence of sets such that $A = \bigcup_{n \in \mathbb{N}} A_n$, and $(B_n)_{n \in \mathbb{N}}$ is a *decreasing* sequence of sets such that $B = \bigcap_{n \in \mathbb{N}} B_n$.

The σ-additivity has the following important characterizations.

Proposition 1.1.32 *[!] Let $\mathscr{A}$ be an algebra on Ω and*

$$\mu : \mathscr{A} \longrightarrow [0, +\infty]$$

an additive function. The following properties are equivalent:

 (i) *μ is σ-additive;*
 (ii) *μ is σ-subadditive;[4]*
 (iii) *μ is continuous from below, i.e., for every sequence $(A_n)_{n \in \mathbb{N}}$ in $\mathscr{A}$ such that $A_n \nearrow A$, with $A \in \mathscr{A}$, we have*

$$\lim_{n \to \infty} \mu(A_n) = \mu(A).$$

Moreover, if (i) holds then we also have

 (iv) *μ is continuous from above, i.e., for every sequence $(B_n)_{n \in \mathbb{N}}$ in $\mathscr{A}$, such that $\mu(B_1) < \infty$ and $B_n \searrow B \in \mathscr{A}$, we have*

$$\lim_{n \to \infty} \mu(B_n) = \mu(B).$$

Finally, if $\mu(\Omega) < \infty$ then (i), (ii), (iii) and (iv) are equivalent.

Proof Preliminarily, we observe that μ is monotone: this is proved as Proposition 1.1.22-(i).

  [(i) $\Rightarrow$ (ii)] It is the content of Proposition 1.1.22-(ii).

  [(ii) $\Rightarrow$ (iii)] Let $\mathscr{A} \ni A_n \nearrow A \in \mathscr{A}$. By monotonicity, we have

$$\lim_{n \to \infty} \mu(A_n) \leq \mu(A).$$

---

[4] For every $A \in \mathscr{A}$ and for every sequence $(A_n)_{n \in \mathbb{N}}$ of elements of $\mathscr{A}$ such that $A \subseteq \bigcup_{n \in \mathbb{N}} A_n$, we have

$$\mu(A) \leq \sum_{n=1}^{\infty} \mu(A_n).$$

On the other hand, let

$$C_1 = A_1, \qquad C_{n+1} = A_{n+1} \setminus A_n, \quad n \in \mathbb{N}.$$

Then (C_n) is a disjoint sequence in $\mathscr{A}$ and we have

$$\mu(A) = \mu\left(\biguplus_{k \geq 1} C_k \right) \leq$$

(by the σ-subadditivity of μ)

$$\leq \sum_{k=1}^{\infty} \mu(C_k) = \lim_{n \to \infty} \sum_{k=1}^{n} \mu(C_k) =$$

(by the finite additivity of μ)

$$= \lim_{n \to \infty} \mu(A_n).$$

[(iii) $\Rightarrow$ (i)] Let $(A_n)_{n \in \mathbb{N}}$ be a sequence of disjoint elements of $\mathscr{A}$, such that $A := \biguplus_{n \in \mathbb{N}} A_n \in \mathscr{A}$. Let

$$\bar{A}_n = \bigcup_{k=1}^{n} A_k;$$

we have $\bar{A}_n \nearrow A$ and $\bar{A}_n \in \mathscr{A}$ for each n. Then, by the hypothesis of continuity from below of μ, we have

$$\mu(A) = \lim_{n \to \infty} \mu(\bar{A}_n) =$$

(by the finite additivity of μ)

$$= \lim_{n \to \infty} \sum_{k=1}^{n} \mu(A_k) = \sum_{k=1}^{\infty} \mu(A_k),$$

observing that the limit of the partial sums exists, finite or not, since μ has non-negative values.

[(iii) $\Rightarrow$ (iv)] Suppose (iii) holds. If $B_n \searrow B$ then $A_n := B_1 \setminus B_n$ is such that $A_n \nearrow A := B_1 \setminus B$. If $\mu(B_1) < \infty$, by the property (1.1.6) which holds under the sole hypothesis of additivity, we have[5]

$$\mu(B) = \mu(B_1 \setminus A)$$
$$= \mu(B_1) - \mu(A) =$$

(by continuity from below of μ)

$$= \mu(B_1) - \lim_{n\to\infty} \mu(A_n) = \lim_{n\to\infty} (\mu(B_1) - \mu(A_n)) = \lim_{n\to\infty} \mu(B_n).$$

[(iv) $\Rightarrow$ (iii)] Under the hypothesis that $\mu(\Omega) < \infty$, the fact that (iv) implies (iii) is proven as in the previous point by setting $B_n = \Omega \setminus A_n$ and using the fact that if $(A_n)_{n\in\mathbb{N}}$ is increasing then $(B_n)_{n\in\mathbb{N}}$ is decreasing and obviously $\mu(B_1) < \infty$. $\square$

1.2   Finite Spaces and Counting Problems

In this section, we assume that Ω is finite and consider some problems in which the *discrete uniform probability* of Example 1.1.11 is used. These are referred to as *counting problems* because, as stated in (1.1.1), calculating probabilities boils down to determining the cardinality of events.

Combinatorics is the mathematical tool that allows these calculations to be carried out. Although these are problems that have an elementary formulation (given in terms of coins, dice, cards, etc.), often the calculation can be very complicated and can be intimidating at first glance. On this aspect, it is important to downplay because it is a technical complication more than substantial, which should not create an unjustified concern. The discrete uniform probability is definitely a minor and limited aspect of probability theory, constraining its overall importance and appeal. In fact, unless there is a particular interest in the subject matter, *this section can be either skipped or initially read through swiftly.*

---

[5] In detail: we have $B_1 \setminus \bigcup_{n=1}^{\infty} A_n = B_1 \cap \bigcap_{n=1}^{\infty} A_n^c = \bigcap_{n=1}^{\infty} (B_1 \cap A_n^c) = \bigcap_{n=1}^{\infty} B_n.$

1.2.1 Cardinality of Sets

Let us start by recalling some basic notions about the cardinality of finite sets. In the following we use the following

Notation 1.2.1

$$I_n = \{k \in \mathbb{N} \mid k \leq n\} = \{1, 2, \ldots, n\}, \qquad n \in \mathbb{N}.$$

We say that a set A has cardinality $n \in \mathbb{N}$, and write $|A| = n$ or $\sharp A = n$, if there is a bijective function from I_n to A. Moreover, by definition $|A| = 0$ if $A = \emptyset$. We write $A \leftrightarrow B$ if $|A| = |B|$. In this section, we only consider sets with finite cardinality.

Prove as an exercise the following properties:

(i) $|A| = |B|$ if and only if there is a bijective function from A to B;
(ii) if A, B are disjoint then

$$|A \uplus B| = |A| + |B|$$

and more generally, this property extends to the case of a finite disjoint union;
(iii) for every A, B we have

$$|A \times B| = |A||B| \tag{1.2.1}$$

(1.2.1) can be proved using ii) and the fact that

$$A \times B = \biguplus_{x \in A} \{x\} \times B$$

where the union is disjoint and $|\{x\} \times B| = |B|$ for each $x \in A$;
(iii) we denote by A^B the set of functions from B to A. Then we have

$$\left| A^B \right| = |A|^{|B|} \tag{1.2.2}$$

since $A^B \leftrightarrow \underbrace{A \times \cdots \times A}_{|B| \text{ times}}$.

1.2.2 Three Reference Random Experiments: Drawings from an URN

When using combinatorics for the study of a random experiment, the choice of the sample space is important because it can simplify the counting of total and favorable cases. The most convenient choice, from this standpoint, generally depends on the random phenomenon under consideration. However, it is often useful to rethink the random experiment (or, possibly, each sub-random experiment into which it can be decomposed) as an appropriate drawing of balls from an urn (with replacement, without replacement, simultaneous) which we now describe.

Consider an urn containing n balls, labeled with $e_1, e_2, \ldots, e_n$. We draw k balls from the urn in one of the following three ways:

(1) drawing *with replacement*, with $k \in \mathbb{N}$, in which, for the next drawing, the drawn ball is reinserted into the urn;
(2) drawing *without replacement*, with $k \in \{1, \ldots, n\}$, in which the drawn ball is not reinserted into the urn;
(3) *simultaneous* drawing, with $k \in \{1, \ldots, n\}$, in which the k balls are drawn simultaneously.

Note that:

- in the drawing with replacement, the total number of balls in the urn and its composition remain constant in the subsequent drawings; since we draw one ball at a time, we take into account the *order of drawing*; moreover, there may be *repetitions*, i.e., it is possible to draw the same ball multiple times;
- in the drawing without replacement, at each drawing the total number of balls in the urn is reduced by one unit and therefore the composition of the urn itself changes each time; in this case, we also take into account the order of drawing; however, repetitions are no longer possible (in fact, once drawn, the ball is not reinserted into the urn);
- simultaneous drawing corresponds to drawing without replacement in which *we do not* take into account the order of drawing.

We can thus summarize what has been said so far in Table 1.1:

We will return later to the fourth case corresponding to the empty box and, in particular, to why it has not been considered (see Remark 1.2.13). For each of the three types of drawing described above, we want to determine a sample space Ω, with the smallest possible cardinality, which allows us to describe such a random experiment. We will address this issue in Sect. 1.2.4 where we will see that Ω will be given respectively by:

(1) the set $\mathbf{DR}_{n,k}$ of *arrangements with repetition of k elements of $\{e_1, \ldots, e_n\}$*, in the case of drawing with replacement;
(2) the set $\mathbf{D}_{n,k}$ of *simple arrangements of k elements of $\{e_1, \ldots, e_n\}$*, in the case of drawing without replacement;

Table 1.1 Classification of drawings from an urn

REPETITION / ORDER	*Without* repetition	*With* repetition
Order is considered	Drawing *without replacement*	Drawing *with replacement*
Order is *not* considered	*Simultaneous* drawing	–

(3) the set $\mathbf{C}_{n,k}$ of *combinations of k elements of* $\{e_1, \ldots, e_n\}$, in the case of simultaneous drawing.

Before introducing these three fundamental sets, we illustrate a general method that we will use to determine the cardinality of $\mathbf{DR}_{n,k}$, $\mathbf{D}_{n,k}$, $\mathbf{C}_{n,k}$ and other finite sets.

1.2.3   Method of Successive Choices

In this section, we illustrate an algorithm, known as the *method of successive choices* (or *scheme of successive choices* or also *fundamental principle of combinatorics*), which allows us to determine the cardinality of a set once its elements are uniquely characterized by a finite number of successive choices.

Method of Successive Choices *Given a finite set A whose cardinality $|A|$ we want to determine, we proceed as follows:*

*(1) at the first step, we consider a partition of A into $n_1 \in \mathbb{N}$ subsets $A_1, \ldots, A_{n_1}$, all having the **same cardinality**; this partition is obtained by making a "choice", i.e., distinguishing the elements of A based on a property they possess;*

*(2) at the second step, for each $i = 1, \ldots, n_1$, we proceed as in point 1) with the set A_i instead of A, considering a partition $A_{i,1}, \ldots, A_{i,n_2}$ of A_i into n_2 subsets all having the **same cardinality**, with $n_2 \in \mathbb{N}$ that **does not depend on** i;*

(3) we continue this way until, after a finite number $k \in \mathbb{N}$ of steps, the elements of the partition have cardinality equal to 1.

The cardinality of A is then given by

$$|A| = n_1 \, n_2 \, \cdots \, n_k.$$

For example, we apply the method of successive choices to prove the formula

$$\left| A^B \right| = |A|^{|B|}.$$

Let $n = |A|$ be the cardinality of A and denote its elements by $a_1, \ldots, a_n$. Similarly, let $k = |B|$ be the cardinality of B and denote its elements by $b_1, \ldots, b_k$. Since A^B is the set of functions from B to A, we can uniquely characterize each function in A^B through the following $k = |B|$ successive choices:

(1) as the first choice, we fix the value that the functions of A^B take in correspondence with b_1; we have $n = |A|$ possibilities (thus $n_1 = n$), i.e., this first choice determines a partition of A into n subsets (we do not need to write which these subsets are, but only how much n_1 is);

(2) as the second choice, we fix the value that the functions of A^B take in correspondence with b_2; we have $n = |A|$ possibilities (thus $n_2 = n$);

(3) $\cdots$

(4) as the k-th and last choice (with $k = |B|$) we fix the value that the functions of A^B take in correspondence with b_k; we have $n = |A|$ possibilities (thus $n_k = n$).

From the method of successive choices, it follows that

$$\left|A^B\right| = \underbrace{|A| \cdots |A|}_{k = |B| \text{ times}} = |A|^{|B|}.$$

Later, when we apply the method of successive choices, we will follow the procedure outlined in points (1)–(4), indicating each choice made at every step and the number of possibilities (or options) available. We will not explicitly mention the partition resulting from each choice, as it is typically understood.

1.2.4   Arrangements and Combinations

In this section, we consider a set with $n \in \mathbb{N}$ elements

$$E = \{e_1, e_2, \ldots, e_n\}$$

which represents an urn, containing n numbered balls, with which the random experiments of drawing are carried out.

Definition 1.2.2 (Arrangements with Repetition) Given $k \in \mathbb{N}$, we say that

$$\mathbf{DR}_{n,k} := \underbrace{E \times \cdots \times E}_{k \text{ times}} = \{(\omega_1, \ldots, \omega_k) \mid \omega_1, \ldots, \omega_k \in E\}$$

is the set of *arrangements with repetition of k elements of E*. By (1.2.2) we have $|\mathbf{DR}_{n,k}| = n^k$.

The set $\mathbf{DR}_{n,k}$ is the natural sample space for describing the outcome of k drawings *with replacement* from an urn containing n balls: each element $\omega = (\omega_1, \ldots, \omega_k)$ indicates the sequence of balls drawn. More generally, $\mathbf{DR}_{n,k}$ expresses the ways in which we can choose, in an **ordered** and **repeated** manner, k objects taken from a set of n objects.

Example 1.2.3 Let $E = \{a, b, c\}$. Then $|\mathbf{DR}_{3,2}| = 3^2$ and precisely

$$\mathbf{DR}_{3,2} = \{(a, a), (a, b), (a, c), (b, a), (b, b), (b, c), (c, a), (c, b), (c, c)\}.$$

Exercise 1.2.4 Identify the number of potential outcomes of the following random experiments:[6]

(i) a random word (even without meaning) composed of 8 letters of the English alphabet (which has 26 letters) is chosen;
(ii) a football pool ticket is played, in which for each of 13 matches you can choose between 1, 2, or X;
(iii) a fair six-sided die is rolled 10 times.

Definition 1.2.5 (Simple Arrangements) Given $k \leq n$, we say that

$$\mathbf{D}_{n,k} = \{(\omega_1, \ldots, \omega_k) \mid \omega_1, \ldots, \omega_k \in E \text{ with } \omega_i \neq \omega_j \text{ for } i \neq j\}$$

is the set of *simple arrangements of k elements of E*. We have

$$|\mathbf{D}_{n,k}| = n(n-1)\cdots(n-k+1) = \frac{n!}{(n-k)!}. \qquad (1.2.3)$$

The set $\mathbf{D}_{n,k}$ is the natural sample space for describing the outcome of k drawings *without replacement* from an urn containing n balls: again, each element $\omega = (\omega_1, \ldots, \omega_k)$ indicates the sequence of balls drawn. More generally, $\mathbf{D}_{n,k}$ expresses the ways in which we can arrange, in an **ordered** and **non-repeated** manner, a number of k objects chosen from a set of n objects.

Formula (1.2.3) can be proven using the method of successive choices, characterizing the generic element $(\omega_1, \ldots, \omega_k)$ of $\mathbf{D}_{n,k}$ as follows:

(1) as the first choice, we fix ω_1: we have $n = |E|$ possibilities, so $n_1 = n$;
(2) as the second choice, we fix ω_2, different from ω_1: we have $n - 1$ possibilities, so $n_2 = n - 1$;
(3) $\cdots$
(4) as the k-th and last choice, we fix ω_k: we have $n - k + 1$ possibilities, since we have already chosen $\omega_1, \ldots, \omega_{k-1}$, so $n_k = n - k + 1$.

From the method of successive choices, the validity of (1.2.3) is deduced.

Example 1.2.6 Let $E = \{a, b, c\}$. Then $|\mathbf{D}_{3,2}| = \frac{3!}{1!} = 6$ and precisely

$$\mathbf{D}_{3,2} = \{(a, b), (a, c), (b, a), (b, c), (c, a), (c, b)\}.$$

Example 1.2.7 What is the probability of getting a straight five-number sequence (for which the order of drawing matters) in the lottery game (in which five numbers are drawn without replacement from the first ninety natural numbers), assuming to play a single five-number sequence (for example, the ordered sequence 13, 5, 45, 21, 34)? What is the probability of getting a simple five-number sequence instead (for which the order of drawing does not matter)?

---

[6] Solution of Exercise 1.2.4: (i) $|\mathbf{DR}_{26,8}| = 26^8$; (ii) $|\mathbf{DR}_{3,13}| = 3^{13}$; (iii) $|\mathbf{DR}_{6,10}| = 6^{10}$.

Solution The probability of getting a straight five-number sequence is simply $\frac{1}{|\mathbf{D}_{90,5}|} \approx 1.89 \cdot 10^{-10}$.

If instead, we consider a simple five-number sequence, we must first count in how many In different ways, we can order 5 numbers, equal to $|\mathbf{D}_{5,5}| = 5!$. So the probability of a simple five-number combination after 5 draws is $\frac{|\mathbf{D}_{5,5}|}{|\mathbf{D}_{90,5}|} \approx 2.27 \cdot 10^{-8}$.

Definition 1.2.8 (Permutations) We denote by $\mathbf{P}_n := \mathbf{D}_{n,n}$ the set of *permutations* of n objects. We have

$$|\mathbf{P}_n| = n!$$

The set $\mathbf{P}_n$ expresses the ways in which we can reorder, that is, arrange in an **ordered** and **non-repeated** manner, a number n of objects.

Definition 1.2.9 (Combinations) Given $k \leq n$, we denote by $\mathbf{C}_{n,k}$ the set of *combinations of k elements of E*, defined as the family of subsets of E with cardinality k:

$$\mathbf{C}_{n,k} = \{A \subseteq E \mid |A| = k\}.$$

The set $\mathbf{C}_{n,k}$ is the natural sample space for describing the outcome of the simultaneous drawing of k balls from an urn containing n of them: each element $\omega = \{\omega_1, \ldots, \omega_k\}$ indicates the group (set) of k extracted balls. More generally, $\mathbf{C}_{n,k}$ expresses all the groups of k objects chosen from a set of n objects, in a **non-ordered** and **non repeated** manner.

Example 1.2.10 Let $E = \{a, b, c\}$. Then $|\mathbf{C}_{3,2}| = 3$ and precisely

$$\mathbf{C}_{3,2} = \big\{\{a, b\}, \{a, c\}, \{b, c\}\big\}.$$

Proposition 1.2.11 *We have*

$$|\mathbf{C}_{n,k}| = \frac{|\mathbf{D}_{n,k}|}{|\mathbf{P}_k|} = \frac{n!}{k!(n-k)!} = \binom{n}{k}. \tag{1.2.4}$$

Proof Unlike for $|\mathbf{DR}_{n,k}|$ and $|\mathbf{D}_{n,k}|$, it is not possible to decompose the calculation of $|\mathbf{C}_{n,k}|$ into a sequence of successive choices. However, proving (1.2.4) is equivalent to proving the following equality:

$$|\mathbf{D}_{n,k}| = |\mathbf{C}_{n,k}| \, |\mathbf{P}_k|. \tag{1.2.5}$$

We prove (1.2.5) by applying the method of successive choices to the set $\mathbf{D}_{n,k}$, characterizing the generic element $\omega = (\omega_1, \ldots, \omega_k)$ of $\mathbf{D}_{n,k}$ according to the following scheme:

(1) as the first choice, we fix the subset $\{\omega_1, \ldots, \omega_k\}$ of E formed by the components of ω: we have $|\mathbf{C}_{n,k}|$ possibilities and therefore $n_1 = |\mathbf{C}_{n,k}|$;

(2) as the second and last choice, we fix the permutation of the k elements $\omega_1, \ldots, \omega_k$ that describes the order in which they are arranged in ω: we have $|\mathbf{P}_k|$ possibilities and therefore $n_2 = |\mathbf{P}_k|$.

From the method of successive choices, we deduce the validity of (1.2.5) and therefore of (1.2.4). □

The sets $\mathbf{DR}_{n,k}$, $\mathbf{D}_{n,k}$ (and thus also $\mathbf{P}_n = \mathbf{D}_{n,n}$) and $\mathbf{C}_{n,k}$ are important not only because they are the sample spaces of the three random experiments introduced in Sect. 1.2.2, but also because the cardinalities of such sets often correspond to the numbers $n_1, n_2, \ldots, n_k$ of the method of successive choices; for example, for the calculation of $|\mathbf{D}_{n,k}|$ in (1.2.5) we chose $n_1 = |\mathbf{C}_{n,k}|$ and $n_2 = |\mathbf{P}_k|$.

We can complete the table of Sect. 1.2.2, also reporting the sample spaces and their cardinalities (i.e., the "total cases").

Here are some concluding remarks regarding Table 1.2.

Remark 1.2.12 Although three random experiments have been introduced, it suffices to consider only the first two: drawing without replacement and drawing with replacement. Simultaneous drawing can indeed be viewed as a specific case of drawing without replacement where the order is not considered. More precisely, to each element of $\mathbf{C}_{n,k}$, that is, to each subset of k balls chosen among n, correspond $k!$ elements (or k-tuples) of $\mathbf{D}_{n,k}$, therefore we have

$$\frac{\text{favorable cases in } \mathbf{C}_{n,k}}{\text{total cases in } \mathbf{C}_{n,k}} = \frac{k! \, (\text{favorable cases in } \mathbf{C}_{n,k})}{k! \, (\text{total cases in } \mathbf{C}_{n,k})} = \frac{\text{favorable cases in } \mathbf{D}_{n,k}}{\text{total cases in } \mathbf{D}_{n,k}}.$$

Remark 1.2.13 The empty cell in the table above corresponds to the set of so-called "combinations with repetition", that is, the set of all groups, unordered and possibly repeated, of k objects chosen from a set of n objects. The corresponding random experiment is the drawing with replacement in which the order is not considered: this random experiment can also be described by the sample space $\mathbf{DR}_{n,k}$ equipped with the discrete uniform probability. On the contrary, on the space

Table 1.2 Classification of drawings from an urn and relation with arrangements and combinations

ORDER \ REPETITION	Without repetition	With repetition
Order is considered	Drawing *without replacement* $\Omega = \mathbf{D}_{n,k}$ $|\Omega| = \frac{n!}{(n-k)!}$	Drawing *with replacement* $\Omega = \mathbf{DR}_{n,k}$ $|\Omega| = n^k$
Order is *not* considered	*Simultaneous* drawing $\Omega = \mathbf{C}_{n,k}$ $|\Omega| = \frac{|\mathbf{D}_{n,k}|}{k!} = \binom{n}{k}$	–

of combinations with repetition, the probability cannot be the discrete uniform one. In fact, each combination with repetition does not always correspond to the same number of elements of $\mathbf{DR}_{n,k}$ (as it happens in the case of $\mathbf{C}_{n,k}$ and $\mathbf{D}_{n,k}$) and the proportionality constant depends on how many repetitions there are within the combination: combinations with more repetitions are less probable. For this reason, on this space, the formula "favorable cases/total cases" is not valid, that is, combinatorics techniques cannot be used.

Example 1.2.14 Let us reconsider the calculation of the probability of a simple five-number combination in the lottery game: since the order of drawing of the numbers does not matter, we are in the case of simultaneous drawing, so it is natural to consider $\Omega = \mathbf{C}_{90,5}$. In fact, the probability of the five-number combination is $\frac{1}{|C_{90,5}|}$ which coincides with the result we had already found using simple arrangements, that is, $\frac{5!}{|D_{90,5}|}$.

Exercise 1.2.15 Find the probability of obtaining a simple five-number combination after $k \geq 5$ drawings.

Solution Let $\Omega = \mathbf{C}_{90,k}$. Let A be the event we are interested in, that is, the family of sets of k numbers in which 5 are fixed and the remaining $k - 5$ can be any of the remaining 85 numbers. Then we have

$$P(A) = \frac{|\mathbf{C}_{85,k-5}|}{|\mathbf{C}_{90,k}|}.$$

For example, $P(A) \approx 6 \cdot 10^{-6}$ for $k = 10$ and $P(A) \approx 75\%$ for $k = 85$.

Exercise 1.2.16 Consider a deck of 40 cards consisting of four suits: hearts, diamonds, clubs, and spades. Find the probability of event A defined in each of the following ways:

(1) in 5 draws without replacement, 5 hearts are obtained;
(2) in 5 draws with replacement, 5 hearts are obtained;
(3) in 5 draws without replacement, the numbers from 1 to 5 are obtained in order, of any suit, even different from each other.

Solution

(1) The drawing is without replacement, but the event $A =$ "5 hearts are obtained" does not consider the order. Therefore such drawing can also be seen as a simultaneous drawing. So we can choose as sample space $\Omega = \mathbf{C}_{40,5}$ (choosing $\Omega = \mathbf{D}_{40,5}$ would still be fine). The outcome $\omega = \{\omega_1, \omega_2, \omega_3, \omega_4, \omega_5\}$ corresponds to the set of cards drawn. Then $A \leftrightarrow \mathbf{C}_{10,5}$ (the possible choices, unordered and not repeated, of 5 hearts) and therefore

$$P(A) = \frac{\binom{10}{5}}{\binom{40}{5}} \approx 0.04\,\%.$$

(2) This time the drawing is with replacement, so we must consider $\Omega = \mathbf{DR}_{40,5}$ (actually, even in this case the event A does not take into account the order; however, when there is repetition the only space we can choose to use the combinatorics techniques is the space of arrangements with repetition). The outcome ω can be identified with the sequence $(\omega_1, \omega_2, \omega_3, \omega_4, \omega_5)$, ordered and with possible repetitions, of the cards drawn. In this case $A \leftrightarrow \mathbf{DR}_{10,5}$ (the possible choices, ordered and repeated, of 5 hearts) and therefore

$$P(A) = \frac{10^5}{40^5} \approx 0.1\,\%.$$

(3) In this case the drawing is without replacement and the event $A =$ "the numbers from 1 to 5 are obtained in order, of any suit, even different from each other" takes into account the order, so the natural sample space is $\Omega = \mathbf{D}_{40,5}$. We have that $A \leftrightarrow \mathbf{DR}_{4,5}$ (the ordered sequence of the suits of the 5 cards drawn is chosen) and therefore

$$P(A) = \frac{|\mathbf{DR}_{4,5}|}{|\mathbf{D}_{40,5}|} \approx 10^{-3}\,\%.$$

1.2.5   Binomial and Hypergeometric Probability

We present two fundamental examples that we will further explore, as they are closely connected to two significant probability measures: the binomial and hypergeometric distributions. We assume

$$0! = 1 \quad \text{and} \quad 0^0 = 1. \tag{1.2.6}$$

We recall that for $k, n \in \mathbb{N}_0$, with $k \leq n$,

$$\binom{n}{k} = \frac{n!}{k!(n-k)!}.$$

From the definition it follows directly that

$$\binom{n}{k} = \binom{n}{n-k}, \qquad \binom{n}{0} = \binom{n}{n} = 1, \qquad \binom{n}{1} = n.$$

Moreover, for $k, n \in \mathbb{N}$ with $k < n$, we have

$$\binom{n}{k} = \binom{n-1}{k-1} + \binom{n-1}{k}. \tag{1.2.7}$$

As an exercise, using (1.2.7) prove by induction the *binomial formula*[7] (or Newton's formula)

$$(a + b)^n = \sum_{k=0}^{n} \binom{n}{k} a^k b^{n-k}, \qquad a, b \in \mathbb{R}. \qquad (1.2.8)$$

As special cases of (1.2.8):

- if $a = b = 1$ we have

$$\sum_{k=0}^{n} \binom{n}{k} = 2^n. \qquad (1.2.9)$$

Remembering that if $|A| = n$ then $\binom{n}{k} = |C_{n,k}|$ is equal to the number of subsets of A of cardinality k, (1.2.9) shows that $|\mathscr{P}(A)| = 2^n$.

- remembering the convention (1.2.6) for the cases $p = 0$ and $p = 1$, we have

$$\sum_{k=0}^{n} \binom{n}{k} p^k (1 - p)^{n-k} = 1, \qquad p \in [0, 1]. \qquad (1.2.10)$$

In other words, letting for simplicity

$$p_k := \binom{n}{k} p^k (1 - p)^{n-k}, \qquad k = 0, \ldots, n,$$

then $p_0, \ldots, p_n$ are non-negative numbers with sum equal to 1. Therefore, setting $P(\{k\}) = p_k$, by Remark 1.1.13 we define a probability measure on the sample space $\Omega = \{0, \ldots, n\}$, called *binomial probability*.

We give an interpretation of the binomial probability in the following

Example 1.2.17 (Binomial) [!] We consider a urn containing b white balls and r red balls, with $b, r \in \mathbb{N}$, and perform n draws *with* replacement. Then we calculate the probability of the event A_k which consists in drawing exactly k white balls, with $0 \leq k \leq n$.

First, we determine the sample space: a priori the drawing order does not matter, but observing that there is replacement (i.e., the repetition of a possible ball already drawn), we are led to consider $\Omega = \mathbf{DR}_{b+r,n}$. The outcome ω can be identified

---

[7] An alternative proof, of a combinatorial nature, of Newton's formula is the following: the product $(a + b)(a + b) \cdots (a + b)$ of n factors expands into a sum of monomials of degree n of the type $a^{n-k} b^k$ with $0 \leq k \leq n$. How many monomials of a certain type (i.e., with k fixed) are there? The monomial $a^{n-k} b^k$ is obtained by choosing the value b from k of the n factors available in the product $(a + b)(a + b) \cdots (a + b)$ (and, therefore, choosing a from the remaining $n - k$), which is in $\binom{n}{k}$ ways.

with the k-tuple that identifies the sequence, ordered and with possible repetitions, of the balls drawn (assuming that the balls have been numbered to identify them). We characterize the generic outcome $\omega \in A_k$ through the following successive choices:

  (i) we choose the sequence (ordered and with possible repetitions) of the k white balls drawn from the b present in the urn: there are $|\mathbf{DR}_{b,k}|$ possible ways;
 (ii) we choose the sequence (ordered and with possible repetitions) of the $n-k$ red balls drawn from the r present in the urn: there are $|\mathbf{DR}_{r,n-k}|$ possible ways;
(iii) we choose in which of the n drawings the k white balls were drawn; there are $|\mathbf{C}_{n,k}|$ possible ways.[8]

Then we have

$$P(A_k) = |\mathbf{C}_{n,k}| \frac{|\mathbf{DR}_{b,k}||\mathbf{DR}_{r,n-k}|}{|\mathbf{DR}_{b+r,n}|} = \binom{n}{k} \frac{b^k r^{n-k}}{(b+r)^n},$$

or, equivalently,

$$P(A_k) = \binom{n}{k} p^k (1-p)^{n-k}, \qquad k = 0, 1, \ldots, n,$$

where $p = \frac{b}{b+r}$ is the probability of drawing a white ball, according to the uniform probability.

Remark 1.2.18 As we will explain better later, the binomial probability can be interpreted as *the probability of having k successes by repeating n times an experiment that has only two outcomes*: success with probability p and failure with probability $1-p$. For example, the probability of getting exactly k heads by flipping a coin n times is equal to $\binom{n}{k} p^k (1-p)^{n-k}$ with $p = \frac{1}{2}$, that is $\binom{n}{k} \frac{1}{2^n}$.

Example 1.2.19 (Hypergeometric) Consider an urn containing b white balls and r red balls, with $b, r \in \mathbb{N}$. We perform $n \leq b+r$ drawings *without* replacement. Let us calculate the probability of the event A_k which consists in drawing exactly k white balls, with $\max\{0, n-r\} \leq k \leq \min\{n, b\}$. The condition $\max\{0, n-r\} \leq k \leq \min\{n, b\}$ is equivalent to requiring that the following three conditions hold true simultaneously:

* $0 \leq k \leq n$;
* $k \leq b$, i.e., the number of white balls drawn does not exceed b;
* $n - k \leq r$, i.e., the number of red balls drawn does not exceed r.

---

[8] In fact, each subset of cardinality k of I_n identifies k drawings of the n, and vice versa. For example, if $n = 4$ and $k = 2$, the subset $\{2, 3\}$ of $I_4 = \{1, 2, 3, 4\}$ corresponds to the second and third drawing, and vice versa.

First, we determine the sample space: since the drawing order does not matter, we can consider $\Omega = \mathbf{C}_{b+r,n}$ (alternatively, we can choose $\Omega = \mathbf{D}_{b+r,n}$). The outcome ω corresponds to the *set* of balls drawn (assuming that the balls have been numbered to identify them). We characterize the generic outcome $\omega \in A_k$ through the following successive choices:

(i) we choose the k white balls drawn from the b present in the urn: there are $|\mathbf{C}_{b,k}|$ possible ways;

(ii) we choose the $n - k$ red balls drawn from the r present in the urn: there are $|\mathbf{C}_{r,n-k}|$ possible ways.

Then we have

$$P(A_k) = \frac{|\mathbf{C}_{b,k}||\mathbf{C}_{r,n-k}|}{|\mathbf{C}_{b+r,n}|} = \frac{\binom{b}{k}\binom{r}{n-k}}{\binom{b+r}{n}}, \qquad \max\{0, n - r\} \le k \le \min\{n, b\}.$$

1.2.6   Examples

We propose a series of examples useful for becoming acquainted with counting problems.

Example 1.2.20   Consider a group of $k \ge 2$ people born in the same year (of 365 days). Find the probability that at least two people in the group were born on the same day.

Solution   We can reformulate the problem as follows: an urn contains 365 balls numbered from 1 to 365; ball number N corresponds to the N-th day of the year; we draw with replacement k balls; what is the probability of drawing the same number twice? We have thus reduced the problem to drawing with replacement k balls from an urn containing 365 balls. We know that the natural sample space is $\Omega = \mathbf{DR}_{365,k}$. Let A be the event of interest, i.e., $A = $ "at least two people were born on the same day". Then $A^c \leftrightarrow \mathbf{D}_{365,k}$ and therefore

$$P(A) = 1 - P(A^c) = 1 - \frac{|\mathbf{D}_{365,k}|}{|\mathbf{DR}_{365,k}|} = 1 - \frac{365!}{(365 - k)! \cdot 365^k}.$$

One finds that $P(A) \approx 0.507 > \frac{1}{2}$ for $k = 23$ and $P(A) \approx 97\%$ for $k = 50$.

Example 1.2.21   Two cards are drawn (without replacement) from a deck of 40 cards identified by suit (hearts $\heartsuit$, diamonds $\diamondsuit$, clubs $\clubsuit$, spades $\spadesuit$) and number from 1 to 10. Find the probability of event A defined in each of the following ways:

(1)   the two cards are, in order, a heart card and a diamond card;

(2)   the two cards are, in order, a heart card and a 7;

(3)   the two cards are a heart card and a 7, regardless of the order.

Solution

(1) Let $\Omega = \mathbf{D}_{40,2}$. The outcome $\omega = (\omega_1, \omega_2)$ corresponds to the pair of cards drawn. We characterize the generic outcome $\omega = (\omega_1, \omega_2) \in A$ through the following successive choices:

(i) we choose the first card drawn (i.e., ω_1) among the heart cards: there are 10 possible choices;
(ii) we choose the second card drawn (i.e., ω_2) among the diamond cards: there are 10 possible choices.

In conclusion

$$P(A) = \frac{100}{|\mathbf{D}_{40,2}|} = \frac{5}{78} \approx 6.4\%.$$

If instead we did not take into account the order of drawing, we could consider, alternatively, the sample space $\Omega = \mathbf{C}_{40,2}$. In this case, the outcome $\omega = \{\omega_1, \omega_2\}$ corresponds to the set of cards drawn. So, proceeding as before,

$$\frac{100}{|\mathbf{C}_{40,2}|} = \frac{5}{39} = 2P(A).$$

(2) Let $\Omega = \mathbf{D}_{40,2}$. We cannot determine $|A|$ through the two successive choices (i)–(ii) of point (1), as proceeding in this way would also count the pair $(7\heartsuit, 7\heartsuit)$ which must be excluded since the cards are not replaced in the deck. Instead of directly applying the successive choices method to A, we note that A is the disjoint union of $A_1 = \mathbf{D}_{9,1} \times \mathbf{D}_{4,1}$ (the first card is a heart card different from 7 and the second card is one of the four 7s) and $A_2 = \mathbf{D}_{3,1}$ (the first card is the 7 of hearts and the second card is one of the remaining three 7s). Therefore

$$P(A) = P(A_1) + P(A_2) = \frac{9 \cdot 4}{|\mathbf{D}_{40,2}|} + \frac{3}{|\mathbf{D}_{40,2}|} = \frac{1}{40}.$$

(3) Since the order does not matter, $P(A)$ is double compared to case (2), so $P(A) = \frac{1}{20}$.

Example 1.2.22 Divide a deck of 40 cards into two decks of 20. Find the probability of event A defined in each of the following ways:

(1) the first deck contains exactly one 7;
(2) the first deck contains at least one 7.

Solution Let $\Omega = \mathbf{C}_{40,20}$. The outcome ω can be thought of as *the set* of cards in the first deck.

(1) We characterize the generic outcome $\omega \in A$ through the following successive choices:

    (i) we choose the only 7 that belongs to the first deck: there are 4 possible ways;

    (ii) we choose the remaining 19 cards of the first deck, which must not be 7s: there are $|\mathbf{C}_{36,19}|$ possible ways.

In conclusion

$$P(A) = \frac{4|\mathbf{C}_{36,19}|}{|\mathbf{C}_{40,20}|} = \frac{120}{481} \approx 25\%.$$

(2) We have

$$P(A) = 1 - P(A^c) = 1 - \frac{|\mathbf{C}_{36,20}|}{|\mathbf{C}_{40,20}|} \approx 95.7\%. \tag{1.2.11}$$

To better understand, let us see alternative ways to solve the problem: we could try to characterize the generic outcome $\omega \in A$ through the following successive choices:

    (i) we choose the 7 that belongs to the first deck: there are 4 possible ways;

    (ii) we choose the remaining 19 cards of the first deck among the remaining 39: there are $|\mathbf{C}_{39,19}|$ possible ways.

In this case, we would find

$$P(A) = \frac{4|\mathbf{C}_{39,19}|}{|\mathbf{C}_{40,20}|} = 2$$

which is obviously a wrong result. The error lies in the fact that the successive choices do not uniquely identify ω, in the sense that the same ω is "counted" more than once: for example, an ω containing $7\heartsuit$ and $7\diamondsuit$ is identified by choosing $7\heartsuit$ in choice i) and $7\diamondsuit$ in choice ii) but also by swapping the roles of $7\heartsuit$ and $7\diamondsuit$.

If we do not want to use the complementary event, we can alternatively calculate $|A|$ expressing A as the union of disjoint events $A_k =$"the first deck contains exactly a number k of 7s", for $k = 1, 2, 3, 4$. The generic outcome $\omega \in A_k$ is determined uniquely by the following successive choices:

    (i) among the 7s, we choose k that are those that belong to the first deck: there are $|\mathbf{C}_{4,k}|$ possible ways;

    (ii) we choose the remaining $20 - k$ of the first deck, different from 7: there are $|\mathbf{C}_{36,20-k}|$ possible ways.

Thus

$$P(A_k) = \frac{|\mathbf{C}_{4,k}||\mathbf{C}_{36,20-k}|}{|\mathbf{C}_{40,20}|}, \qquad k = 1, 2, 3, 4,$$

and as a final result, we obtain (1.2.11) again.

Example 1.2.23 From an urn containing b white balls and r red balls, with $b, r \in \mathbb{N}$, k balls are drawn without replacement, with $k \leq b + r$. Find the probability of the event B_k consisting of drawing a white ball at the k-th draw.

Solution Let $\Omega = \mathbf{D}_{b+r,k}$. The outcome ω can be identified with the *vector* indicating the ordered sequence and without repetitions of the k draws (assuming the balls have been numbered to identify them). Then

$$B_k \leftrightarrow \{(\omega_1, \ldots, \omega_k) \mid \omega_k \text{"white"}\}.$$

To determine $|B_k|$ we use the method of successive choices, characterizing a generic k-tuple $(\omega_1, \ldots, \omega_k)$ through the following scheme:

 (i) we choose the white ball of the k-th draw, i.e., ω_k: there are b possible ways;
 (ii) we choose the sequence (ordered and without repetitions) of the $k - 1$ previous draws: there are $|\mathbf{D}_{b+r-1,k-1}|$ possible ways.

Then, with $b + r = n$, we have

$$P(B_k) = \frac{b|\mathbf{D}_{n-1,k-1}|}{|\mathbf{D}_{n,k}|} = \frac{b\frac{(n-1)!}{(n-k)!}}{\frac{n!}{(n-k)!}} = \frac{b}{n}.$$

Thus $P(B_k) = \frac{b}{b+r}$ coincides with the probability of drawing a white ball at the first draw, i.e., $P(B_k) = P(B_1)$. This fact can be explained by observing that B_k is in bijective correspondence with the set $\{(\omega_1, \ldots, \omega_k) \mid \omega_1 \text{"white"}\}$.

Example 1.2.24 Consider a deck of 40 cards, from which k cards are drawn without replacement, with $k \leq 40$. Find the probability that a heart card is drawn at the k-th draw.

Solution This example is similar to the previous one: letting $\Omega = \mathbf{D}_{40,k}$ and $A_k = $ "a heart card is drawn at the k-th draw", the probability of A_k is given by

$$P(A_k) = \frac{10|\mathbf{D}_{39,k-1}|}{|\mathbf{D}_{40,k}|} = \frac{1}{4}.$$

Example 1.2.25 From an urn containing b white balls and r red balls, 2 balls are drawn with replacement. Find the probability of the event A defined in each of the following ways:

(1) the two balls have the same color;
(2) at least one of the two balls is red.

Solution Let $\Omega = \mathbf{DR}_{b+r,2}$. The outcome ω can be identified with the pair (ω_1, ω_2) indicating the ordered sequence (and with possible repetition) of the two draws (assuming the balls have been numbered to identify them).

(1) We have that A is the disjoint union of $A_1 = \mathbf{DR}_{b,2}$ (the two balls are white) and $A_2 = \mathbf{DR}_{r,2}$ (the two balls are red). Therefore

$$P(A) = P(A_1) + P(A_2) = \frac{|\mathbf{DR}_{b,2}|}{|\mathbf{DR}_{b+r,2}|} + \frac{|\mathbf{DR}_{r,2}|}{|\mathbf{DR}_{b+r,2}|} = \frac{b^2 + r^2}{(b+r)^2}.$$

(2) We have $P(A) = 1 - P(A^c)$ with $A^c = \mathbf{DR}_{b,2}$ (the two balls are white) and thus

$$P(A) = 1 - \frac{b^2}{(b+r)^2}.$$

Example 1.2.26 Consider a deck of poker cards with 52 cards, identified by the suit (hearts $\heartsuit$, diamonds $\diamondsuit$, clubs $\clubsuit$, spades $\spadesuit$) and the type (a number from 2 to 10 or J, Q, K, A). Find the probability of getting a served three-of-a-kind, i.e., of receiving from the dealer 5 cards of which 3 are of the same type, while the other two are of different types from each other and from the first three.

Solution Let $\Omega = \mathbf{C}_{52,5}$ and $A =$ "get a served three-of-a-kind". We characterize the generic outcome $\omega \in A$ through the following successive choices:

  (i) we choose the type of the cards that form the three-of-a-kind: there are 13 possible types;
 (ii) we choose the three suits of the three-of-a-kind: there are $|\mathbf{C}_{4,3}|$ possible choices;
(iii) we choose the types of the other 2 cards from the remaining 12 possible types: there are $|\mathbf{C}_{12,2}|$ possible choices;
(iv) we choose the suit of the other 2 cards from the 4 possible suits: there are $4 \cdot 4 = 16$ possible ways.

Then we have

$$P(A) = \frac{13 \cdot 4 \cdot |\mathbf{C}_{12,2}| \cdot 16}{|\mathbf{C}_{52,5}|} \approx 2.11\%.$$

As we said earlier, although most of the random experiments described by the discrete uniform probability can be formulated on one of the three sample spaces $\mathbf{DR}_{n,k}$, $\mathbf{D}_{n,k}$, $\mathbf{C}_{n,k}$, there are cases where this is not possible. However, it is always possible to decompose the random phenomenon into suitable sub-experiments, which can then be formulated using $\mathbf{DR}_{n,k}$, $\mathbf{D}_{n,k}$ or $\mathbf{C}_{n,k}$: this allows the initial random event to be described within their Cartesian product. Let's examine in more detail how to proceed with the following three examples.

Example 1.2.27 Consider a deck of 30 cards (for example, hearts, diamonds, and clubs). After dividing it into three decks of 10 cards each, calculate the probability of the event A defined in each of the following ways:

(1) the three aces are in different decks;
(2) the three aces are in the same deck.

Solution Let $\Omega = \mathbf{C}_{30,10} \times \mathbf{C}_{20,10}$: the outcome $\omega = (\omega_1, \omega_2)$ can be thought of as *the pair* in which ω_1 is the set of cards in the first deck and ω_2 is the set of cards in the second deck.

(1) Characterize the generic outcome $\omega \in A$ through the following successive choices:

   (i) choose the decks in which the aces are: there are $|\mathbf{P}_3| = 6$ possible ways;
   (ii) choose the remaining 9 cards of the first deck, which must not be aces: there are $|\mathbf{C}_{27,9}|$ possible ways;
   (iii) choose the remaining 9 cards of the second deck, which must not be aces: there are $|\mathbf{C}_{18,9}|$ possible ways.

   Then we have

$$P(A) = \frac{6|\mathbf{C}_{27,9}||\mathbf{C}_{18,9}|}{|\mathbf{C}_{30,10}||\mathbf{C}_{20,10}|} = \frac{50}{203} \approx 24.6\%.$$

(2) Similarly, characterize the generic outcome $\omega \in A$ through the following successive choices:

   (i) choose the deck in which the aces are: there are 3 possible ways;
   (ii) choose the remaining 7 cards of the deck in which the aces are, which must not be aces: there are $|\mathbf{C}_{27,7}|$ possible ways;
   (iii) choose the 10 cards of a second deck, which must not be aces: there are $|\mathbf{C}_{20,10}|$ possible ways.

   Then we have

$$P(A) = \frac{3|\mathbf{C}_{27,7}||\mathbf{C}_{20,10}|}{|\mathbf{C}_{30,10}||\mathbf{C}_{20,10}|} = \frac{18}{203} \approx 8.8\%.$$

Example 1.2.28 A fair coin is flipped ten times; then a ten-sided die, featuring integers from 1 to 10, is rolled. Find the probability of the event

$A = $ "the coin toss, determined by the outcome of the die, resulted in heads".

In other words, event A occurs if, after randomly choosing one of the 10 tosses (through the roll of the die), the result of that toss is heads.

Solution Intuitively the probability is $\frac{1}{2}$. Consider $\Omega = \mathbf{DR}_{2,10} \times I_{10}$ (note that instead of the set I_{10} it is possible to use indifferently $\mathbf{DR}_{10,1}$, $\mathbf{D}_{10,1}$ or $\mathbf{C}_{10,1}$, since $|I_{10}| = |\mathbf{DR}_{10,1}| = |\mathbf{D}_{10,1}| = |\mathbf{C}_{10,1}|$). The outcome $\omega = (\omega_1, \ldots, \omega_{10}, k)$ corresponds to the sequence $\omega_1, \ldots, \omega_{10}$ of the toss results and the choice k of the toss among the 10 made. Characterize the generic outcome $\omega \in A$ through the following successive choices:

(i) choose the number k of the toss: there are 10 possible values;
(ii) choose the result of the other 9 tosses: there are $|\mathbf{DR}_{2,9}|$ possible ways.

Then we have

$$P(A) = \frac{10|\mathbf{DR}_{2,9}|}{|\mathbf{DR}_{2,10} \times I_{10}|} = \frac{10 \cdot 2^9}{10 \cdot 2^{10}} = \frac{1}{2}.$$

Example 1.2.29

(i)   How many ways are there to arrange 3 coins (denote them as c_1, c_2 and c_3) in 10 boxes, where each box can hold only one coin?
(ii) Once the coins are arranged, what is the probability that the first box contains a coin?
(iii) Answer the previous questions when each box can contain at most 2 coins.

Solution

(1) We can imagine that the experiment takes place as follows: an urn contains 10 balls numbered from 1 to 10; each ball corresponds to a box (assuming that the boxes have also been numbered from 1 to 10); then three balls are drawn without replacement: the number of the i-th ball drawn indicates the box in which the coin c_i will be placed, with $i = 1, 2, 3$. We have thus reduced the experiment to drawing 3 balls without replacement from an urn containing 10. We know that the natural sample space is $\Omega = \mathbf{D}_{10,3}$. Question 1) asks to calculate the "total cases", that is, $|\mathbf{D}_{10,3}| = \frac{10!}{7!} = 720$.

(2) Intuitively (?) the probability is $\frac{3}{10}$. To prove it, let A be the event for which we want to calculate the probability, that is,

$A = $ "the first box contains a coin" $=$ "ball number 1 has been drawn".

We have that

$$P(A) = \frac{|A|}{|\mathbf{D}_{10,3}|} = \frac{|A|}{720}$$

or, alternatively,

$$P(A) = 1 - P(A^c) = 1 - \frac{|A^c|}{|\mathbf{D}_{10,3}|} = 1 - \frac{|A^c|}{720}.$$

It remains to determine $|A|$ or $|A^c|$. Note that A^c is the event in which the three coins are not placed in the first box and therefore is equivalent to placing the 3 coins in the remaining 9 boxes (equivalently, in the three draws from the urn, ball number 1 does not come out), that is, $A^c \leftrightarrow \mathbf{D}_{9,3}$. Therefore, $|A^c| = |\mathbf{D}_{9,3}|$, from which

$$P(A) = 1 - \frac{|\mathbf{D}_{9,3}|}{|\mathbf{D}_{10,3}|} = 1 - \frac{7}{10} = \frac{3}{10}.$$

Alternatively, $|A|$ can be determined by the method of successive choices proceeding as follows:

- choose the coin to put in the first box: 3 possible choices;
- choose where to place the remaining two coins in the remaining nine boxes: $|\mathbf{D}_{9,2}|$ possible ways.

Therefore, $|A| = 3|\mathbf{D}_{9,2}|$, so

$$P(A) = \frac{3|\mathbf{DR}_{9,2}|}{720} = \frac{3}{10}.$$

(3) Let $\Omega = \Omega_1 \uplus \Omega_2$, where:

- Ω_1 contains the "total cases" in which the first two coins are in the same box, and, consequently, the third coin is in one of the remaining nine boxes: there are $10 \cdot 9$ total cases of this type, so $|\Omega_1| = 10 \cdot 9$;
- Ω_2 contains the "total cases" in which the first two coins are in different boxes, while the third coin is in any of the ten boxes: there are $|\mathbf{D}_{10,2}| \cdot 10$ total cases of this type, so $|\Omega_2| = |\mathbf{D}_{10,2}| \cdot 10$.

Since $\Omega = \Omega_1 \uplus \Omega_2$, we have that

$$|\Omega| = |\Omega_1| + |\Omega_2| = 10 \cdot 9 + |\mathbf{D}_{10,2}| \cdot 10 = 990.$$

In summary, this section delves into discrete uniform probability, essentially defined as the ratio between "favorable cases" and "total cases". The computation of uniform probability boils down to a counting problem, solvable using combinatorial tools. Within this framework, the "method of successive choices" proves to be a valuable algorithm for tallying both "favorable cases" and "total cases". The most common errors made in using this method are:

- counting outcomes that do not exist (see Example 1.2.21);
- counting the same outcome more than once (see Example 1.2.22);
- not counting all outcomes.

We have also seen that, in the case of discrete uniform probability, it is often useful to rethink the random phenomenon as an experiment (or, possibly, a sequence of experiments) in which, from an urn that contains n distinct balls, k balls are drawn with replacement, without replacement or simultaneously. In this context, we have ultimately presented two noteworthy examples of probability: the so-called binomial and the hypergeometric distributions.

1.3   Conditional Probability and Independence of Events

The concepts of independence and conditional probability are fundamental to Probability theory. Up to this point, we have primarily reviewed some concepts from combinatorics and measure theory, interpreting them through a probabilistic lens. However, the introduction of independence and conditional probability brings forth entirely new and distinctive concepts within Probability theory: they allow us to analyze how the information regarding the occurrence of an event *affects the probability* of another event.

1.3.1   *Conditional Probability*

As already explained, Probability theory deals with phenomena whose outcome is uncertain: now uncertainty about a fact essentially means "partial or total lack of knowledge" of the fact itself. In other words, uncertainty is due to a *lack of information* about the phenomenon because it will happen in the future (for example, the price of a stock tomorrow) or because it has already happened but it was not possible to observe it (for example, the draw of a card that is not shown to us or the trajectory of an electron). Clearly, it can happen that some information becomes available and in this case the probability space that describes the phenomenon must be "updated" to take account of it. To this end, the concept of conditional probability is introduced. Let us first consider the following

Example 1.3.1 [!] From an urn containing 2 white balls and 2 black balls, two balls are drawn in sequence and without replacement:

  (i) calculate the probability that the second ball is white;
 (ii) knowing that the first ball drawn is black, calculate the probability that the second ball is white;
(iii) knowing that the second ball drawn is black, calculate the probability that the first ball is white.

Using combinatorics, it is quite easy to solve question (i). Let us consider the sample space $\Omega = \mathbf{D}_{4,2}$ of possible drawings, taking into account the order. Then $|\Omega| = |\mathbf{D}_{4,2}| = 12$ and the event $A =$ "the second ball is white" has 6 elements, so $P(A) = \frac{1}{2}$.

Question (ii) is elementary from an intuitive standpoint: since we have the information that the first ball drawn is black, in the second draw the urn is composed of two white balls and one black ball, and therefore the probability sought is $\frac{2}{3}$.

Conditionally on the information given, the event A now has a probability greater than $\frac{1}{2}$.

On the contrary, the last question does not seem to have an intuitive solution. One might think that the second draw does not affect the first because it occurs later, but this is not correct. Since we are given information about the second draw, we must think that the two draws have already taken place and in this case *the information about the outcome of the second draw affects the probability of the outcome of the first*: in fact, knowing that the second drawn ball is a black ball, it is as if in the first draw that black ball had been "reserved" and could not be drawn; so there are two chances out of three to draw a white ball. In fact, even using combinatorics, it is easy to prove that the probability sought is $\frac{2}{3}$.

Let us formalize the previous ideas.

Definition 1.3.2 (Conditional Probability) In a probability space $(\Omega, \mathscr{F}, P)$ let B be a non-negligible event, i.e., such that $P(B) > 0$. The *conditional probability of A given B* is defined by

$$P(A \mid B) := \frac{P(A \cap B)}{P(B)}, \qquad A \in \mathscr{F}. \tag{1.3.1}$$

Remark 1.3.3 Definition 1.3.2 is motivated in the following way: if we know that event B has occurred then the sample space is "reduced" from Ω to B and, conditionally on such information, it is natural to define the probability of A as in (1.3.1) since:

(i) only the outcomes of A that are also in B can occur;
(ii) since the new sample space is B, we must divide by $P(B)$ so that $P(B \mid B) = 1$.

Proposition 1.3.4 *In the probability space $(\Omega, \mathscr{F}, P)$ let B be a non-negligible event. We have:*

(i) *$P(\cdot \mid B)$ is a probability measure on $(\Omega, \mathscr{F})$;*
(ii) *if $A \cap B = \emptyset$ then $P(A \mid B) = 0$;*
(iii) *if $A \subseteq B$ then $P(A \mid B) = \frac{P(A)}{P(B)}$ and consequently $P(A \mid B) \geq P(A)$;*
(iv) *if $B \subseteq A$ then $P(A \mid B) = 1$;*
(v) *if $P(A) = 0$ then $P(A \mid B) = 0$.*

Proof The properties follow directly from Definition 1.3.2: proving the details is a very useful and instructive exercise.   □

Example 1.3.5 [!] Let us take up point (ii) of Example 1.3.1 and consider the events $B =$"the first ball drawn is black" and $A =$"the second ball drawn is white". Intuitively, we said that the conditional probability of A given B is equal to $\frac{2}{3}$: now we calculate $P(A \mid B)$ using Definition 1.3.2. Clearly $P(B) = \frac{1}{2}$, while in the sample space $\mathbf{D}_{4,2}$ there are 4 possible draws in which the first ball is black and the second is white, so $P(A \cap B) = \frac{4}{12} = \frac{1}{3}$. It follows that

$$P(A \mid B) = \frac{P(A \cap B)}{P(B)} = \frac{2}{3}$$

which confirms the intuitive result.

Now let us solve point (i) of Example 1.3.1 using the concept of conditional probability to avoid the use of combinatorics. The difficulty of the question lies in the fact that the result of the second draw depends on the result of the first draw and the latter is unknown: for this reason, at first glance, it seems impossible[9] to calculate the probability of event A. The idea is to partition the sample space and consider separately the cases in which B occurs or not to exploit the definition of conditional probability: we have already proved that $P(A \mid B) = \frac{2}{3}$ and similarly it can be seen that $P(A \mid B^c) = \frac{1}{3}$. Then we have

$$P(A) = P(A \cap B) + P(A \cap B^c)$$
$$= P(A \mid B)P(B) + P(A \mid B^c)P(B^c)$$
$$= \frac{2}{3} \cdot \frac{1}{2} + \frac{1}{3} \cdot \frac{1}{2} = \frac{1}{2}$$

which confirms what we have already seen.

Proposition 1.3.6 (Law of Total Probability) *[!] For every event B such that $0 < P(B) < 1$, we have*

$$P(A) = P(A \mid B)P(B) + P(A \mid B^c)(1 - P(B)), \qquad A \in \mathscr{F}. \qquad (1.3.2)$$

More generally, if $(B_i)_{i \in I}$ is a finite or countable partition[10] of Ω, with $P(B_i) > 0$ for each $i \in I$, then we have

$$P(A) = \sum_{i \in I} P(A \mid B_i)P(B_i), \qquad A \in \mathscr{F} \qquad (1.3.3)$$

Proof We prove (1.3.3), of which (1.3.2) is a particular case. Since

$$A = \biguplus_{i \in I}(A \cap B_i),$$

---

[9] A survey carried out in the fourth year of some high schools in Italy has highlighted a significant number of students who, faced with this question, have answered that *it is not possible* to calculate the probability of event A. To challenge this kind of conviction, students can be pointed out that there is no reason why black balls have a higher probability of being drawn second and therefore intuitively $P(A)$ must be equal to $\frac{1}{2}$.

[10] That is, $(B_i)_{i \in I}$ is a family of events that are pairwise disjoint, and their union is equal to Ω. Sometimes $(B_i)_{i \in I}$ is called a *system of alternatives*.

by the σ-additivity of P we have

$$P(A) = \sum_{i \in I} P(A \cap B_i) = \sum_{i \in I} P(A \mid B_i) P(B_i).$$

$\square$

We consider a typical example of application of the Law of total probability.

Example 1.3.7 We have two urns: urn α contains 3 white balls and 1 red ball; urn β contains 1 white ball and 1 red ball. Find the probability that, choosing an urn at random and drawing a ball, it is white.

First Solution Let A be the event for which we want to calculate the probability and B the event in which urn α is chosen. It seems natural to set

$$P(B) = \frac{1}{2}, \qquad P(A \mid B) = \frac{3}{4}, \qquad P(A \mid B^c) = \frac{1}{2}.$$

Then, by (1.3.2) we obtain

$$P(A) = \frac{3}{4} \cdot \frac{1}{2} + \frac{1}{2} \cdot \frac{1}{2} = \frac{5}{8}.$$

Note that we have formally calculated $P(A)$ without even specifying the probability space!

Second Solution We give another, more detailed solution: let

$$\Omega = \{\alpha b_1, \alpha b_2, \alpha b_3, \alpha r, \beta b, \beta r\}$$

where αb_1 is the outcome in which the first urn is chosen and the first white ball is drawn, and the other outcomes are defined similarly. Clearly

$$A = \{\alpha b_1, \alpha b_2, \alpha b_3, \beta b\}$$

and in this case the correct probability to use is not the uniform one on Ω. In fact, B, the event in which urn α is chosen, must have probability $\frac{1}{2}$ and the elements of B are equiprobable: it follows that $P(\{\omega\}) = \frac{1}{8}$ for each $\omega \in B$. Similarly $P(B^c) = \frac{1}{2}$ and the elements of B^c are equiprobable, so

$$P(\{\beta b\}) = P(\{\beta r\}) = \frac{1}{4}.$$

We can then calculate

$$P(A) = P(\{\alpha b_1\}) + P(\{\alpha b_2\}) + P(\{\alpha b_3\}) + P(\{\beta b\}) = \frac{5}{8}$$

in agreement with what was previously found.

Exercise 1.3.8 A die is rolled and then a coin is flipped a number of times equal to the result of the die roll. What is the probability of getting exactly two heads?

Example 1.3.9 An urn contains 6 white balls and 4 black balls. Drawing 2 balls without replacement, what is the probability that both are white (event A)?

We can interpret the question as a counting problem, using the uniform probability P on the space $\Omega = \mathbf{C}_{10,2}$ of combinations of two balls drawn from the 10 available. Then we have

$$P(A) = \frac{|\mathbf{C}_{6,2}|}{|\mathbf{C}_{10,2}|} = \frac{\frac{6!}{2!4!}}{\frac{10!}{2!8!}} = \frac{6 \cdot 5}{10 \cdot 9}. \tag{1.3.4}$$

Now note that $\frac{6}{10} = P(A_1)$ where A_1 is the event "the first ball drawn is white". On the other hand, if A_2 is the event "the second ball drawn is white", then $\frac{5}{9}$ is the conditional probability of A_2 given A_1, that is $\frac{5}{9} = P(A_2 \mid A_1)$. Thus, observing also that $A = A_1 \cap A_2$, (1.3.4) is equivalent to

$$P(A_1 \cap A_2) = P(A_1)P(A_2 \mid A_1)$$

and so we find precisely formula (1.3.1) that defines the conditional probability.

More generally, from the definition of conditional probability, we directly obtain the following useful result.

Proposition 1.3.10 (Multiplication Rule) *[!] Let $A_1, \ldots, A_n$ be events such that $P(A_1 \cap \cdots \cap A_{n-1}) > 0$. Then we have*

$$P(A_1 \cap \cdots \cap A_n) = P(A_1)P(A_2 \mid A_1) \cdots P(A_n \mid A_1 \cap \cdots \cap A_{n-1}) \tag{1.3.5}$$

Exercise 1.3.11 Use formula (1.3.5) to calculate the probability that, when drawing 3 cards from a deck of 40, the value of each card is not greater than 5.

Solution Let A_i, $i = 1, 2, 3$, denote the event "the i-th card drawn is less than or equal to 5". The sought-after probability is equal to

$$P(A_1 \cap A_2 \cap A_3) = P(A_1)P(A_2 \mid A_1)P(A_3 \mid A_1 \cap A_2) = \frac{20}{40} \cdot \frac{19}{39} \cdot \frac{18}{38}.$$

Solving the exercise as a counting problem, we would find the equivalent solution $\frac{|\mathbf{C}_{20,3}|}{|\mathbf{C}_{40,3}|}$.

Example 1.3.12   We calculate the probability of getting a pair in the lottery with the numbers 1 and 3 (event A), knowing that the drawing has already taken place and three of the five drawn numbers are odd (event B).

Solution   Let $\Omega = \mathbf{C}_{90,5}$: the outcome $\omega = \omega_1, \ldots, \omega_5$ can be thought of as the *set* of the drawn numbers. We have $\omega \in A$ if $1, 3 \in \omega$, hence $A \leftrightarrow \mathbf{C}_{88,3}$. Moreover, $B \leftrightarrow \mathbf{C}_{45,3} \times \mathbf{C}_{45,2}$ (corresponding to the choice of three odd numbers and two even numbers from 90), and $A \cap B \leftrightarrow \mathbf{C}_{43,1} \times \mathbf{C}_{45,2}$ (corresponding to the choice of the third odd number, in addition to 1 and 3, and two even numbers from 90). Therefore, we have

$$P(A) = \frac{|\mathbf{C}_{88,3}|}{|\mathbf{C}_{90,5}|} \approx 0.25\% \qquad \text{and} \qquad P(A \mid B) = \frac{43|\mathbf{C}_{45,2}|}{|\mathbf{C}_{45,3}||\mathbf{C}_{45,2}|} \approx 0.3\%.$$

Remark 1.3.13   According to the Law of total probability (1.3.2), if $0 < P(B) < 1$, *we can uniquely determine $P(A)$ from $P(B)$, $P(A \mid B)$, and $P(A \mid B^c)$.* We also note that Eq. (1.3.2) implies that $P(A)$ *belongs to the interval with endpoints $P(A \mid B)$ and $P(A \mid B^c)$*: therefore, regardless of the knowledge of $P(B)$, $P(A \mid B)$ and $P(A \mid B^c)$ provide estimates of the value of $P(A)$. In particular, if $P(A \mid B) = P(A \mid B^c)$, then we also have $P(A) = P(A \mid B)$ or equivalently $P(A \cap B) = P(A)P(B)$.

We now consider the problem of assessment of students' opinions on the quality of teaching, carried out at some universities. We define the following random events:

- A = "a professor receives a positive evaluation in the survey of students' opinions";
- B = "a professor is "good" (assuming we know what that means) at teaching".

Generally, events A and B do not coincide. Therefore, we can interpret the conditional probabilities $P(A \mid B)$ and $P(B \mid A)$ as follows:

- $P(A \mid B)$ is the probability that a "good" professor receives a positive evaluation;
- $P(B \mid A)$ is the probability that a professor who receives a positive evaluation is "good".

Reflecting carefully on the meaning of these two conditional probabilities, it becomes clear that sometimes we may be interested in deriving one from the knowledge of the other. Actually, we may have a general estimate (based on historical data) of $P(A \mid B)$ and be interested in knowing $P(B \mid A)$ based on the recently conducted survey. A solution to this problem is provided by the classic Bayes' theorem.

Theorem 1.3.14 (Bayes' Formula) [!]   *For any non-negligible events A, B we have*

$$P(B \mid A) = \frac{P(A \mid B)P(B)}{P(A)} \qquad\qquad (1.3.6)$$

Proof Equation (1.3.6) is equivalent to

$$P(B \mid A)P(A) = P(A \mid B)P(B)$$

and directly follows from the definition of conditional probability.   □

Example 1.3.15 Let us revisit Example 1.3.7: knowing that a white ball has been drawn, what is the probability that urn α was chosen?

Solution As before, we denote by A the event "a white ball is drawn" and by B the event "urn α is chosen". We had already calculated $P(A) = \frac{5}{8}$, while we assume $P(A \mid B) = \frac{3}{4}$ and $P(B) = \frac{1}{2}$. Then, by Bayes' formula, we have

$$P(B \mid A) = \frac{P(A \mid B)P(B)}{P(A)} = \frac{3}{5}.$$

Exercise 1.3.16 Assuming $P(A \mid B) \neq P(A \mid B^c)$, prove that

$$P(B) = \frac{P(A) - P(A \mid B^c)}{P(A \mid B) - P(A \mid B^c)}, \tag{1.3.7}$$

and thus it is possible to uniquely determine $P(B)$ from $P(A)$, $P(A \mid B)$, and $P(A \mid B^c)$.

Exercise 1.3.17 (Teaching Assessment) Suppose that on a historical basis "good" professors receive a positive evaluation in 95% of cases, while "less good" professors receive a positive evaluation in 10% of cases (some professors are sly...). If 80% of evaluations are positive, what is the probability that:

 (i) professors who received a positive evaluation are truly "good"?
(ii) professors who received a negative evaluation are actually "good"?

Notice that by combining Bayes' theorem with Eq. (1.3.7), we obtain

$$P(B \mid A) = \frac{P(A \mid B)P(B)}{P(A)} = \frac{P(A \mid B)\,(P(A) - P(A \mid B^c))}{P(A)\,(P(A \mid B) - P(A \mid B^c))}.$$

Solution As before, we set

- $A =$ "a professor receives a positive evaluation from students";
- $B =$ "a professor is good at teaching".

We are given the following probabilities: $P(A \mid B) = 0.95$, $P(A \mid B^c) = 0.10$ and $P(A) = 0.80$.

(i) We want to find the probability that a professor who received a positive evaluation is actually "good", which is $P(B \mid A)$. Using Bayes' formula, we have:

$$P(B \mid A) = \frac{P(A \mid B)P(B)}{P(A)}.$$

We don't have the value of $P(B)$, but we have

$$P(B) = \frac{P(A) - P(A \mid B^c)}{P(A \mid B) - P(A \mid B^c)} = \frac{0.80 - 0.10}{0.95 - 0.10} = \frac{0.70}{0.85} \approx 0.8235.$$

So we get

$$P(B \mid A) = \frac{P(A \mid B)P(B)}{P(A)} = \frac{0.95 \cdot 0.8235}{0.80} \approx 0.9768.$$

(ii) We want to find the probability that a professor who received a negative evaluation is actually "good", which is $P(B \mid A^c)$. Using Bayes' formula, we have:

$$P(B \mid A^c) = \frac{P(A^c \mid B)P(B)}{P(A^c)}.$$

We can infer $P(A^c \mid B)$ and $P(A^c)$ from the given probabilities:

$$P(A^c \mid B) = 1 - P(A \mid B) = 1 - 0.95 = 0.05,$$
$$P(A^c) = 1 - P(A) = 1 - 0.80 = 0.20.$$

Now we can find

$$P(B \mid A^c) = \frac{P(A^c \mid B)P(B)}{P(A^c)} = \frac{0.05 \cdot 0.8235}{0.20} \approx 0.2059.$$

Observe that, combining Bayes' formula with formula (1.3.7), we obtain

$$P(B \mid A) = \frac{P(A \mid B)P(B)}{P(A)} = \frac{P(A \mid B)\,(P(A) - P(A \mid B^c))}{P(A)\,(P(A \mid B) - P(A \mid B^c))}.$$

1.3.2   Independence

Definition 1.3.18 In a probability space $(\Omega, \mathscr{F}, P)$, we say that two events A, B are independent under P if

$$P(A \cap B) = P(A)P(B). \tag{1.3.8}$$

The concept of independence *depends on the given* probability measure.[11] It expresses the fact that *the information on the occurrence of event B does not influence the probability of A:* in fact, if $P(B) > 0$, then (1.3.8) is equivalent to

$$P(A \mid B) = P(A),$$

that is

$$\frac{P(A \cap B)}{P(B)} = \frac{P(A)}{P(\Omega)}$$

which can be interpreted as a proportionality relationship

$$P(A \cap B) : P(B) = P(A) : P(\Omega).$$

Similarly, if

$$P(A \cap B) > P(A)P(B) \tag{1.3.9}$$

then A, B are said to be *positively correlated under P* since (1.3.9) implies[12]

$$P(A \mid B) > P(A), \qquad P(B \mid A) > P(B),$$

that is, the probability of A *increases conditionally on the information on the occurrence of B* and vice versa.

Remark 1.3.19 Clearly, the fact that A, B are independent does not mean that they are disjoint, on the contrary: if $P(A) > 0$, $P(B) > 0$ and (1.3.8) holds, then also $P(A \cap B) > 0$ and therefore $A \cap B \neq \emptyset$. On the other hand, if $P(A) = 0$ then also $P(A \cap B) = 0$ (due to (1.1.5) and the fact that $A \cap B \subseteq A$) and therefore (1.3.8) holds for every B, that is, A is independent of every event B.

---

[11] Sometimes it is necessary to explicitly declare the probability measure being considered. In fact, in applications, several probability measures may be involved simultaneously: two events that are independent under one probability measure may not necessarily be independent under another measure.

[12] When A, B are not negligible under P.

Remark 1.3.20 We have defined the concept of independence but not that of *dependence*. If two events A, B are not independent *we do not say that they are dependent*: we will later define a concept of dependence that is well distinct and somehow unrelated to that of independence.

Example 1.3.21 Two athletes have probability of 70% and 80% of breaking a record in a race, respectively. What is the probability that at least one of the two breaks the record?

If A is the event "the first athlete breaks the record", B is the event "the second athlete breaks the record" and we assume that A and B are independent, then we have

$$P(A \cup B) = P(A) + P(B) - P(A \cap B) =$$

(due to independence)

$$= P(A) + P(B) - P(A)P(B)$$

$$= 150\% - 70\% \cdot 80\% = 94\%.$$

Example 1.3.22 The fact that two events are independent does not mean that "they have nothing to do with each other". Consider the throw of two dice and the events "the sum of the throws is 7" (event A) and "the result of the first throw is 3" (event B). Then A and B are independent under the uniform probability.

Example 1.3.23 We will soon see that the concept of independence is natural for describing an experiment that is repeated in such a way that each repetition does not influence the probability of the other repetitions (for example, a sequence of rolls of a die or coin tosses). In this case, it is natural to use a sample space that is a Cartesian product. For example, let $\Omega = \Omega_1 \times \Omega_2$ be finite, endowed with the uniform probability P: consider $A = E_1 \times \Omega_2$ and $B = \Omega_1 \times E_2$ with $E_i \subseteq \Omega_i$, $i = 1, 2$. Then

$$P(A \cap B) = P(E_1 \times E_2) = \frac{|E_1||E_2|}{|\Omega|} = \frac{|E_1 \times \Omega_2||\Omega_1 \times E_2|}{|\Omega|^2} = P(A)P(B)$$

and therefore A and B are independent under P. We will deepen the link between the concepts of independence and product of measures starting from Sect. 2.3.

Exercise 1.3.24 At the cinema, two people α, β decide which film to watch, among two available, independently and with the following probabilities:

$$P(\alpha_1) = \frac{1}{3}, \qquad P(\beta_1) = \frac{1}{4}$$

where α_1 indicates the event "α chooses the first film". Find the probability that α and β watch the same film (event A).

First Solution We have

$$P(A) = P(\alpha_1 \cap \beta_1) + P(\alpha_2 \cap \beta_2) =$$

(by independence and since $P(\alpha_2) = 1 - P(\alpha_1)$)

$$= P(\alpha_1)P(\beta_1) + P(\alpha_2)P(\beta_2) = \frac{7}{12}.$$

This simple example shows that it is possible to calculate the probability of an event that depends on independent events, starting from the knowledge of the probabilities of the individual events, and most importantly, without the need to explicitly construct the probability space.

Second Solution It is also useful to proceed in the "classical" way, solving the exercise as a counting problem: in this case we must first construct the sample space

$$\Omega = \{(1, 1), (1, 2), (2, 1), (2, 2)\}$$

where (i, j) indicates the outcome "α chooses movie i and β chooses movie j" with $i, j = 1, 2$. By assumption, we know the probabilities of the events

$$\alpha_1 = \{(1, 1), (1, 2)\}, \qquad \beta_1 = \{(1, 1), (2, 1)\},$$

however, this is not enough to uniquely determine the probability P, that is, to determine the probabilities of the individual outcomes. In fact, to do this, it is necessary to use also the hypothesis of independence (under P) of α_1 and β_1, from which we obtain, for example,

$$P(\{(1, 1)\}) = P(\alpha_1 \cap \beta_1) = P(\alpha_1)P(\beta_1) = \frac{1}{12}.$$

Similarly, we can calculate all the probabilities of the outcomes and consequently solve the problem. We note that this counting-based procedure is more involved and less intuitive.

Proposition 1.3.25 *If A, B are independent then A, B^c are also independent.*

Proof We have

$$P(A \cap B^c) = P(A \setminus B) = P(A \setminus (A \cap B)) =$$

(by (1.1.6))

$$= P(A) - P(A \cap B) =$$

(by the hypothesis of independence of A, B)

$$= P(A) - P(A)P(B) = P(A)P(B^c).$$

$\square$

Exercise 1.3.26 At the cinema, two people α, β decide which movie to watch among three available, in the following way:

(i) α chooses a movie at random with the following probabilities

$$P(\alpha_1) = \frac{1}{2}, \qquad P(\alpha_2) = \frac{1}{3}, \qquad P(\alpha_3) = \frac{1}{6}$$

where α_i indicates the event "α chooses the i-th movie" for $i = 1, 2, 3$;

(ii) β flips a coin and if the result is "heads" then he chooses the same movie as α, otherwise he chooses a movie at random, independently of α.

Find the probability $P(A)$ where A is the event "α and β watch the same movie".

Solution Let T denote the event "the result of the coin toss is heads". We have $P(T) = \frac{1}{2}$ and by hypothesis $P(A \mid T) = 1$ and $P(\beta_i \mid T^c) = \frac{1}{3}$ for $i = 1, 2, 3$. Moreover, since $P(\cdot \mid T^c)$ is a probability measure, we have

$$P(A \mid T^c) = \sum_{i=1}^{3} P(\alpha_i \cap \beta_i \mid T^c) =$$

$$= \sum_{i=1}^{3} P(\alpha_i \mid T^c)P(\beta_i \mid T^c)$$

$$= \frac{1}{3} \sum_{i=1}^{3} P(\alpha_i \mid T^c) = \frac{1}{3},$$

since $\sum_{i=1}^{3} P(\alpha_i \mid T^c) = 1$ being $P(\cdot \mid T^c)$ a probability measure. Then, by (1.3.2), we have

$$P(A) = P(A \mid T)P(T) + P(A \mid T^c)(1 - P(T)) = 1 \cdot \frac{1}{2} + \frac{1}{3} \cdot \frac{1}{2} = \frac{2}{3}.$$

We leave the reader to calculate the probability that α and β choose the first movie, that is, $P(\alpha_1 \cap \beta_1)$.

Next, we consider the case of more than two events.

Definition 1.3.27 Let $(A_i)_{i \in I}$ be a family of events. We say that these events are independent if

$$P\left(\bigcap_{j \in J} A_j\right) = \prod_{j \in J} P(A_j)$$

for every $J \subseteq I$, with J finite.

Consider three events A, B, C: Exercises 1.3.41 and 1.3.42 show that in general *there is no implication* between the property

$$P(A \cap B \cap C) = P(A)P(B)P(C) \qquad\qquad (1.3.10)$$

and the properties

$$P(A \cap B) = P(A)P(B), \quad P(A \cap C) = P(A)P(C), \quad P(B \cap C) = P(B)P(C). \qquad (1.3.11)$$

In particular, a family of pairwise independent events is not generally a family of independent events.

We conclude the section with a remarkable result. Given a sequence of events $(A_n)_{n \geq 1}$, we set[13]

$$(A_n \text{ i.o.}) := \bigcap_{n \geq 1} \bigcup_{k \geq n} A_k.$$

Note that

$$(A_n \text{ i.o.}) = \{\omega \in \Omega \mid \forall n \in \mathbb{N} \ \exists k \geq n \text{ such that } \omega \in A_k\},$$

that is, $(A_n \text{ i.o.})$ is the event consisting of all $\omega \in \Omega$ that *belong to an infinite number* of A_n.

Lemma 1.3.28 (Borel-Cantelli) *[!] Let $(A_n)_{n \geq 1}$ be a sequence of events in the space $(\Omega, \mathscr{F}, P)$:*

(i) if

$$\sum_{n \geq 1} P(A_n) < +\infty$$

then $P(A_n \text{ i.o.}) = 0$;

---

[13] i.o. stands for *infinitely often*.

(ii) if the sets A_n are independent and

$$\sum_{n \geq 1} P(A_n) = +\infty$$

then $P\,(A_n\ i.o.) = 1$.

Proof By the continuity from above of P we have

$$P\,(A_n\ i.o.) = \lim_{n \to \infty} P\left(\bigcup_{k \geq n} A_k\right) \leq$$

(by σ-subadditivity, Proposition 1.1.22-(ii))

$$\leq \lim_{n \to \infty} \sum_{k \geq n} P\,(A_k) = 0$$

by hypothesis. This proves the first part of the thesis.

As for (ii), we prove that

$$P\left(\bigcup_{k \geq n} A_k\right) = 1 \tag{1.3.12}$$

for every $n \in \mathbb{N}$, from which the thesis will follow. Fixed n, N with $n \leq N$, we have

$$P\left(\bigcup_{k=n}^{N} A_k\right) = 1 - P\left(\bigcap_{k=n}^{N} A_k^c\right) =$$

(by independence)

$$= 1 - \prod_{k=n}^{N} (1 - P(A_k)) \geq$$

(by the elementary inequality $1 - x \leq e^{-x}$ valid for $x \in \mathbb{R}$)

$$\geq 1 - \exp\left(-\sum_{k=n}^{N} P(A_k)\right).$$

Then (1.3.12) follows by taking the limit as $N \to \infty$. $\square$

In conclusion, *conditional probability* and *independence* are the first truly new concepts exclusive to Probability theory and not encountered in other mathematically "related" theories like measure theory or combinatorics.

The purpose of both concepts is to express the probability $P(A \cap B)$ in terms of probabilities of the individual events A and B. This is obviously possible if A, B are independent under P since in this case we have

$$P(A \cap B) = P(A)P(B),$$

while in general we have

$$P(A \cap B) = P(A \mid B)P(B).$$

Many problems are solved much more easily using the previous identities (and other useful formulas such as the Law of total probability, the multiplication rule, and Bayes' formula) instead of combinatorics.

1.3.3   *Repeated Independent Trials*

Definition 1.3.29 [!] In a probability space $(\Omega, \mathscr{F}, P)$, let $(C_h)_{h=1,\dots,n}$ be a finite family of independent and equiprobable events, that is, such that $P(C_h) = p \in [0, 1]$ for each $h = 1, \dots, n$. Then we say that $(C_h)_{h=1,\dots,n}$ is a *family of n repeated independent trials with probability p*.

Intuitively, we can imagine repeating an experiment n times that can have two outcomes, success or failure: C_h represents the event "the h-th experiment is successful". For example, in a sequence of n coin tosses, C_h can represent the event "at the h-th toss we get *heads*".

For each $n \in \mathbb{N}$ and $p \in [0, 1]$, it is always possible to construct *a discrete space* (Ω, P) on which a family $(C_h)_{h=1,\dots,n}$ of n repeated independent trials with probability p is defined. The following result also shows that on a discrete probability space *it is not possible to define a sequence* $(C_h)_{h\in\mathbb{N}}$ *of* repeated independent trials unless it is trivial, that is, with $p = 0$ or $p = 1$.

Proposition 1.3.30 *For each* $n \in \mathbb{N}$ *and* $p \in [0, 1]$, *there exists a discrete space* (Ω, P) *on which a family* $(C_h)_{h=1,\dots,n}$ *of n repeated independent trials with probability p is canonically defined.*

If $(C_h)_{h\in\mathbb{N}}$ *is a sequence of independent events on a discrete space* (Ω, P), *such that* $P(C_h) = p \in [0, 1]$ *for each* $h \in \mathbb{N}$, *then necessarily* $p = 0$ *or* $p = 1$.

Proof See Sect. 1.5.1.   □

We examine two significant examples.

Example 1.3.31 (Probability of First Success at Trial k) [!] Let $(C_h)_{h=1,\dots,n}$ be a family of n repeated independent trials with probability p. The event "the first success is at the k-th trial" is defined as

$$A_k := C_1^c \cap C_2^c \cap \cdots \cap C_{k-1}^c \cap C_k, \qquad 1 \le k \le n.$$

By independence, we have

$$P(A_k) = (1-p)^{k-1}p, \qquad 1 \le k \le n. \tag{1.3.13}$$

For example, A_k represents the event in which, in a sequence of n coin tosses, heads are obtained for the first time at the k-th toss. We note that $P(A_k)$ in (1.3.13) *does not depend on n*: intuitively, A_k depends only on what happened up to the k-th trial and is independent of the total number n of trials.

Example 1.3.32 (Probability of k Successes in n Trials) [!] Consider a family $(C_h)_{h=1,\dots,n}$ of n repeated independent trials with probability p. We calculate the probability of the event A_k "exactly k trials are successful".

First Method Referring to the canonical space of Proposition 1.3.30 and in particular to formula (1.5.1), we have $A_k = \Omega_k$. Therefore

$$P(A_k) = \sum_{\omega \in \Omega_k} P(\{\omega\}) = |\Omega_k| p^k (1-p)^{n-k} = \binom{n}{k} p^k (1-p)^{n-k}, \qquad 0 \le k \le n.$$

We will see that $P(A_k)$ is related to the concept of *binomial distribution* in Example 1.4.17.

Second Method The event A_k is of the type

$$C_{i_1} \cap \cdots \cap C_{i_k} \cap C_{i_{k+1}}^c \cdots \cap C_{i_n}^c$$

as $\{i_1, \dots, i_k\}$, family of indices of I_n, varies: the possible choices of such indices are exactly $|\mathbf{C}_{n,k}|$. Moreover, by independence, we have

$$P\left(C_{i_1} \cap \cdots \cap C_{i_k} \cap C_{i_{k+1}}^c \cdots \cap C_{i_n}^c\right) = p^k (1-p)^{n-k}$$

and thus we find

$$P(A_k) = \binom{n}{k} p^k (1-p)^{n-k}, \qquad 0 \le k \le n. \tag{1.3.14}$$

Remark 1.3.33 Reconsider Example 1.2.17 related to calculating the probability of drawing (with replacement) exactly k white balls from an urn containing b white and r red balls. If C_h is the event "the ball of the h-th draw is white" then $p =$

$P(C_h) = \frac{b}{b+r}$ and (1.3.14) provides the desired probability, in agreement with what we had obtained in Example 1.2.17 through combinatorics.

Note that in the approach based on combinatorics, the *uniform probability* is used, as always in counting problems. Instead, in the approach based on the family of repeated independent trials, we implicitly use the canonical space of Proposition 1.3.30 without, however, the need to explicitly declare the sample space and the probability measure (which, in any case, is not the uniform one).

1.3.4 Examples

We propose some summary examples and review exercises on the concepts of independence and conditional probability.

Example 1.3.34

- Mr. Smith has two children: what is the probability that both children are male (event A)?

  Considering the sample space

$$\Omega = \{(M, M), (M, F), (F, M), (F, F)\} \tag{1.3.15}$$

with obvious meaning of the symbols, it is clear that $P(A) = \frac{1}{4}$. The situation is summarized in the following table where the cells represent the four possible cases and the corresponding probabilities are indicated inside the circles: we have $A = \{(M, M)\}$.

	Male	Female
Male	(M, M) $\left(\frac{1}{4}\right)$	(M, F) $\left(\frac{1}{4}\right)$
Female	(F, M) $\left(\frac{1}{4}\right)$	(F, F) $\left(\frac{1}{4}\right)$

- Mr. Smith has two children. Knowing that one of them is male (event B), what is the probability that both children are male?

  The "intuitive" answer (the probability is equal to $\frac{1}{2}$) is unfortunately wrong. To realize this, it suffices to consider again the sample space Ω: now, having the information that (F, F) is not possible (i.e., it has zero probability "conditionally" on the given information that is the occurrence of event B) and assuming that the outcomes $(M, M), (M, F), (F, M)$ are equally likely, it follows that the desired probability is equal to $\frac{1}{3}$. The following table shows how the probability is redistributed conditionally on the information that B occurs.

	Maschio	Femmina
Maschio	$(M, M)\left(\frac{1}{3}\right)$	$(M, F)\left(\frac{1}{3}\right)$
Femmina	$(F, M)\left(\frac{1}{3}\right)$	$(F, F)\left(0\right)$

- Mr. Smith has two children. Knowing that the firstborn is male (event C, different from B of the previous point), what is the probability that both children are male? The "intuitive" answer (the probability is equal to $\frac{1}{2}$) is correct because in this case FM and FF both have zero probability (conditionally on the given information that the event C occurs). In other words, knowing that the firstborn is male, everything depends on whether the second child is male or female, i.e., on two equiprobable events with probability equal to $\frac{1}{2}$. The following table shows how the probability is redistributed conditionally on the information that C occurs.

	Male	Female
Male	$(M, M)\left(\frac{1}{2}\right)$	$(M, F)\left(\frac{1}{2}\right)$
Female	$(F, M)\left(0\right)$	$(F, F)\left(0\right)$

Let P denote the uniform probability on Ω in (1.3.15). We have

$$P(A) = P(\{MM\}) = \frac{1}{4}, \quad P(B) = P(\{MM, MF, FM\}) = \frac{3}{4},$$

$$P(C) = P(\{MM, MF\}) = \frac{1}{2},$$

and therefore, according to Definition 1.3.2, we have

$$P(A \mid B) = \frac{P(A)}{P(B)} = \frac{1}{3}, \qquad P(A \mid C) = \frac{P(A)}{P(C)} = \frac{1}{2},$$

in agreement with what we had conjectured above intuitively.

Exercise 1.3.35 Prove Proposition 1.3.4.

Exercise 1.3.36 Using Bayes' formula, prove that

$$P(B \mid A) = \frac{P(A \mid B)P(B)}{P(A \mid B)P(B) + P(A \mid B^c)(1 - P(B))} \tag{1.3.16}$$

and therefore it is possible to uniquely determine $P(B \mid A)$ from $P(B)$, $P(A \mid B)$ and $P(A \mid B^c)$.

Exercise 1.3.37 We know that 4% of a certain population α is sick. Performing an experimental test to detect whether an individual of α is sick, we see that the test has the following reliability:

 (i) if the individual is sick, the test gives a positive result in 99% of cases;
 (ii) if the individual is healthy, the test gives a positive result in 2% of cases.

Based on this data, what is the probability that an individual of α, positive to the test, is really sick? Suppose then to use the test on another population β: considering valid the reliability estimates (i) and (ii), and observing that the test gives a positive result on 6% of the population β, what is the probability that an individual of β is sick?

Solution Let T be the event "the test on an individual gives a positive result" and M be the event "the individual is sick". By assumption, $P(M) = 4\%$, $P(T \mid M) = 99\%$ and $P(T \mid M^c) = 2\%$. Then, by (1.3.16) with $B = M$ and $A = T$ we have

$$P(M \mid T) \approx 67.35\%$$

and therefore there is a high number of "false positives". This is due to the fact that the percentage of sick people is relatively low: we note that in general

$$P(M \mid T) = \frac{P(T \mid M)P(M)}{P(T \mid M)P(M) + P(T \mid M^c)(1 - P(M))} \longrightarrow 0^+ \quad \text{as } P(M) \to 0^+$$

while $P(M \mid T) \to 1^-$ as $P(M) \to 1^-$. We observe that based on the data we can also calculate, through (1.3.2), the percentage of positive tests

$$P(T) = P(T \mid M)P(M) + P(T \mid M^c)(1 - P(M)) \approx 5.88\%.$$

Regarding the second question, we have that by hypothesis $P(T \mid M) = 99\%$ and $P(T \mid M^c) = 2\%$. If the observed data is that $P(T) = 6\%$ then from (1.3.7) we obtain

$$P(M) = \frac{P(T) - P(T \mid M^c)}{P(T \mid M) - P(T \mid M^c)} \approx 4.12\%$$

The result can be interpreted by saying that, taken as valid the reliability estimates (i) and (ii) of the test, we have that out of 6% of positive tests about 33% are false positives.

Exercise 1.3.38 Prove in detail what was stated in Example 1.3.22.

Exercise 1.3.39 In relation to Exercise 1.3.24, construct a probability measure Q on Ω, different from P, such that we still have

$$Q(\alpha_1) = \frac{1}{3}, \qquad Q(\beta_1) = \frac{1}{4}$$

but α_1 and β_1 are not independent under Q.

Exercise 1.3.40 Consider a deck of 40 cards: verify that, with respect to the uniform probability,

  (i) the events "drawing an odd card" (event A) and "drawing a 7" (event B) are not independent;
 (ii) the events "drawing an odd card" (event A) and "drawing a diamond card" (event B) are independent.

Exercise 1.3.41 ((1.3.11) **Does not Imply** (1.3.10)) Consider the roll of three dice and the events A_{ij} defined by "the result of the i-th die is equal to that of the j-th die". Then A_{12}, A_{13}, A_{23} are pairwise independent but not independent.

Exercise 1.3.42 ((1.3.10) **Does not Imply** (1.3.11)) Consider the roll of two dice and, setting $\Omega = I_6 \times I_6$, the events

$$A = \{(\omega_1, \omega_2) \mid \omega_2 \in \{1, 2, 5\}\}, \quad B = \{(\omega_1, \omega_2) \mid \omega_2 \in \{4, 5, 6\}\},$$
$$C = \{(\omega_1, \omega_2) \mid \omega_1 + \omega_2 = 9\}.$$

Then (1.3.10) is true, but (1.3.11) is not.

Exercise 1.3.43 Suppose that n objects are randomly placed in r boxes, with $r \geq 1$. Find the probability that "exactly k objects are placed in the first box" (event A_k).

Solution If C_h is the event "the h-th object is placed in the first box" then $p = P(C_h) = \frac{1}{r}$. Moreover, $P(A_k)$ is given by (1.3.14).

1.4   Distributions

In this section, we deal with the construction and characterization of measures on the Euclidean space, with particular attention to probability measures on $\mathbb{R}^d$, called *distributions*. The fundamental result in this direction is the Carathéodory's theorem, which we state in Sect. 1.4.7 and will often use later. The idea is to define a distribution first on a particular family $\mathscr{A}$ of subsets of the sample space Ω (for example, the family of intervals in the case $\Omega = \mathbb{R}$) and then extend it to an appropriate σ-algebra containing $\mathscr{A}$. The problem of choosing such σ-algebra is related to the cardinality of Ω: if Ω is finite or countable, giving a probability on Ω is equivalent to assigning the probabilities of the individual outcomes (cf. Remark 1.1.13); therefore, it is natural to assume $\mathscr{P}(\Omega)$ as the σ-algebra of events.

The general case, as we have already seen in Example 1.1.30, is considerably more complex; in fact, the cardinality of $\mathscr{P}(\Omega)$ can be "too large" to define a probability measure on it.[14]

1.4.1 Completion of a Probability Space

Consider a generic non-empty set Ω. We observe that if $(\mathscr{F}_i)_{i \in I}$ is a family (not necessarily countable) of σ-algebras on Ω, then the intersection

$$\bigcap_{i \in I} \mathscr{F}_i$$

is still a σ-algebra. This justifies the following

Definition 1.4.1 Given a family $\mathscr{A}$ of subsets of Ω, we denote by $\sigma(\mathscr{A})$ the intersection of all the σ-algebras containing $\mathscr{A}$. Since $\sigma(\mathscr{A})$ *is the smallest σ-algebra that contains $\mathscr{A}$*, we say that $\mathscr{A}$ is the σ-*algebra generated by $\mathscr{A}$*.

Example 1.4.2 When $\mathscr{A} = \{A\}$ consists of a single set $A \subseteq \Omega$, we write $\sigma(A)$ instead of $\sigma(\{A\})$. Note that

$$\sigma(A) = \{\emptyset, \Omega, A, A^c\}.$$

The intersection of σ-algebras is still a σ-algebra, but a similar result is not true for the union: given two σ-algebras $\mathscr{F}_1$ and $\mathscr{F}_2$, we have $\mathscr{F}_1 \cup \mathscr{F}_2 \subseteq \sigma(\mathscr{F}_1 \cup \mathscr{F}_2)$ and the inclusion can be strict.

In general, it is difficult to give an explicit representation of the σ-algebra generated by a family $\mathscr{A}$: clearly $\sigma(\mathscr{A})$ must contain the complements and countable unions of elements of $\mathscr{A}$ but, as we will see in the next section, there are cases in which these operations do not yield all the elements of $\sigma(\mathscr{A})$. For this reason, it is useful to introduce techniques that allow proving that if a certain property is true for the elements of a family $\mathscr{A}$ then it is true also for all the elements of $\sigma(\mathscr{A})$: this type of results are the subject of Appendix A.

Remark 1.4.3 (Completion of a Probability Space) We recall that a probability space $(\Omega, \mathscr{F}, P)$ is complete if $\mathscr{N} \subseteq \mathscr{F}$, i.e., the negligible sets (and the almost certain ones) are events: a subset N of Ω is P-negligible if there exists $B \in \mathscr{F}$ such that $N \subseteq B$ and $P(B) = 0$. One can always "complete" a space $(\Omega, \mathscr{F}, P)$

---

[14] If the cardinality of Ω is finite, say $|\Omega| = n$, then $\mathscr{P}(\Omega) = 2^n$ and if Ω has countable cardinality then $\mathscr{P}(\Omega)$ has the cardinality of the continuum (of $\mathbb{R}$). However, if $\Omega = \mathbb{R}$, by Cantor's theorem the cardinality of $\mathscr{P}(\mathbb{R})$ is strictly greater than the cardinality of $\mathbb{R}$.

extending P to the σ-algebra $\sigma(\mathscr{F} \cup \mathscr{N})$ in the following way. First of all, notice that[15] $\sigma(\mathscr{F} \cup \mathscr{N}) = \bar{\bar{\mathscr{F}}}$ where

$$\bar{\bar{\mathscr{F}}} := \{A \cup N \mid A \in \mathscr{F}, \ N \in \mathscr{N}\}.$$

Then, we extend P to $\bar{\bar{\mathscr{F}}}$, setting $\bar{P}(A \cup N) := P(A)$; to check that this definition is well-posed, we prove that if $A_1 \cup N_1 = A_2 \cup N_2$, with $A_1, A_2 \in \mathscr{F}$ and $N_1, N_2 \in \mathscr{N}$, then $P(A_1) = P(A_2)$. Indeed, setting $B = B_1 \cup B_2$ where B_i are negligible events such that $N_i \subseteq B_i$ for $i = 1, 2$, we have

$$A_1 \cup B = (A_1 \cup N_1) \cup B = (A_2 \cup N_2) \cup B = A_2 \cup B$$

which implies $P(A_1) = P(A_2)$. It is also easy to verify that $\bar{P}$ is a probability measure on $\bar{\bar{\mathscr{F}}}$. Clearly, the completion of a space depends on the σ-algebra and the probability measure considered (cf. Exercise 1.4.14).

1.4.2 Borel σ-Algebra

We introduce the σ-algebra that we will systematically use when the sample space is $\mathbb{R}^d$. In fact, since it does not involve any additional difficulty and will be convenient later, we consider the case where the sample space is a generic *metric space* $(\mathbb{M}, \varrho)$. Beyond Euclidean spaces, a noteworthy example is $\mathbb{M} = C[0, 1]$, the space of continuous functions on the interval $[0, 1]$, equipped with the maximum distance

$$\varrho_{\max}(f, g) = \max_{t \in [0,1]} |f(t) - g(t)|, \qquad f, g \in C[0, 1].$$

This space is significant in the study of stochastic processes.

Another useful example is $\mathbb{M} = \bar{\mathbb{R}} := \mathbb{R} \cup \{-\infty, +\infty\}$, the extended real line equipped with the metric $\varrho(x, y) = |\delta(x) - \delta(y)|$ where $\delta(x) = \frac{x}{1+|x|}$ maps $\bar{\mathbb{R}}$ into $[-1, 1]$: this space proves essential for modeling random variables assuming extended real values, such as random times.

In a metric space $(\mathbb{M}, \varrho)$, the Borel σ-algebra $\mathscr{B}_\varrho$ is the σ-algebra generated by the topology (the family of open sets) induced by ϱ.

---

[15] Since $\mathscr{F} \cup \mathscr{N} \subseteq \bar{\bar{\mathscr{F}}}$, it suffices to prove that $\bar{\bar{\mathscr{F}}}$ is a σ-algebra: it is clearly non-empty and σ-$\cup$-closed. To show that $\bar{\bar{\mathscr{F}}}$ is closed under complement, we take $B \in \mathscr{F}$ such that $P(B) = 0$ and $N \subseteq B$, and notice that, for $A \cup N \in \bar{\bar{\mathscr{F}}}$,

$$(A \cup N)^c = (A \cup B)^c \cup (B \setminus (A \cup N)).$$

Definition 1.4.4 (Borel σ-Algebra) The *Borel σ-algebra* $\mathscr{B}_\varrho$ is the smallest σ-algebra containing the open sets of $(\mathbb{M}, \varrho)$. The elements of $\mathscr{B}_\varrho$ are called Borel sets.

Notation 1.4.5 Hereafter, we will denote by $\mathscr{B}_d$ the Borel σ-algebra in the Euclidean space $\mathbb{R}^d$. It is known that $\mathscr{B}_d$ is strictly contained in the σ-algebra $\mathscr{L}$ of Lebesgue measurable sets.[16] In the case $d = 1$, we simply write $\mathscr{B}$ instead of $\mathscr{B}_1$.

Remark 1.4.6 [!] By definition, $\mathscr{B}_\varrho$ contains all subsets of $\mathbb{M}$ that are obtained from open sets by operations of taking complements and countable unions: for example, singletons are Borel sets,[17] that is, $\{x\} \in \mathscr{B}_\varrho$ for every $x \in \mathbb{M}$.

However, *not all elements of $\mathscr{B}_\varrho$ can be obtained using only the operations of taking complements and countable unions.* Even more, in [6] it is shown that even a countable sequence of operations of taking complements and countable unions is not sufficient to attain $\mathscr{B}_\varrho$. More precisely, given a family $\mathscr{H}$ of subsets of a space Ω, we denote by $\mathscr{H}^*$ the family containing the elements of $\mathscr{H}$, the complements of the elements of $\mathscr{H}$ and the countable unions of elements of $\mathscr{H}$. We also define $\mathscr{H}_0 = \mathscr{H}$ and, by recursion, the increasing sequence of families

$$\mathscr{H}_n = \mathscr{H}^*_{n-1}, \qquad n \in \mathbb{N}.$$

By induction, it can be seen that $\mathscr{H}_n \subseteq \sigma(\mathscr{H})$ for every $n \in \mathbb{N}$; however (cf. [6] p. 30) when $\Omega = \mathbb{R}$ and $\mathscr{H}$ is as in Exercise 1.4.7-(ii), we have that

$$\bigcup_{n=0}^{\infty} \mathscr{H}_n$$

is strictly included in $\mathscr{B} = \sigma(\mathscr{H})$.

Exercise 1.4.7 Let $\mathscr{B}$ the Borel σ-algebra in $\mathbb{R}$. Prove that $\mathscr{B} = \sigma(\mathscr{H})$ where $\mathscr{H}$ is any of the following families of subsets of $\mathbb{R}$:

  (i) $\mathscr{H} = \{\,]a, b] \mid a, b \in \mathbb{R}, \ a < b\}$;
 (ii) $\mathscr{H} = \{\,]a, b] \mid a, b \in \mathbb{Q}, \ a < b\}$ (note that $\mathscr{H}$ is countable and therefore we say that $\mathscr{B}$ is *countably generated*);
(iii) $\mathscr{H} = \{\,]-\infty, a] \mid a \in \mathbb{R}\}$.

---

[16] $(\mathbb{R}^d, \mathscr{L}_d, \mathrm{Leb}_d)$ is the completion (cf. Remark 1.4.3) with respect to the Lebesgue measure Leb_d of $(\mathbb{R}^d, \mathscr{B}_d, \mathrm{Leb}_d)$.

[17] Indeed,

$$\{x\} = \bigcap_{n \geq 1} D(x, 1/n)$$

where the disks $D(x, 1/n) := \{y \in \mathbb{M} \mid \varrho(x, y) < 1/n\} \in \mathscr{B}_\varrho$ being open by definition.

An analogous result holds in dimensions greater than one, considering multi-intervals.

1.4.3 Distributions

Let $\mathscr{B}_\varrho$ be the Borel σ-algebra on a metric space $(\mathbb{M}, \varrho)$. Clearly, the Euclidean case $\mathbb{M} = \mathbb{R}^d$ is of particular interest and should always be kept as a reference point.

Definition 1.4.8 (Distribution) A *distribution* is a probability measure on $(\mathbb{M}, \mathscr{B}_\varrho)$.

To fix ideas, it is good to give the following "physical" interpretation of the concept of distribution μ. We think of the sample space $\mathbb{R}^d$ as the set of possible positions in space of a particle that cannot be observed with precision: then $H \in \mathscr{B}_d$ is interpreted as the event according to which "the particle is in the Borel set H" and $\mu(H)$ is the probability that the particle is in H.

> The concept of distribution will be fully understood only after we introduce random variables. At this point, we do not yet possess sufficient knowledge to fully appreciate distributions. Therefore, we will limit ourselves to mentioning a few examples, which we will revisit in more detail later, in Chap. 2.

We begin by proving some general properties of distributions.

Proposition 1.4.9 (Internal and External Regularity) *Let μ be a distribution on $(\mathbb{M}, \mathscr{B}_\varrho)$. For every $H \in \mathscr{B}_\varrho$ we have*

$$\mu(H) = \sup\{\mu(C) \mid C \subseteq H, \ C \text{ closed}\}$$
$$= \inf\{\mu(A) \mid A \supseteq H, \ A \text{ open}\}.$$

The proof of Proposition 1.4.9 is postponed to Sect. 1.5.2. An immediate consequence is the following

Corollary 1.4.10 *Two distributions μ_1 and μ_2 on $(\mathbb{M}, \mathscr{B}_\varrho)$ are equal if and only if $\mu_1(H) = \mu_2(H)$ for every open set H (or for every closed set H).*

Remark 1.4.11 If μ is a distribution on $(\mathbb{M}, \mathscr{B}_\varrho)$ then

$$A := \{x \in \mathbb{M} \mid \mu(\{x\}) > 0\}$$

is *finite or at most countable*. In fact, let

$$A_n = \{x \in \mathbb{M} \mid \mu(\{x\}) > 1/n\}, \qquad n \in \mathbb{N}.$$

Then, for each $x_1, \ldots, x_k \in A_n$ we have

$$1 = \mu(\mathbb{M}) \geq \mu(\{x_1, \ldots, x_k\}) \geq \frac{k}{n}$$

and consequently A_n has at most n elements. Then the thesis follows from the fact that $A = \bigcup_{n \geq 1} A_n$ where the union is finite or countable.

The "extreme" case in which μ concentrates all the mass in a single point is illustrated in the following example.

Example 1.4.12 Given $x_0 \in \mathbb{R}^d$, the *Dirac delta distribution* δ_{x_0} centered at x_0, is defined by

$$\delta_{x_0}(H) = \begin{cases} 1 & \text{if } x_0 \in H, \\ 0 & \text{if } x_0 \notin H, \end{cases} \qquad H \in \mathcal{B}_d.$$

Note in particular that $\delta_{x_0}(\{x_0\}) = 1$ and think about the "physical" interpretation of this fact.

Before considering other remarkable examples of distributions, we observe that by combining appropriately distributions we still obtain a distribution.

Proposition 1.4.13 *Let* $(\mu_n)_{n \in \mathbb{N}}$ *be a sequence of distributions on* $(\mathbb{M}, \mathcal{B}_\varrho)$ *and* $(p_n)_{n \in \mathbb{N}}$ *be a sequence of real numbers such that*

$$\sum_{n=1}^{\infty} p_n = 1 \quad \text{and} \quad p_n \geq 0, \ n \in \mathbb{N}. \tag{1.4.1}$$

Then μ *defined by*

$$\mu(H) := \sum_{n=1}^{\infty} p_n \mu_n(H), \qquad H \in \mathcal{B}_\varrho,$$

is a distribution.

Proof It is easy to verify that $\mu(\emptyset) = 0$ and $\mu(\mathbb{M}) = 1$. It remains to prove the σ-additivity: we have

$$\mu\left(\biguplus_{k \in \mathbb{N}} H_k\right) = \sum_{n=1}^{\infty} p_n \mu_n\left(\biguplus_{k \in \mathbb{N}} H_k\right) =$$

(by the σ-additivity of μ_n)

$$= \sum_{n=1}^{\infty} p_n \sum_{k=1}^{\infty} \mu_n(H_k) =$$

(reordering the terms, since they are non-negative)

$$= \sum_{k=1}^{\infty} \sum_{n=1}^{\infty} p_n \mu_n(H_k) = \sum_{k=1}^{\infty} \mu(H_k).$$

$\square$

Exercise 1.4.14 Recall the notion of completion of a space, given in Remark 1.4.3. On $\mathbb{R}$ consider the Dirac delta distribution δ_x centered at $x \in \mathbb{R}$, the trivial σ-algebra $\{\emptyset, \mathbb{R}\}$ and the Borel σ-algebra $\mathscr{B}$. Prove that the space $(\mathbb{R}, \{\emptyset, \mathbb{R}\}, \delta_x)$ is complete while the space $(\mathbb{R}, \mathscr{B}, \delta_x)$ is not complete. The completion of $(\mathbb{R}, \mathscr{B}, \delta_x)$ is the space $(\mathbb{R}, \mathscr{P}(\mathbb{R}), \delta_x)$.

1.4.4 Discrete Distributions

From now on, we focus on the case $\mathbb{M} = \mathbb{R}^d$.

Definition 1.4.15 A *discrete distribution* is a distribution of the form

$$\mu(H) := \sum_{n=1}^{\infty} p_n \delta_{x_n}(H), \qquad H \in \mathscr{B}_d, \tag{1.4.2}$$

where (x_n) is a sequence of distinct points in $\mathbb{R}^d$ and (p_n) satisfies the properties in (1.4.1).

Remark 1.4.16 To a discrete distribution of the form (1.4.2), it is natural to associate the *function*

$$\bar{\mu} : \mathbb{R}^d \longrightarrow [0, 1],$$

defined by

$$\bar{\mu}(x) = \mu(\{x\}), \qquad x \in \mathbb{R}^d,$$

or more explicitly

$$\bar{\mu}(x) = \begin{cases} p_n & \text{if } x = x_n, \\ 0 & \text{otherwise.} \end{cases}$$

Since

$$\mu(H) = \sum_{x \in H \cap \{x_n \mid n \in \mathbb{N}\}} \bar{\mu}(x), \qquad H \in \mathscr{B}_d, \qquad (1.4.3)$$

the distribution μ is uniquely associated with the function $\bar{\mu}$, which is sometimes called the *distribution function of* μ. As we will see in the following examples, it is generally much easier to assign the distribution function $\bar{\mu}$ than the distribution μ itself: in fact, μ is a measure (i.e., a set function) unlike $\bar{\mu}$, which is a function on $\mathbb{R}^d$.

Let us consider some noteworthy examples of discrete distributions.

Example 1.4.17

(i) **(Bernoulli)** The *Bernoulli distribution with parameter* $p \in [0, 1]$ is denoted by Be_p and is defined as a linear combination of two Dirac deltas:

$$\mathrm{Be}_p = p\delta_1 + (1 - p)\delta_0.$$

Explicitly, we have

$$\mathrm{Be}_p(H) = \begin{cases} 0 & \text{if } 0, 1 \notin H, \\ 1 & \text{if } 0, 1 \in H, \\ p & \text{if } 1 \in H, \ 0 \notin H, \\ 1 - p & \text{if } 0 \in H, \ 1 \notin H. \end{cases} \qquad H \in \mathscr{B},$$

and the distribution function is simply

$$\bar{\mu}(x) = \begin{cases} p & \text{if } x = 1, \\ 1 - p & \text{if } x = 0, \\ 0 & \text{otherwise.} \end{cases}$$

(ii) **(Discrete uniform)** Let $H = \{x_1, \ldots, x_n\}$ be a finite subset of $\mathbb{R}^d$. The *discrete uniform distribution on* H is denoted by Unif_H and is defined by

$$\mathrm{Unif}_H = \frac{1}{n} \sum_{k=1}^{n} \delta_{x_k},$$

that is,

$$\mathrm{Unif}_H(\{x\}) = \begin{cases} \frac{1}{n} & \text{if } x \in H, \\ 0 & \text{otherwise.} \end{cases}$$

(iii) **(Binomial)** Let $n \in \mathbb{N}$ and $p \in [0, 1]$. The *binomial distribution with parameters n and p* is defined on $\mathbb{R}$ by

$$\text{Bin}_{n,p} = \sum_{k=0}^{n} \binom{n}{k} p^k (1 - p)^{n-k} \delta_k,$$

that is, the distribution function is

$$\bar{\mu}(k) = \text{Bin}_{n,p}(\{k\}) = \begin{cases} \binom{n}{k} p^k (1 - p)^{n-k} & \text{for } k = 0, 1, \ldots, n, \\ 0 & \text{otherwise.} \end{cases}$$

Example 1.2.17 gives an interpretation of the binomial distribution.

(iv) **(Geometric)** Given $p \in {]0, 1]}$, the *geometric distribution with parameter p* is defined by

$$\text{Geom}_p = \sum_{k=1}^{\infty} p(1 - p)^{k-1} \delta_k.$$

The distribution function is

$$\bar{\mu}(k) = \text{Geom}_p(\{k\}) = \begin{cases} p(1 - p)^{k-1} & \text{for } k \in \mathbb{N}, \\ 0 & \text{otherwise.} \end{cases}$$

Note that

$$\sum_{k=1}^{\infty} p(1 - p)^{k-1} = p \sum_{h=0}^{\infty} (1 - p)^h =$$

(since by hypothesis $0 < p \leq 1$)

$$= \frac{p}{1 - (1 - p)} = 1.$$

Example 1.3.31 gives an interpretation of the geometric distribution.

(iv) **(Poisson)** The *Poisson distribution with parameter $\lambda > 0$, centered at $x \in \mathbb{R}$,* is defined by

$$\text{Poisson}_{x,\lambda} := e^{-\lambda} \sum_{k=0}^{\infty} \frac{\lambda^k}{k!} \delta_{x+k}.$$

In the case $x = 0$, we simply speak of the Poisson distribution with parameter $\lambda > 0$ and indicate it with Poisson$_\lambda$: in this case, the distribution function is

$$\bar{\mu}(k) = \text{Poisson}_\lambda(\{k\}) = \begin{cases} \frac{e^{-\lambda}\lambda^k}{k!} & \text{for } k \in \mathbb{N}_0, \\ 0 & \text{otherwise.} \end{cases}$$

1.4.5 Absolutely Continuous Distributions

Consider a $\mathscr{B}_d$-measurable[18] function

$$\gamma : \mathbb{R}^d \longrightarrow [0, +\infty[\quad \text{such that} \quad \int_{\mathbb{R}^d} \gamma(x)dx = 1. \tag{1.4.4}$$

Then μ defined as

$$\mu(H) = \int_H \gamma(x)dx, \qquad H \in \mathscr{B}_d, \tag{1.4.5}$$

is a distribution. Indeed, it is obvious that $\mu(\emptyset) = 0$ and $\mu(\mathbb{R}^d) = 1$. Moreover, if $(H_n)_{n \in \mathbb{N}}$ is a sequence of disjoint Borel sets, then, due to the properties of the Lebesgue integral,[19] we have

$$\mu\left(\biguplus_{n \geq 1} H_n\right) = \int_{\biguplus_{n \geq 1} H_n} \gamma(x)dx = \sum_{n \geq 1} \int_{H_n} \gamma(x)dx = \sum_{n \geq 1} \mu(H_n),$$

which proves that μ is σ-additive.

Definition 1.4.18 (Absolutely Continuous Distribution) A $\mathscr{B}_d$-measurable function γ that satisfies the properties in (1.4.4) is called a *density function* (or, simply, a *density* or a *PDF*[20]). We say that a distribution μ on $\mathbb{R}^d$ is *absolutely continuous*, and we write $\mu \in AC$, if there exists a density γ for which (1.4.5) holds.

Note the analogy between the properties (1.4.4) of a density γ and the properties (1.4.1).

---

[18] That is, $\gamma^{-1}(H) \in \mathscr{B}_d$ for any $H \in \mathscr{B}$.

[19] In particular, here we use Beppo Levi's theorem.

[20] PDF stands for "Probability Density Function" and is also the command used in the software Mathematica® for density functions.

Remark 1.4.19 [!] The PDF of a distribution $\mu \in AC$ is not uniquely determined: it is determined only up to Borel sets that have Lebesgue measure equal to zero; in fact, the value of the integral in (1.4.5) does not change by modifying γ on a set of null Lebesgue measure.

Actually, if γ_1, γ_2 are PDFs of $\mu \in AC$, then $\gamma_1 = \gamma_2$ a.e. (with respect to the Lebesgue measure). In fact, let

$$A_n = \{x \mid \gamma_1(x) - \gamma_2(x) \geq 1/n\} \in \mathscr{B}_d, \qquad n \in \mathbb{N}.$$

Then

$$\frac{\text{Leb}(A_n)}{n} \leq \int_{A_n} (\gamma_1(x) - \gamma_2(x)) \, dx = \int_{A_n} \gamma_1(x)dx - \int_{A_n} \gamma_2(x)dx$$

$$= \mu(A_n) - \mu(A_n) = 0,$$

from which $\text{Leb}(A_n) = 0$ for each $n \in \mathbb{N}$. It follows that also

$$\{x \mid \gamma_1(x) > \gamma_2(x)\} = \bigcup_{n=1}^{\infty} A_n$$

has Lebesgue measure zero, that is, $\gamma_1 \leq \gamma_2$ a.e. Similarly, we have $\gamma_1 \geq \gamma_2$ a.e.

Remark 1.4.20 [!] Unless otherwise specified, when considering a Lebesgue integral, we will always assume that the integrand function is $\mathscr{B}$-measurable (and therefore, in particular, Lebesgue measurable). Thus, unless explicitly indicated, "measurable" means "$\mathscr{B}$-measurable" and also in the definition of L^p space (the space of integrable functions of order p) the $\mathscr{B}$-measurability is implicitly assumed. This is convenient for many reasons: for example, the composition of $\mathscr{B}$-measurable functions is still $\mathscr{B}$-measurable (a fact not necessarily true for Lebesgue measurable functions).

Remark 1.4.21 [!] If μ on $\mathbb{R}^d$ is absolutely continuous, then μ assigns zero probability to Borel sets that are negligible with respect to the Lebesgue measure: precisely, we have

$$\text{Leb}_d(H) = 0 \quad \Longrightarrow \quad \mu(H) = \int_H \gamma(x)dx = 0. \tag{1.4.6}$$

In particular, if H is finite or countable, then $\mu(H) = 0$. In a certain sense, the distributions in AC are "complementary" to discrete distributions (but be careful because of Remark 1.4.23 below!): in fact, the latter assign positive probability precisely to individual points or countable subsets of $\mathbb{R}^d$. Condition (1.4.6) is

necessary[21] for $\mu \in$ AC and provides a very useful and practical test to verify
that μ does *not* admit density: if there exists $H \in \mathscr{B}_d$ such that $\mathrm{Leb}_d(H) = 0$ and
$\mu(H) > 0$ then $\mu \notin$ AC. A typical applications is illustrated in Example 2.1.47.

Every density function identifies a distribution: in practice, assigning a density
is the simplest and most commonly used way to define an absolutely continuous
distribution, as shown by the following remarkable examples.

Example 1.4.22

(i) **(Uniform)** The *uniform distribution* Unif_K on K, where $K \in \mathscr{B}_d$ has Lebesgue
measure $0 < \mathrm{Leb}_d(K) < \infty$, is the distribution with density

$$\gamma = \frac{1}{\mathrm{Leb}_d(K)} \mathbb{1}_K .$$

Then

$$\mathrm{Unif}_K(H) = \int_{H \cap K} \frac{1}{\mathrm{Leb}_d(K)} dx = \frac{\mathrm{Leb}_d(H \cap K)}{\mathrm{Leb}_d(K)}, \qquad H \in \mathscr{B}_d .$$

What happens if $\mathrm{Leb}_d(K) = \infty$? Is it possible to define a uniform probability
on $\mathbb{R}^d$?

(ii) **(Exponential)** The *exponential distribution* Exp_λ *with parameter* $\lambda > 0$ is the
distribution with density

$$\gamma(x) = \begin{cases} \lambda e^{-\lambda x} & \text{if } x \geq 0, \\ 0 & \text{if } x < 0. \end{cases}$$

Then

$$\mathrm{Exp}_\lambda(H) = \lambda \int_{H \cap [0, +\infty[} e^{-\lambda x} dx, \qquad H \in \mathscr{B} .$$

Note that $\mathrm{Exp}_\lambda(\mathbb{R}) = \mathrm{Exp}_\lambda(\mathbb{R}_{\geq 0}) = 1$.

(iii) **(Normal)** The *real normal distribution* $\mathscr{N}_{\mu,\sigma^2}$ *with parameters* $\mu \in \mathbb{R}$ *and*
$\sigma > 0$ is the distribution on $\mathscr{B}$ with density

$$\gamma(x) = \frac{1}{\sqrt{2\pi\sigma^2}} e^{-\frac{1}{2}\left(\frac{x-\mu}{\sigma}\right)^2}, \qquad x \in \mathbb{R} .$$

---

[21] Actually, by the Radon-Nikodym Theorem B.1.3, (1.4.6) is a necessary and sufficient condition
for absolute continuity.

Then

$$\mathcal{N}_{\mu,\sigma^2}(H) = \frac{1}{\sqrt{2\pi\sigma^2}} \int_H e^{-\frac{1}{2}\left(\frac{x-\mu}{\sigma}\right)^2} dx, \qquad H \in \mathcal{B}.$$

Distribution $\mathcal{N}_{0,1}$, corresponding to $\mu = 0$ and $\sigma = 1$, is called the *standard normal distribution*.

Remark 1.4.23 **[!]** Not all distributions are of the type analyzed so far (i.e., discrete or absolutely continuous). For example, consider the segment in $\mathbb{R}^2$

$$I = \{(x,0) \mid 0 \le x \le 1\}$$

and the distribution

$$\mu(H) = \mathrm{Leb}_1(H \cap I), \qquad H \in \mathcal{B}_2,$$

where Leb_1 denotes the 1-dimensional Lebesgue measure (or more precisely the 1-dimensional Hausdorff[22] measure in $\mathbb{R}^2$). Clearly, $\mu \notin AC$ since $\mu(I) = 1$ and I has zero Lebesgue measure in $\mathbb{R}^2$; on the other hand, μ is not a discrete distribution because $\mu(\{(x,y)\}) = 0$ for every $(x,y) \in \mathbb{R}^2$.

The idea is that a distribution can concentrate probability on subsets of $\mathbb{R}^d$ with dimension (in the sense of Hausdorff[23]) *less* than d: for example, a spherical surface (which has Hausdorff dimension equal to 2) in $\mathbb{R}^3$. Things can get even more complicated since the Hausdorff dimension can be fractional (cf. Example 1.4.36).

1.4.6 Cumulative Distribution Functions (CDF)

The concept of density enables the identification of a *distribution* (that is, a probability measure) through a *function* defined on $\mathbb{R}^d$ (which, from a mathematical standpoint, is more tractable than directly dealing with a measure). Naturally, this approach is feasible when the distribution is absolutely continuous. A similar result holds for discrete distributions (cf. Remark 1.1.13).

In this section, we present a much more general approach and introduce the concept of *cumulative* distribution function that will allow us to identify a generic distribution through a function. For now, we limit ourselves to considering the one-dimensional case: in Sect. 1.4.9 we will deal with the multi-dimensional case.

---

[22] See, for example, Chapter 2 in [27].

[23] Cf. Chapter 2.5 in [27].

Definition 1.4.24 The *cumulative* distribution function F_μ of a distribution μ on $(\mathbb{R}, \mathscr{B})$ is defined as

$$F_\mu : \mathbb{R} \longrightarrow [0, 1],$$

$$x \longrightarrow F_\mu(x) := \mu(]-\infty, x]).$$

We use the abbreviation CDF for cumulative distribution functions.

Example 1.4.25

(i) The CDF of the Dirac delta δ_{x_0} is

$$F(x) = \begin{cases} 0 & \text{if } x < x_0, \\ 1 & \text{if } x \geq x_0. \end{cases}$$

(ii) The CDF of the discrete distribution $\text{Unif}_n := \frac{1}{n} \sum_{k=1}^{n} \delta_k$ is

$$F(x) = \begin{cases} 0 & \text{if } x < 1, \\ \frac{k}{n} & \text{if } k \leq x < k+1, \text{ for } 1 \leq k \leq n-1, \\ 1 & \text{if } x \geq n. \end{cases} \tag{1.4.7}$$

See Fig. 1.1 for the case $n = 5$.

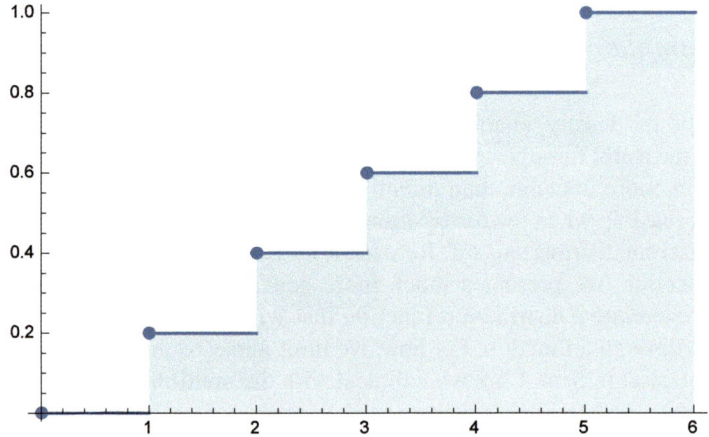

Fig. 1.1 Graph of the CDF of a r.v. with distribution Unif_5

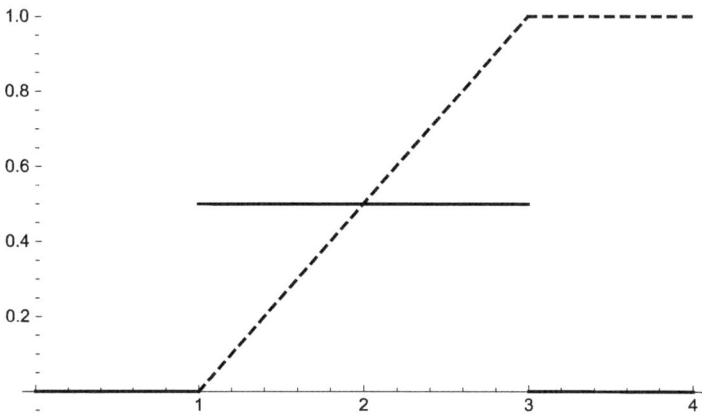

Fig. 1.2 Density function (solid line) and distribution (dashed line) of the distribution $\text{Unif}_{[1,3]}$

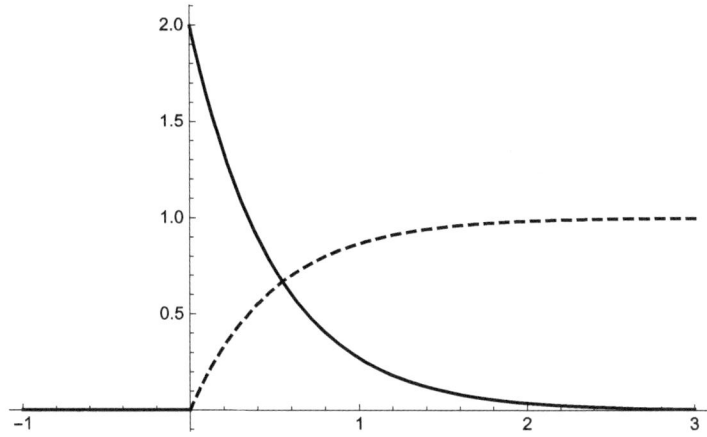

Fig. 1.3 Density function (solid line) and CDF (dashed line) of the distribution Exp_2

(iii) As shown in Fig. 1.2, the density and CDF of the distribution $\text{Unif}_{[1,3]}$ are respectively

$$\gamma = \frac{1}{2}\mathbb{1}_{[1,3]} \qquad \text{and} \qquad F(x) = \begin{cases} 0 & x \leq 1, \\ \frac{x-1}{2} & 1 < x \leq 3, \\ 1 & x > 3. \end{cases}$$

(iv) As shown in Fig. 1.3 with $\lambda = 2$, the density and CDF of the distribution Exp_λ are respectively

$$\gamma(x) = \lambda e^{-\lambda x} \qquad \text{and} \qquad F(x) = 1 - e^{-\lambda x}, \qquad x \geq 0, \qquad (1.4.8)$$

and are null for $x < 0$.

(v) The CFD of $\mathscr{N}_{\mu,\sigma^2}$ is

$$F(x) = \frac{1}{\sqrt{2\pi\sigma^2}} \int_{-\infty}^{x} e^{-\frac{1}{2}\left(\frac{t-\mu}{\sigma}\right)^2} dt, \qquad x \in \mathbb{R}.$$

For the standard normal, we have

$$F(x) = \frac{1}{2}\left(\mathrm{erf}\left(\frac{x}{\sqrt{2}}\right) + 1\right), \qquad x \in \mathbb{R},$$

where

$$\mathrm{erf}(x) = \frac{2}{\sqrt{\pi}} \int_{0}^{x} e^{-t^2} dt, \qquad x \in \mathbb{R},$$

is the *error function*. Figure 1.4 shows the density and CDF of the standard normal distribution.

Theorem 1.4.26 *[!] The CDF F_μ of a distribution μ has the following properties:*

(i) *F_μ is monotone (weakly) increasing;*
(ii) *F_μ is right continuous, i.e.*

$$F_\mu(x) = F_\mu(x+) := \lim_{y \to x^+} F_\mu(y), \qquad x \in \mathbb{R};$$

(iii) *we have*

$$\lim_{x \to -\infty} F_\mu(x) = 0 \qquad and \qquad \lim_{x \to +\infty} F_\mu(x) = 1;$$

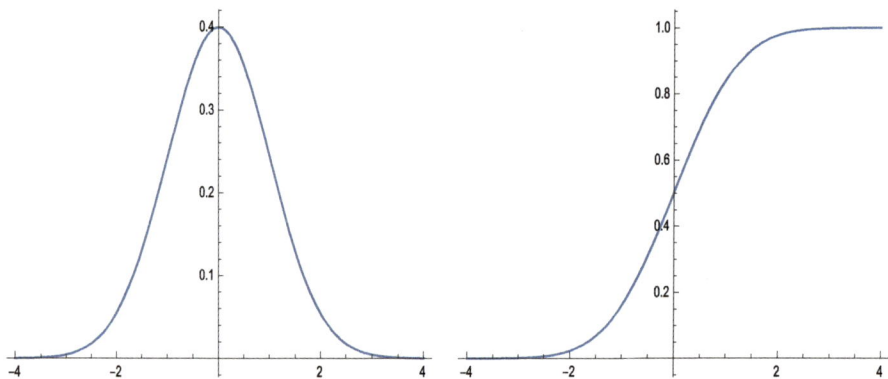

Fig. 1.4 On the left: plot of the standard normal density. On the right: plot of the standard normal CDF. Note the different scale on the ordinate axis

Proof If $x \leq y$ then $]-\infty, x] \subseteq]-\infty, y]$ and therefore, by the monotonicity of μ, $F_\mu(x) \leq F_\mu(y)$, which proves i). Next, consider a decreasing sequence $(x_n)_{n\in\mathbb{N}}$ that converges x as $n \to \infty$: we have

$$]-\infty, x] = \bigcap_{n\in\mathbb{N}}]-\infty, x_n]$$

and therefore, by the continuity from above of μ (cf. Proposition 1.1.32-(iii))

$$F_\mu(x) = \mu(]-\infty, x]) = \lim_{n\to\infty} \mu(]-\infty, x_n]) = \lim_{n\to\infty} F_\mu(x_n).$$

Then ii) follows from the arbitrariness of the sequence $(x_n)_{n\in\mathbb{N}}$. Property iii) follows from the continuity from above and below of μ respectively. □

Remark 1.4.27 [!]   Under the assumptions of the previous proposition, given the monotonicity of F_μ, also the left limit exists

$$F_\mu(x-) := \lim_{y\to x^-} F_\mu(y),$$

but in general, we only have

$$F_\mu(x-) \leq F_\mu(x), \qquad x \in \mathbb{R}.$$

In fact, for every *increasing* sequence $(x_n)_{n\in\mathbb{N}}$ that converges x as $n \to \infty$, we have

$$]-\infty, x_n] \nearrow]-\infty, x[$$

and by the continuity from below of P (cf. Proposition 1.1.32-ii)) we have

$$F_\mu(x-) = \mu(]-\infty, x[) \qquad \text{and} \qquad \mu(\{x\}) = \Delta F_\mu(x) := F_\mu(x) - F_\mu(x-).$$
$$(1.4.9)$$

Thus μ *assigns positive probability at points where F_μ is discontinuous and at such points the probability is equal to the size of the jump of F_μ.* By the way, it is easy to see that *a monotone increasing function*

$$F : \mathbb{R} \longrightarrow \mathbb{R}$$

admits at most a countable infinity of discontinuity points. In fact, let

$$A_n = \{x \in \mathbb{R} \mid |x| \leq n, \ \Delta F(x) \geq \frac{1}{n}\}, \qquad n \in \mathbb{N},$$

the cardinality $|A_n|$ is finite since

$$\frac{|A_n|}{n} \leq \sum_{x \in A_n} \Delta F(x) \leq F(n) - F(-n) < \infty.$$

Since the set of discontinuity points of F is equal to the union of A_n, $n \in \mathbb{N}$, this confirms Remark 1.4.11 according to which, *for every distribution μ, the set of points such that $\mu(\{x\}) > 0$ is at most countable.*

Exercise 1.4.28 Prove that the CDF of the normal distribution $\mathscr{N}_{\mu,\sigma^2}$ is strictly increasing.

1.4.7 Carathéodory's Extension Theorem

Recall the concept of measure on an algebra (Definitions 1.1.21 and 1.1.19). One of the results on which the entire theory of probability is based is the following

Theorem 1.4.29 (Carathéodory's theorem) *[!!!] Let μ be a σ-finite measure on an algebra $\mathscr{A}$. There exists a unique σ-finite measure that extends μ to the σ-algebra generated by $\mathscr{A}$.*

Proof The proof is long and articulated; in Sect. 1.5.3 we prove a more general version of Theorem 1.4.29, which will be easier to apply later.   □

Carathéodory's theorem establishes the *existence* and *uniqueness* of the extension of μ from $\mathscr{A}$ to $\sigma(\mathscr{A})$. It is remarkable that no assumption is required on Ω which is any non-empty set: in fact, the proof is based on purely set-theoretic arguments.

1.4.8 From CDFs to Distributions

The construction of a probabilistic model on $\mathbb{R}$ (representing a random phenomenon, be it the position of a particle in a physics model or the price of a risky asset in a financial model or the temperature in a meteorological model) often consists in assigning a particular distribution. From a practical and intuitive perspective, the first step is to establish how the distribution assigns probability to the *intervals* which are the simplest events to think about: we had done so in Example 1.1.30, when we defined the uniform distribution. In fact, we know (from

Corollary 1.4.10) that a real distribution is identified by how it acts on intervals or equivalently, since

$$\mu(]a, b]) = F_\mu(b) - F_\mu(a),$$

from the cumulative distribution function. Then it seems natural to ask if, *given a function F that satisfies the properties that a CDF must have, there exists a distribution μ that has F as CDF.*

The answer is affirmative and is contained in the following Theorem 1.4.33 which we prove as a corollary of Carathéodory's Theorem 1.4.29. First, we make some preliminary reminders.

Definition 1.4.30 (Absolutely Continuous Function (AC)) A function F is *absolutely continuous*[24] on $[a, b]$ (in symbols, $F \in AC[a, b]$) if it is written in the form

$$F(x) = F(a) + \int_a^x \gamma(t)dt, \qquad x \in [a, b], \tag{1.4.10}$$

for some $\gamma \in L^1([a, b])$.

The following result, whose proof is given in the appendix (see Proposition B.3.3), states that absolutely continuous functions are almost everywhere differentiable.

Proposition 1.4.31 *Let $F \in AC[a, b]$ be of the form* (1.4.10). *Then F is a.e. differentiable with $F' = \gamma$: as a consequence, we have*

$$F(x) = F(a) + \int_a^x F'(t)dt, \qquad x \in [a, b]. \tag{1.4.11}$$

In other words, *absolutely continuous functions comprise the class of functions for which the fundamental theorem of calculus holds*, meaning, in simple terms, functions that are equal to the integral of their own derivative. It is worth noting that even if F is a.e. differentiable with $F' \in L^1([a, b])$, this does not ensure the validity of formula (1.4.11). A trivial counter-example is given by the function $F = \mathbb{1}_{[1/2,1]}$: we have $F' = 0$ a.e. on $[0, 1]$ but

$$1 = F(1) - F(0) \neq \int_0^1 F'(x)dx = 0.$$

We will see in Example 1.4.36, that F can also be continuous, a.e. differentiable with $F' \in L^1([a, b])$ and this still does not ensure the validity of formula (1.4.11).

---

[24] The true definition of absolutely continuous function is given in Appendix B.4: actually, Definition 1.4.30 is an equivalent characterization of absolute continuity.

Exercise 1.4.32 Check that the function

$$F(x) = \begin{cases} 0 & x \le 0, \\ \sqrt{x} & 0 < x < 1, \\ 1 & x \ge 1, \end{cases}$$

is absolutely continuous on $[0, 1]$.

The main result of this section is the following

Theorem 1.4.33 *[!!]* *Let $F : \mathbb{R} \longrightarrow \mathbb{R}$ be an increasing[25] and right continuous function (that is, F satisfies properties (i) and (ii) of Theorem 1.4.26). Then:*

(i) there exists a unique measure μ_F on $(\mathbb{R}, \mathcal{B})$ that is σ-finite and satisfies

$$\mu_F(]a, b]) = F(b) - F(a), \qquad a, b \in \mathbb{R}, \ a < b; \tag{1.4.12}$$

(ii) if F also verifies

$$\lim_{x \to -\infty} F(x) = 0 \qquad and \qquad \lim_{x \to +\infty} F(x) = 1,$$

(that is, F satisfies property (iii) of Theorem 1.4.26) then μ_F is a distribution;
(iii) finally, F is absolutely continuous if and only if $\mu_F \in AC$: in this case, F' is a density of μ_F.

Proof See Sect. 1.5.4.   □

Remark 1.4.34 It is worth emphasizing that Theorem 1.4.33 also contains a *uniqueness* result, based on which each CDF is associated with a unique measure for which Eq. (1.4.12) holds. For example, the measure associated with the function $F(x) = x$ is the Lebesgue measure and the same holds taking $F(x) = x + c$ for every $c \in \mathbb{R}$.

Remark 1.4.35 There are two particularly important cases in applications:

(1) if F is piecewise constant and we denote by x_n the points of discontinuity of F (which, by Remark 1.4.27, are at most a countable quantity) then, by (1.4.9), μ_F is the discrete distribution

$$\mu_F = \sum_n \Delta F(x_n) \delta_{x_n}$$

where $\Delta F(x_n)$ denotes the size of the jump of F at x_n;
(2) if F is absolutely continuous then $\mu_F \in AC$ with a density equal to F'.

---

[25] Not necessarily *strictly* increasing.

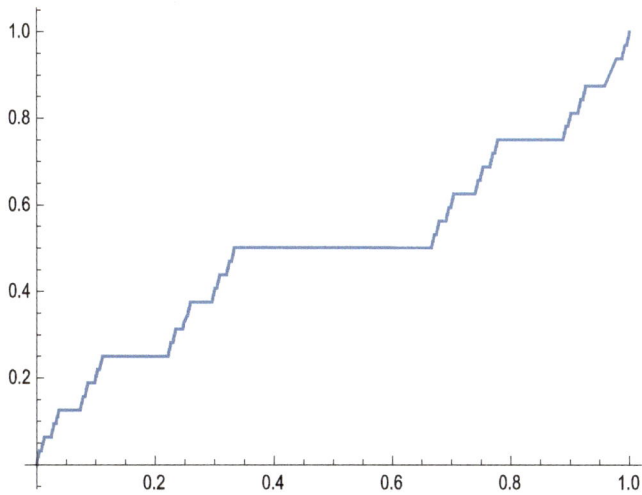

Fig. 1.5 Graph of the Cantor-Vitali function

Example 1.4.36 The Cantor-Vitali function

$$V : \mathbb{R} \longrightarrow [0, 1]$$

is a continuous and increasing function such that $V(x) = 0$ for $x \leq 0$, $V(x) = 1$ for $x \geq 1$ and with first derivative V' that exists almost everywhere and is equal to zero: for a construction of the Cantor-Vitali function see, for example, [27] page 192. Since V satisfies the assumptions of Theorem 1.4.33, there exists a unique distribution μ_V such that $\mu_V(]a, b]) = V(b) - V(a)$. Figure 1.5 depicts the graph of the Cantor-Vitali function.

Since V is continuous, we have $\mu_V(\{x\}) = 0$ for every $x \in [0, 1]$ (cf. (1.4.9)) and therefore μ_V *is not a discrete distribution*. If $\mu_V \in AC$ there would exist a density γ such that

$$V(x) = \mu_V([0, x]) = \int_0^x \gamma(y)dy, \qquad x \geq 0.$$

From Proposition 1.4.31 we infer $\gamma = V' = 0$ a.e. but this is absurd. Hence μ_V *is not even an absolutely continuous distribution*, although its CDF V is continuous and almost everywhere differentiable.

For those who want to deepen the issue, the fact is that μ_V assigns probability 1 to the *Cantor set* (for more details see [27], p. 37) which is a subset of the interval $[0, 1]$, with null Lebesgue measure and Hausdorff dimension equal to $\frac{\log 2}{\log 3}$.

Exercise 1.4.37 Verify that the function

$$F(x) := \begin{cases} 0 & \text{for } x < 0, \\ \frac{x}{3} & \text{for } 0 \le x < 1, \\ 1 & \text{for } x \ge 1, \end{cases}$$

is a CDF. If μ_F is the associated distribution, calculate $\mu_F([0, 1])$, $\mu_F([0, 1[)$ and $\mu_F(\mathbb{Q})$. Finally, verify that $\mu_F = \frac{2}{3}\delta_1 + \frac{1}{3}\text{Unif}_{[0,1]}$.

Exercise 1.4.38 For each $n \in \mathbb{N}$ let

$$F_n(x) = \begin{cases} 0 & \text{for } x < 0, \\ x^n & \text{for } 0 \le x < 1, \\ 1 & \text{for } x \ge 1. \end{cases}$$

Prove that F_n is an absolutely continuous CDF and determine the density γ_n of the associated distribution μ_n. Verify that

$$F(x) := \lim_{n\to\infty} F_n(x)$$

is a CDF and determine the associated distribution. Is the function

$$\gamma(x) := \lim_{n\to\infty} \gamma_n(x)$$

a density?

Exercise 1.4.39 Given a numbering $(q_n)_{n\in\mathbb{N}}$ of the rationals in $[0, 1]$, we define the distribution

$$\mu(\{x\}) = \begin{cases} 2^{-n} & \text{if } x = q_n, \\ 0 & \text{otherwise.} \end{cases}$$

Is the CDF F_μ continuous at point 1? Determine $F_\mu(1)$ and $F_\mu(1-)$.

Solution If $\bar{n} \in \mathbb{N}$ is such that $q_{\bar{n}} = 1$ then $\Delta F_\mu(1) = \frac{1}{2^{\bar{n}}}$. Since $F_\mu(1) = 1$ then $F_\mu(1-) = 1 - \frac{1}{2^{\bar{n}}}$.

1.4.9 Cumulative Distribution Functions on $\mathbb{R}^d$

The multi-dimensional case is analogous to the scalar case with some small differences.

Fig. 1.6 Graph of the bivariate Dirac CDF centered at $(1, 1)$

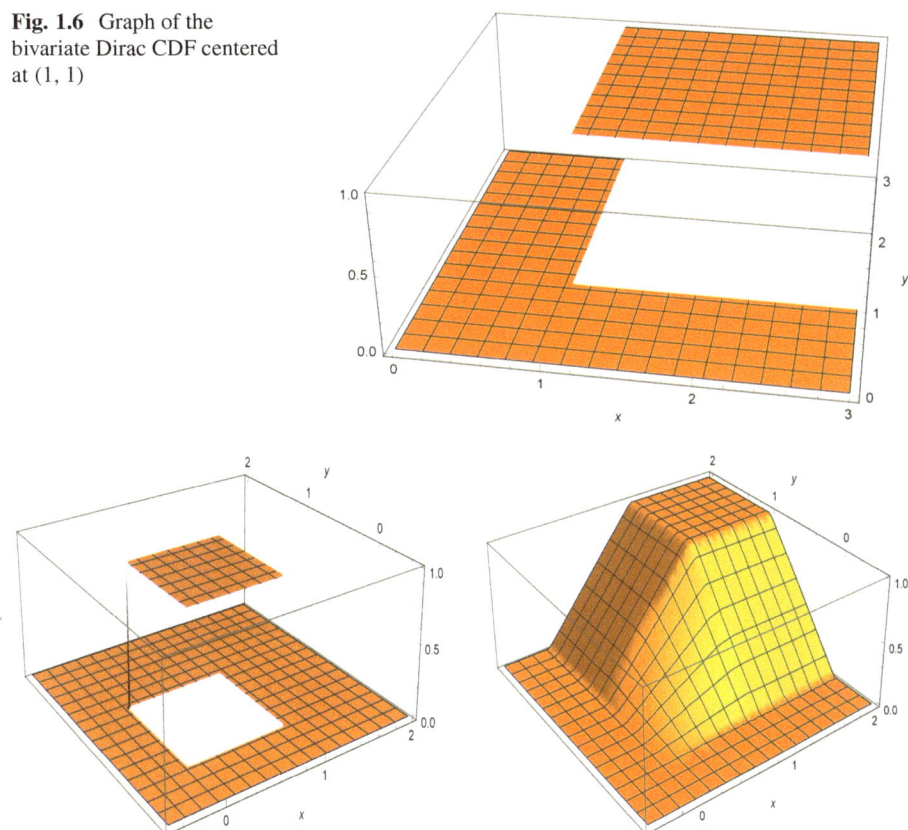

Fig. 1.7 Distribution $\text{Unif}_{[0,1]\times[0,1]}$: plot of the density (on the left) and of the CDF (on the right)

Definition 1.4.40 The *cumulative distribution function* of a distribution μ on $(\mathbb{R}^d, \mathcal{B}_d)$ is defined by

$$F_\mu(x) := \mu(]-\infty, x_1] \times \cdots \times]-\infty, x_d]), \qquad x = (x_1, \ldots, x_d) \in \mathbb{R}^d. \tag{1.4.13}$$

Example 1.4.41 We showcase the plots of various two-dimensional CDFs:

(i) Dirac distribution centered at $(1, 1)$ in Fig. 1.6;
(ii) Uniform distribution on the square $[0, 1] \times [0, 1]$ in Fig. 1.7. The density is the indicator function $\gamma = \mathbb{1}_{[0,1]\times[0,1]}$;
(iii) Standard bivariate normal distribution in Fig. 1.8, with density

$$\gamma(x, y) = \frac{e^{-\frac{x^2}{2} - \frac{y^2}{2}}}{2\pi}, \qquad (x, y) \in \mathbb{R}^2.$$

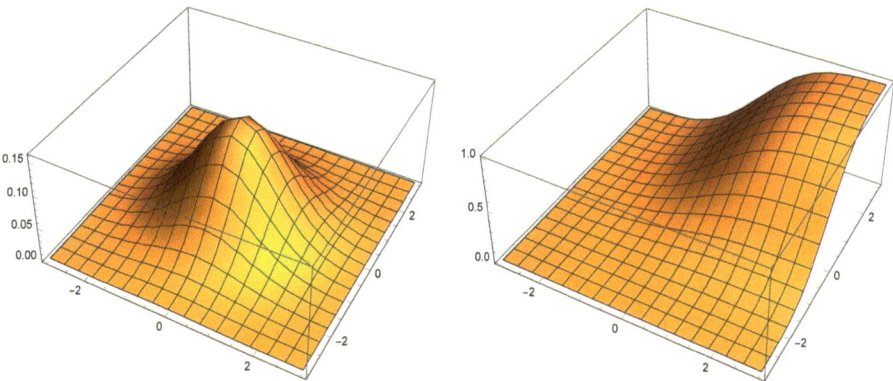

Fig. 1.8 Standard bivariate normal distribution: plot of the density (on the left) and of the CDF (on the right)

Example 1.4.42 Consider the bivariate CDF

$$F(x, y) = \left(1 - e^{-y} + \frac{e^{-y(x+1)} - 1}{x + 1} \right) \mathbb{1}_{\mathbb{R}_{\geq 0} \times \mathbb{R}_{\geq 0}}(x, y),$$

and suppose we know that F is absolutely continuous, i.e.,

$$F(x, y) = \int_{-\infty}^{x} \int_{-\infty}^{y} \gamma(\xi, \eta) d\xi d\eta$$

for some $\gamma \in m\mathscr{B}^{+}$. Then, as in the one-dimensional case (cf. Theorem 1.4.33-(iii)), a density for F is obtained simply by differentiating:

$$\partial_x \partial_y F(x, y) = y e^{-xy} \mathbb{1}_{\mathbb{R}_{\geq 0} \times \mathbb{R}_{\geq 0}}(x, y).$$

Now we state a theorem that naturally extends the results established in one dimension. We first observe that, given $k \in \{1, \ldots, d\}$, $a \leq b$ reals and $x \in \mathbb{R}^d$, we have

$$\mu(] - \infty, x_1] \times \cdots \times] - \infty, x_{k-1}] \times]a, b] \times] - \infty, x_{k+1}] \times \cdots \times] - \infty, x_d])$$
$$= F_\mu(x_1, \ldots, x_{k-1}, b, x_{k+1}, \ldots, x_d) - F_\mu(x_1, \ldots, x_{k-1}, a, x_{k+1}, \ldots, x_d)$$
$$=: \Delta_{]a,b]}^{(k)} F_\mu(x),$$

and more generally

$$\mu(]a_1, b_1] \times \cdots \times]a_d, b_d]) = \Delta_{]a_1,b_1]}^{(1)} \cdots \Delta_{]a_d,b_d]}^{(d)} F_\mu(x). \tag{1.4.14}$$

Theorem 1.4.43 *The CDF F_μ of a d-dimensional distribution μ has the following properties:*

(i) monotonicity: for each choice of $b_k > a_k \geq -\infty$, $1 \leq k \leq d$, we have

$$\Delta^{(1)}_{]a_1,b_1]} \cdots \Delta^{(d)}_{]a_d,b_d]} F_\mu(x) \geq 0; \tag{1.4.15}$$

(ii) right continuity: for each $x \in \mathbb{R}^d$

$$\lim_{y \to x^+} F_\mu(y) = F_\mu(x),$$

where $y \to x^+$ means that $y_k \to x_k^+$ for each $k = 1, \ldots, d$;
(iii) if $x_k \to -\infty$ for a $k = 1, \ldots, d$ then $F_\mu(x) \to 0$ and if $x_k \to +\infty$ for each $k = 1, \ldots, d$ then $F_\mu(x) \to 1$.

Conversely, if

$$F : \mathbb{R}^d \longrightarrow [0, 1]$$

is a function satisfying properties (i), (ii) and (iii) then there exists a distribution on $\mathbb{R}^d$ such that $F = F_\mu$, i.e., (1.4.13) holds.

Proof The proof follows a completely analogous approach to that used in the one-dimensional case. We only note that (1.4.15) follows directly from (1.4.14), since μ has non-negative values. $\qquad\square$

Remark 1.4.44 The monotonicity property (1.4.15) is not trivial: for $d = 2$, it becomes

$$0 \leq \Delta^{(1)}_{]a_1,b_1]}\Delta^{(2)}_{]a_2,b_2]} F(x) = F(b_1, b_2) - F(b_1, a_2) - (F(a_1, b_2) - F(a_1, a_2))$$

$$= F(b_1, b_2) - F(a_1, b_2) - (F(b_1, a_2) - F(a_1, a_2))$$

$$= \Delta^{(2)}_{]a_2,b_2]}\Delta^{(1)}_{]a_1,b_1]} F(x).$$

For example, the function

$$F(x_1, x_2) = \begin{cases} 1 & \text{if } x_1, x_2 \geq 1, \\ 2/3 & \text{if } x_1 \geq 1 \text{ and } 0 \leq x_2 < 1, \\ 2/3 & \text{if } x_2 \geq 1 \text{ and } 0 \leq x_1 < 1, \\ 0 & \text{otherwise,} \end{cases}$$

although being "monotone in every direction", does not satisfy (1.4.15). In fact,

$$\Delta^{(1)}_{]1/2,1]}\Delta^{(2)}_{]1/2,1]} F(x) = -1/3,$$

and therefore, if the distribution corresponding to F existed, it would assign negative probability to the square $]1/2, 1] \times]1/2, 1]$ which is obviously absurd.

Exercise 1.4.45 Let $I := [0, 1] \times \{0\} \subseteq \mathbb{R}^2$ and μ be the uniform distribution on I, defined by

$$\mu(H) = \mathrm{Leb}_1(H \cap I), \qquad H \in \mathscr{B}_2,$$

where Leb_1 denotes the one-dimensional Lebesgue measure.[26] Determine the CDF of μ.

1.4.10   Recap

How can we construct and define a probability measure? The first general tool from measure theory is Carathéodory's theorem, according to which every measure defined on an algebra $\mathscr{A}$ extends uniquely to the σ-algebra generated by $\mathscr{A}$. For example, based on this theorem, the measure defined for each interval $[a, b]$ as the length $b - a$ extends uniquely to the Lebesgue measure on the Borel σ-algebra.

Probability measures defined on $(\mathbb{R}^d, \mathscr{B}_d)$, also called *distributions*, play a particularly important role. Among them, *discrete* distributions are linear combinations (even countable) of Dirac deltas: notable examples are Bernoulli, discrete uniform, binomial, and Poisson distributions. Other important distributions are the *absolutely continuous* ones, i.e., those that can be represented in terms of the Lebesgue integral of a certain function, called *density*: notable examples are uniform, exponential, and normal distributions (but we will see many others...).

Discrete and absolutely continuous distributions are defined in terms of *real functions*: the distribution function in the former case and the density in the latter. This is a significant fact because it is much easier to handle a function of a real variable (or, in general, on $\mathbb{R}^d$) than a distribution (which is a measure, that is a function of Borel sets). On the other hand, there are distributions that are neither discrete nor absolutely continuous.

To characterize a generic distribution in terms of a real function, we have introduced the concept of *cumulative distribution function* (or CDF). A CDF has some general properties: in the one-dimensional case, a CDF is monotone increasing (and therefore a.e. differentiable), right continuous, and has limits at $-\infty$ and $+\infty$ equal to 0 and 1, respectively. We have shown that *assigning a distribution or its CDF is equivalent*.

---

[26] A bit improperly, given $A \in \mathscr{B}$, we are identifying $\mathrm{Leb}_1(A)$ with $\mathrm{Leb}_1(A \times \{0\})$.

Finally, the fact that a distribution μ has a density is equivalent to the fact that its CDF F is absolutely continuous, i.e., to the fact that

$$\mu(]a, x]) = F(x) - F(a) = \int_a^x F'(t)dt, \qquad a < x,$$

and in this case F' is a density of μ.

1.5   Appendix

1.5.1   Proof of Proposition 1.3.30

Proposition 1.3.30 For every $n \in \mathbb{N}$ and $p \in [0, 1]$, there exists a discrete space (Ω, P) on which a family $(C_h)_{h=1,\dots,n}$ of n repeated independent trials with probability p is canonically defined.

If $(C_h)_{h \in \mathbb{N}}$ is a sequence of independent events on a *discrete* space (Ω, P), such that $P(C_h) = p \in [0, 1]$ for every $h \in \mathbb{N}$, then necessarily $p = 0$ or $p = 1$.

Proof Let

$$\Omega = \{\omega = (\omega_1, \dots, \omega_n) \mid \omega_i \in \{0, 1\}\}$$

and consider the *partition*

$$\Omega = \bigcup_{k=0}^n \Omega_k, \qquad \Omega_k := \{\omega \in \Omega \mid \omega_1 + \cdots + \omega_n = k\}. \tag{1.5.1}$$

Clearly, each ω belongs to one and only one Ω_k: thus $\Omega_k \cap \Omega_h = \emptyset$ for $k \neq h$, and notice that $\Omega_k \leftrightarrow C_{n,k}$ (the element $(\omega_1, \dots, \omega_n)$ of Ω_k is uniquely determined by the choice of the k among n components that are equal to 1) that is

$$|\Omega_k| = \binom{n}{k}, \qquad k = 0, \dots, n. \tag{1.5.2}$$

We define P by setting

$$P(\{\omega\}) = p^k (1 - p)^{n-k} \qquad \omega \in \Omega_k, \ k = 0, \dots, n.$$

Then P is a probability since

$$P(\Omega) = \sum_{k=0}^{n} P(\Omega_k) = \sum_{k=0}^{n} \sum_{\omega \in \Omega_k} P(\{\omega\}) = \sum_{k=0}^{n} \binom{n}{k} p^k (1-p)^{n-k} = 1,$$

by (1.2.10).

We prove that the events

$$C_h = \{\omega \in \Omega \mid \omega_h = 1\}, \qquad h = 1, \ldots, n,$$

form a *family of n repeated independent trials* with probability p. Indeed, let $r \in \mathbb{N}$, $r \le n$, and $h_1, \ldots, h_r \in I_n$ be distinct. We have[27]

$$P\left(\bigcap_{i=1}^{r} C_{h_i}\right) = \sum_{k=r}^{n} P\left(\Omega_k \cap \left(\bigcap_{i=1}^{r} C_{h_i}\right)\right)$$

$$= \sum_{k=r}^{n} \left| \Omega_k \cap \left(\bigcap_{i=1}^{r} C_{h_i}\right) \right| p^k (1-p)^{n-k} =$$

(observing that, similarly to (1.5.2), the cardinality of $\Omega_k \cap \left(\bigcap_{i=1}^{r} C_{h_i}\right)$ is exactly equal to $\binom{n-r}{k-r}$)

$$= \sum_{k=r}^{n} \binom{n-r}{k-r} p^k (1-p)^{n-k} =$$

(by the change of index $j = k - r$)

$$= p^r \sum_{j=0}^{n-r} \binom{n-r}{j} p^j (1-p)^{n-j-r} = p^r.$$

Hence we have proved that, for $r = 1$,

$$P(C_h) = p, \qquad h = 1, \ldots, n,$$

---

[27] Note that the index in the summation starts from r because $\Omega_k \cap \left(\bigcap_{i=1}^{r} C_{h_i}\right) = \emptyset$ if $k < r$ (why?).

and for $1 < r \leq n$ we have

$$P\left(\bigcap_{i=1}^{r} C_{h_i}\right) = p^r = \prod_{i=1}^{r} P\left(C_{h_i}\right).$$

Thus $(C_h)_{h=1,\dots,n}$ is a *family of n repeated independent trials* with probability p.

As for the second part of the statement: let $(C_k)_{k\in\mathbb{N}}$ be a sequence of independent events on a discrete space (Ω, P), such that $P(C_k) = p \in [0, 1]$ for every $k \in \mathbb{N}$. It is not restrictive to assume $p \geq \frac{1}{2}$ because otherwise it suffices to consider the sequence of complementary events. In this case we prove that necessarily $p = 1$. In fact, suppose by contradiction that $p < 1$. Fix a generic outcome $\omega \in \Omega$: for every $n \in \mathbb{N}$ we set $\bar{C}_n = C_n$ or $\bar{C}_n = C_n^c$ depending on whether $\omega \in C_n$ or $\omega \in C_n^c$. Note that $P(\bar{C}_n) \leq P(C_n)$ since we have assumed $P(C_n) = p \geq \frac{1}{2}$. For every $n \in \mathbb{N}$ the events $\bar{C}_1, \dots, \bar{C}_n$ are independent and

$$\{\omega\} \subseteq \bigcap_{k=1}^{n} \bar{C}_k$$

so that

$$P(\{\omega\}) \leq \prod_{k=1}^{n} P(\bar{C}_k) \leq p^n.$$

By taking the limit as n approaches infinity, we arrive at $P(\{\omega\}) = 0$. However, since $\omega \in \Omega$ is arbitrary, this contradicts the assumption of the discreteness of Ω. $\qquad\square$

1.5.2 Proof of Proposition 1.4.9

Proposition 1.4.9 Let μ be a distribution on a metric space $(\mathbb{M}, \mathscr{B}_\varrho)$. For every $H \in \mathscr{B}_\varrho$ we have

$$\mu(H) = \sup\{\mu(C) \mid C \subseteq H, \ C \text{ closed}\} \qquad (1.5.3)$$

$$= \inf\{\mu(A) \mid A \supseteq H, \ A \text{ open}\}. \qquad (1.5.4)$$

We say that every Borel set is *internally* (by (1.5.3)) and *externally* (by (1.5.4)) μ-regular.

Proof Let $\mathscr{R}$ denote the family of (internally and externally) μ-regular Borel sets. It is clear that $H \in \mathscr{R}$ if and only if for every $\varepsilon > 0$ there exist a closed set C and an open set A such that

$$C \subseteq H \subseteq A, \qquad \mu(A \setminus C) < \varepsilon.$$

First, let us prove that $\mathscr{R}$ is a σ-algebra:

- since the empty set is open and closed, we have $\emptyset \in \mathscr{R}$;
- if $H \in \mathscr{R}$ then for every $\varepsilon > 0$ there exist a closed set C_ε and an open set A_ε such that $C_\varepsilon \subseteq H \subseteq A_\varepsilon$ and $\mu(A_\varepsilon \setminus C_\varepsilon) < \varepsilon$. Taking the complement, we have $A_\varepsilon^c \subseteq H^c \subseteq C_\varepsilon^c$, with A_ε^c closed, C_ε^c open and $C_\varepsilon^c \setminus A_\varepsilon^c = A_\varepsilon \setminus C_\varepsilon$. This proves that $H^c \in \mathscr{R}$;
- let $(H_n)_{n \in \mathbb{N}}$ be a sequence in $\mathscr{R}$ and $H = \bigcup_{n \geq 1} H_n$. Then, for every $\varepsilon > 0$ there exist two sequences, $(C_{n,\varepsilon})_{n \in \mathbb{N}}$ of closed sets and $(A_{n,\varepsilon})_{n \in \mathbb{N}}$ of open sets, such that $C_{n,\varepsilon} \subseteq H_n \subseteq A_{n,\varepsilon}$ and $\mu(A_{n,\varepsilon} \setminus C_{n,\varepsilon}) < \frac{\varepsilon}{3^n}$. Setting $A_\varepsilon = \bigcup_{n \geq 1} A_{n,\varepsilon}$, we have that A_ε is open and $H \subseteq A_\varepsilon$. On the other hand, by the continuity from below of μ (cf. Proposition 1.1.32), there exists $k \in \mathbb{N}$ such that $\mu(C \setminus C_\varepsilon) \leq \frac{\varepsilon}{2}$ where

$$C := \bigcup_{n=1}^{\infty} C_{n,\varepsilon}, \qquad C_\varepsilon := \bigcup_{n=1}^{n} A_{n,\varepsilon}.$$

Clearly, C_ε is closed and $C_\varepsilon \subseteq H$. Finally, we have

$$\mu(A_\varepsilon \setminus C_\varepsilon) \leq \mu(A_\varepsilon \setminus C) + \mu(C \setminus C_\varepsilon) \leq \sum_{n=1}^{\infty} \mu(A_{n,\varepsilon} \setminus C_{n,\varepsilon})$$

$$+ \frac{\varepsilon}{2} \leq \sum_{n=1}^{\infty} \frac{\varepsilon}{3^n} + \frac{\varepsilon}{2} = \varepsilon.$$

This proves that $\mathscr{R}$ is a σ-algebra. Now let us prove that $\mathscr{R}$ contains all the closed sets: given C closed, let $\varrho(x, C) = \inf_{y \in C} \varrho(x, y)$ and

$$A_n = \{x \in \mathbb{M} \mid \varrho(x, C) < 1/n\}, \qquad n \in \mathbb{N}.$$

Then A_n is open and $A_n \searrow C$: indeed, if $x \in \bigcap_{n \geq 1} A_n$ then $\varrho(x, C) = 0$ and therefore $x \in C$, being C closed. Then, due to the continuity from above of μ, we have $\lim_{n \to \infty} \mu(A_n) = \mu(C)$.

The thesis follows from the fact that $\mathscr{B}_\varrho$ is the smallest σ-algebra containing the open (and closed) sets, and therefore $\mathscr{B}_\varrho \subseteq \mathscr{R}$. $\qquad\qquad\square$

1.5.3 Proof of Carathéodory's Theorem 1.4.29

We provide a slightly more general (and definitely more convenient to apply) version of Theorem 1.4.29: in this section we follow the presentation in [24]. We introduce the definition of pre-measure on a generic family of subsets of Ω.

Definition 1.5.1 (Pre-measure) Let $\mathscr{A}$ be a family of subsets of Ω such that $\emptyset \in \mathscr{A}$. A pre-measure on $\mathscr{A}$ is a function

$$\mu : \mathscr{A} \longrightarrow [0, +\infty]$$

such that

(i) $\mu(\emptyset) = 0$;
(ii) μ is *additive on* $\mathscr{A}$ in the sense that for every $A, B \in \mathscr{A}$, disjoint and such that $A \cup B \in \mathscr{A}$, we have

$$\mu(A \uplus B) = \mu(A) + \mu(B);$$

(iii) μ is *σ-sub-additive on* $\mathscr{A}$ in the sense that for every $A \in \mathscr{A}$ and $(A_n)_{n\in\mathbb{N}}$ sequence of elements in $\mathscr{A}$, we have

$$A \subseteq \bigcup_{n\in\mathbb{N}} A_n \quad \Longrightarrow \quad \mu(A) \leq \sum_{n\in\mathbb{N}} \mu(A_n).$$

We say that μ is *σ-finite* if there exists a sequence $(A_n)_{n\in\mathbb{N}}$ in $\mathscr{A}$ such that $\Omega = \bigcup_{n\in\mathbb{N}} A_n$ and $\mu(A_n) < \infty$ for every $n \in \mathbb{N}$.

Definition 1.5.2 (Semi-ring) A family $\mathscr{A}$ of subsets of Ω is a *semi-ring* if:

(i) $\emptyset \in \mathscr{A}$
ii) $\mathscr{A}$ is $\cap$-closed;
(iii) for every $A, B \in \mathscr{A}$ the difference $B \setminus A$ is a finite disjoint union of sets in $\mathscr{A}$.

Example 1.5.3 [!] The family $\mathscr{A}$ of bounded intervals of the type

$$]a, b], \qquad a, b \in \mathbb{R}, \ a \leq b,$$

is a semi-ring (but not an algebra). The family of *finite unions* of intervals (even unbounded) of the type

$$]a, b], \qquad -\infty \leq a \leq b \leq +\infty,$$

is an algebra (but not a σ-algebra). Such families generate the Borel σ-algebra of $\mathbb{R}$.

We recall that a measure μ is a σ-additive function such that $\mu(\emptyset) = 0$ (cf. Definition 1.1.21). We observe that, by Proposition 1.1.32, μ is a pre-measure on an algebra $\mathscr{A}$ if and only if μ is a measure on $\mathscr{A}$. Moreover, the following lemma provides a natural result whose proof, which we postpone to the end of the section, is not that simple.

Lemma 1.5.4 *If μ is a measure on a semi-ring $\mathscr{A}$ then μ is a pre-measure on $\mathscr{A}$.*

Theorem 1.5.5 (Carathéodory's Theorem—General Version) *Let μ be a σ-finite pre-measure on a semi-ring $\mathscr{A}$. There exists a unique σ-finite measure that extends μ to $\sigma(\mathscr{A})$.*

Remark 1.5.6 Theorem 1.4.29 is a corollary of Theorem 1.5.5: in fact, every algebra is a semi-ring and, by Lemma 1.5.4, every measure on a semi-ring is a pre-measure.

Proof of Theorem 1.5.5 Uniqueness is a corollary of Dynkin's Theorem A.0.3: for details, see Corollary A.0.5 and Remark A.0.6. Here we prove the existence of the extension: in this proof we do not use the assumption that μ is σ-finite; on the other hand, if μ is σ-finite then its extension is also σ-finite. We divide the proof into several steps.

Step 1   We introduce the family of coverings of $B \subseteq \Omega$ that are finite or countable and consist of elements of $\mathscr{A}$:

$$\mathscr{U}(B) := \{\mathscr{R} \subseteq \mathscr{A} \mid \mathscr{R} \text{ at most countable and } B \subseteq \bigcup_{A \in \mathscr{R}} A\}.$$

We define

$$\mu^* : \mathscr{P}(\Omega) \longrightarrow [0, +\infty]$$

by setting

$$\mu^*(B) = \inf_{\mathscr{R} \in \mathscr{U}(B)} \sum_{A \in \mathscr{R}} \mu(A), \qquad (1.5.5)$$

with the convention $\inf \emptyset = +\infty$.

$\square$

Lemma 1.5.7 *μ^* is an outer measure that is, it verifies the following properties:*

(i) $\mu^*(\emptyset) = 0$;
(ii) μ^* is monotone;
(iii) μ^* is σ-sub-additive.

Moreover, $\mu^*(A) = \mu(A)$ for every $A \in \mathscr{A}$.

Proof Since $\emptyset \in \mathscr{A}$, (i) is obvious. If $B \subseteq C$, then $\mathscr{U}(C) \subseteq \mathscr{U}(B)$, hence it follows that $\mu^*(B) \leq \mu^*(C)$, and this proves (ii). Finally, given a sequence $(B_n)_{n \in \mathbb{N}}$ of subsets of Ω and setting $B = \bigcup_{n \in \mathbb{N}} B_n$, we prove that

$$\mu^*(B) \leq \sum_{n \in \mathbb{N}} \mu^*(B_n).$$

It is sufficient to consider the case $\mu^*(B_n) < \infty$ for every $n \in \mathbb{N}$, hence it follows in particular that $\mathscr{U}(B_n) \neq \emptyset$. Then, fixed $\varepsilon > 0$, for each $n \in \mathbb{N}$ there exists $\mathscr{R}_n \in \mathscr{U}(B_n)$ such that

$$\sum_{A \in \mathscr{R}_n} \mu(A) \leq \mu^*(B_n) + \frac{\varepsilon}{2^n}.$$

Now $\mathscr{R} := \bigcup_{n \in \mathbb{N}} \mathscr{R}_n \in \mathscr{U}(B)$ and therefore

$$\mu^*(B) \leq \sum_{A \in \mathscr{R}} \mu(A) \leq \sum_{n \in \mathbb{N}} \sum_{A \in \mathscr{R}_n} \mu(A) \leq \sum_{n \in \mathbb{N}} \mu^*(B_n) + \varepsilon$$

hence the thesis follows due to the arbitrariness of ε.

Finally, we prove that μ^* coincides with μ on $\mathscr{A}$. For every $A \in \mathscr{A}$, we have $\mu^*(A) \leq \mu(A)$ by definition. Conversely, since μ is σ-sub-additive on $\mathscr{A}$, for every $\mathscr{R} \in \mathscr{U}(A)$, we have

$$\mu(A) \leq \sum_{B \in \mathscr{R}} \mu(B)$$

and then also $\mu(A) \leq \mu^*(A)$. $\square$

Step 2 We denote by $\mathscr{M}(\mu^*)$ the family of $A \subseteq \Omega$ such that

$$\mu^*(E) = \mu^*(E \cap A) + \mu^*(E \cap A^c), \qquad \forall E \subseteq \Omega.$$

The elements of $\mathscr{M}(\mu^*)$ are called μ^*-measurable. We will prove that $\mathscr{M}(\mu^*)$ is a σ-algebra and μ^* is a measure on $\mathscr{M}(\mu^*)$. We begin with the following partial result.

Lemma 1.5.8 $\mathscr{M}(\mu^*)$ *is an algebra.*

Proof Clearly $\emptyset \in \mathcal{M}(\mu^*)$ and $\mathcal{M}(\mu^*)$ is closed under complement. We prove that the union of $A, B \in \mathcal{M}(\mu^*)$ belongs to $\mathcal{M}(\mu^*)$: for every $E \subseteq \Omega$, we have

$$\mu^*(E) = \mu^*(E \cap A) + \mu^*(E \cap A^c)$$

$$= \underbrace{\mu^*(E \cap A \cap B) + \mu^*(E \cap A \cap B^c) + \mu^*(E \cap A^c \cap B)}_{\geq \mu^*(E \cap A \cup B)}$$

$$+ \underbrace{\mu^*(E \cap A^c \cap B^c)}_{= \mu^*(E \cap (A \cup B)^c)}$$

since

$$(E \cap A \cup B) \subseteq (E \cap A \cap B) \cup (E \cap A \cap B^c) \cup (E \cap A^c \cap B).$$

This proves that

$$\mu^*(E) \geq \mu^*(E \cap (A \cup B)) + \mu^*(E \cap (A \cup B)^c).$$

On the other hand, μ^* is sub-additive and therefore $A \cup B \in \mathcal{M}(\mu^*)$. $\square$

Lemma 1.5.9 μ^* *is a measure on* $\mathcal{M}(\mu^*)$.

Proof It is sufficient to prove that μ^* is σ-additive on $\mathcal{M}(\mu^*)$. For every $A, B \in \mathcal{M}(\mu^*)$ with $A \cap B = \emptyset$, we have

$$\mu^*(A \uplus B) = \mu^*((A \uplus B) \cap A) + \mu^*((A \uplus B) \cap A^c) = \mu^*(A) + \mu^*(B).$$

Hence, μ^* is additive on $\mathcal{M}(\mu^*)$. Moreover, we already know from Step 1 that μ^* is σ-sub-additive and therefore the thesis follows from Proposition 1.1.32. $\square$

Lemma 1.5.10 $\mathcal{M}(\mu^*)$ *is a* σ-*algebra*.

Proof We already know that $\mathcal{M}(\mu^*)$ is $\cap$-closed. If we verify that $\mathcal{M}(\mu^*)$ is a monotone family (cf. Definition A.0.1), the thesis will follow from Lemma A.0.2. To this end, it is sufficient to prove that if $(A_n)_{n \in \mathbb{N}}$ is a sequence in $\mathcal{M}(\mu^*)$ and $A_n \nearrow A$, then $A \in \mathcal{M}(\mu^*)$. Thanks to the sub-additivity of μ^*, it suffices to prove that

$$\mu^*(E) \geq \mu^*(E \cap A) + \mu^*(E \cap A^c), \qquad E \subseteq \Omega. \tag{1.5.6}$$

Let $A_0 = \emptyset$ and observe that

$$\mu^*(E \cap A_n) = \mu^*((E \cap A_n) \cap A_{n-1}) + \mu^*((E \cap A_n) \cap A_{n-1}^c)$$

$$= \mu^*(E \cap A_{n-1}) + \mu^*(E \cap (A_n \setminus A_{n-1})).$$

As a consequence, we have

$$\mu^*(E \cap A_n) = \sum_{k=1}^{n} \mu^*(E \cap (A_k \setminus A_{k-1})) \tag{1.5.7}$$

and, by the monotonicity of μ^*,

$$\mu^*(E) = \mu^*(E \cap A_n) + \mu^*(E \cap A_n^c)$$
$$\geq \mu^*(E \cap A_n) + \mu^*(E \cap A^c) =$$

(by (1.5.7))

$$= \sum_{k=1}^{n} \mu^*(E \cap (A_k \setminus A_{k-1})) + \mu^*(E \cap A^c).$$

Sending n to infinity and using the σ-sub-additivity of μ^*, we get

$$\mu^*(E) \geq \sum_{k=1}^{\infty} \mu^*(E \cap (A_k \setminus A_{k-1})) + \mu^*(E \cap A^c) \geq \mu^*(E \cap A) + \mu^*(E \cap A^c),$$

which proves (1.5.6) and concludes the proof.   □

Step 3   As a final step, we prove that

$$\sigma(\mathscr{A}) \subseteq \mathscr{M}(\mu^*).$$

Since $\mathscr{M}(\mu^*)$ is a σ-algebra, it is sufficient to prove that $\mathscr{A} \subseteq \mathscr{M}(\mu^*)$: moreover, since μ^* is sub-additive, it is enough to prove that for every $A \in \mathscr{A}$ and $E \subseteq \Omega$, with $\mu^*(E) < \infty$, we have

$$\mu^*(E) \geq \mu^*(E \cap A) + \mu^*(E \cap A^c). \tag{1.5.8}$$

Given $\varepsilon > 0$, there exists a covering $(A_n)_{n \in \mathbb{N}}$ of E formed by elements of $\mathscr{A}$ and such that

$$\sum_{n \in \mathbb{N}} \mu(A_n) \leq \mu^*(E) + \varepsilon. \tag{1.5.9}$$

Since $\mathscr{A}$ is a semi-ring, we have $A_n \cap A \in \mathscr{A}$ and therefore, by Lemma 1.5.7,

$$\mu^*(A_n \cap A) = \mu(A_n \cap A). \tag{1.5.10}$$

On the other hand, still due to the fact that $\mathscr{A}$ is a semi-ring, for each $n \in \mathbb{N}$ there exist $B_1^{(n)}, \ldots, B_{k_n}^{(n)} \in \mathscr{A}$ such that

$$A_n \cap A^c = A_n \setminus A = \biguplus_{j=1}^{k_n} B_j^{(n)}.$$

Then

$$\mu^*(A_n \cap A^c) = \mu^*\left(\biguplus_{j=1}^{k_n} B_j^{(n)}\right) \leq$$

(since μ^* is sub-additive)

$$\leq \sum_{j=1}^{k_n} \mu^*(B_j^{(n)}) =$$

(since $\mu^* = \mu$ on $\mathscr{A}$ by Lemma 1.5.7)

$$= \sum_{j=1}^{k_n} \mu(B_j^{(n)}) =$$

(since μ is additive)

$$= \mu(A_n \cap A^c). \tag{1.5.11}$$

Now we prove (1.5.8): due to the σ-sub-additivity of μ^*, we have

$$\mu^*(E \cap A) + \mu^*(E \cap A^c) \leq \sum_{n \in \mathbb{N}} \left(\mu^*(A_n \cap A) + \mu^*(A_n \cap A^c)\right) \leq$$

(by (1.5.10) and (1.5.11))

$$\leq \sum_{n \in \mathbb{N}} \left(\mu(A_n \cap A) + \mu(A_n \cap A^c)\right) = \sum_{n \in \mathbb{N}} \mu(A_n) \leq$$

(by (1.5.9))

$$\leq \mu^*(E) + \varepsilon.$$

The thesis follows from the arbitrariness of ε. This concludes the proof of Theorem 1.5.5.

We now prove that the σ-algebra $\mathcal{M}(\mu^*)$, constructed in Step 2 of the proof of Carathéodory's theorem, contains the family of negligible sets. We note that in general $\mathcal{M}(\mu^*)$ is strictly larger than $\sigma(\mathcal{A})$: this is the case for the Lebesgue measure if $\mathcal{A}$ is the family of bounded intervals of the type

$$]a, b], \qquad a, b \in \mathbb{R}, \ a \le b.$$

In this case, $\sigma(\mathcal{A})$ is the Borel σ-algebra and $\mathcal{M}(\mu^*)$ is the Lebesgue σ-algebra. On the other hand, we also show that the elements of $\mathcal{M}(\mu^*)$ differ from those of $\sigma(\mathcal{A})$ only for μ^*-negligible sets.

Corollary 1.5.11 [!] *Under the assumptions of Carathéodory's theorem, in the measure space $(\Omega, \mathcal{M}(\mu^*), \mu^*)$ we have:*

(i) if $\mu^*(M) = 0$ then $M \in \mathcal{M}(\mu^*)$ and therefore $(\Omega, \mathcal{M}(\mu^*), \mu^*)$ is a complete measure space;
(ii) for every $M \in \mathcal{M}(\mu^*)$, such that $\mu^*(M) < \infty$, there exists $A \in \sigma(\mathcal{A})$ such that $M \subseteq A$ and $\mu^*(A \setminus M) = 0$.

Proof Due to the sub-additivity and monotonicity of μ^*, if $\mu^*(M) = 0$ and $E \subseteq \Omega$ we have

$$\mu^*(E) \le \mu^*(E \cap M) + \mu^*(E \cap M^c) = \mu^*(E \cap M^c) \le \mu^*(E),$$

and this proves (i).

It is clear that, by definition of μ^*, for every $n \in \mathbb{N}$ there exists $A_n \in \sigma(\mathcal{A})$ such that $M \subseteq A_n$ and

$$\mu^*(A_n) \le \mu^*(M) + \frac{1}{n}. \tag{1.5.12}$$

Setting $A = \bigcap_{n \in \mathbb{N}} A_n \in \sigma(\mathcal{A})$, we have $M \subseteq A$ and, passing to the limit in (1.5.12) and thanks to the continuity from above of μ^* on $\mathcal{M}(\mu^*)$, we have $\mu^*(A) = \mu^*(M)$. Then, since $M \in \mathcal{M}(\mu^*)$, we have

$$\mu^*(A) = \mu^*(A \cap M) + \mu^*(A \cap M^c) = \mu^*(M) + \mu^*(A \setminus M)$$

and therefore $\mu^*(A \setminus M) = 0$. $\square$

We conclude the section by giving the

Proof of Lemma 1.5.4 If μ is a measure on the semi-ring $\mathcal{A}$ then properties (i) and (ii) of pre-measure are obvious. Let us prove that μ is monotone: if $A, B \in \mathcal{A}$ with $A \subseteq B$ then, for property (iii) of semi-ring, there exist $C_1, \ldots, C_n \in \mathcal{A}$ such that

$$B \setminus A = \biguplus_{k=1}^{n} C_k.$$

Thus we have

$$\mu(B) = \mu(A \uplus (B \setminus A)) = \mu(A \uplus C_1 \uplus \cdots \uplus C_n)$$

(by the finite additivity of μ)

$$= \mu(A) + \sum_{k=1}^{n} \mu(C_k) \geq \mu(A),$$

from which the monotonicity of μ.

The proof of property (iii), i.e., the σ-sub-additivity of μ, is a slightly more complicated version of the proof of Proposition 1.1.22-(ii): all the complication is due to the fact that μ is defined on a semi-ring (instead of an algebra as in Proposition 1.1.22) and this limits the set operations we can use. Let $A \in \mathscr{A}$ and $(A_n)_{n \in \mathbb{N}}$ be a sequence in $\mathscr{A}$ such that

$$A \subseteq \bigcup_{n \in \mathbb{N}} A_n.$$

We set $\widetilde{A}_1 = A_1$ and

$$\widetilde{A}_n = A_n \setminus \bigcup_{k=1}^{n-1} A_k = \bigcap_{k=1}^{n-1} (A_n \setminus (A_n \cap A_k)), \qquad n \geq 2. \tag{1.5.13}$$

Then, by properties (ii) and (iii) of semi-ring, there exist $J_n \in \mathbb{N}$ and $C_1^{(n)}, \ldots, C_{J_n}^{(n)} \in \mathscr{A}$ such that

$$\widetilde{A}_n = \biguplus_{j=1}^{J_n} C_j^{(n)}.$$

Now, $\widetilde{A}_n \subseteq A_n$ and therefore, by monotonicity and additivity, we have

$$\mu(A_n) \geq \mu(\widetilde{A}_n) = \sum_{j=1}^{J_n} \mu(C_j^{(n)}). \tag{1.5.14}$$

Moreover, by (1.5.13),

$$A \subseteq \bigcup_{n \in \mathbb{N}} A_n = \biguplus_{n \in \mathbb{N}} \widetilde{A}_n = \biguplus_{n \in \mathbb{N}} \biguplus_{j=1}^{J_n} C_j^{(n)}$$

and therefore

$$\mu(A) = \mu\left(\biguplus_{n\in\mathbb{N}} \biguplus_{j=1}^{J_n} \left(A \cap C_j^{(n)}\right)\right) =$$

(since $A \cap C_j^{(n)} \in \mathscr{A}$ and, by hypothesis, μ is a measure and therefore, in particular, σ-additive)

$$= \sum_{n\in\mathbb{N}} \sum_{j=1}^{J_n} \mu\left(A \cap C_j^{(n)}\right) \le$$

(by monotonicity)

$$\le \sum_{n\in\mathbb{N}} \sum_{j=1}^{J_n} \mu\left(C_j^{(n)}\right) =$$

(by (1.5.14))

$$\le \sum_{n\in\mathbb{N}} \mu(A_n)$$

and this concludes the proof.   □

1.5.4 Proof of Theorem 1.4.33

Theorem 1.4.33 [!!] Let $F : \mathbb{R} \longrightarrow \mathbb{R}$ be an increasing and right continuous function (i.e., F satisfies properties (i) and (ii) of Theorem 1.4.26). Then:

(i) there exists a unique measure μ_F on $(\mathbb{R}, \mathscr{B})$ that is σ-finite and satisfies

$$\mu_F(]a, b]) = F(b) - F(a), \qquad a, b \in \mathbb{R}, \ a < b;$$

(ii) if F also satisfies

$$\lim_{x\to-\infty} F(x) = 0 \qquad \text{and} \qquad \lim_{x\to+\infty} F(x) = 1,$$

(i.e., F satisfies property iii) of Theorem 1.4.26) then μ_F is a distribution;

(iii) finally, F is absolutely continuous if and only if $\mu_F \in$ AC: in this case, F' is a density of μ_F.

Proof

[Part (i)] Consider the semi-ring $\mathscr{A}$ in Example 1.5.3, consisting of bounded intervals of the type

$$]a, b], \qquad a, b \in \mathbb{R}, \ a \leq b.$$

On $\mathscr{A}$ we define μ_F by

$$\mu_F(]a, b]) = F(b) - F(a).$$

The thesis follows from Carathéodory's Theorem 1.5.5 once we have proved that μ_F is a σ-finite pre-measure (cf. Definition 1.5.1). By definition, $\mu_F(\emptyset) = 0$ and clearly μ_F is σ-finite. Moreover, μ_F is additive since, if $]a, b],]c, d]$ are disjoint intervals such that their union is an interval then necessarily[28] $b = c$, so

$$\mu_F\left(]a, b] \uplus]b, d]\right) = \mu_F\left(]a, d]\right)$$
$$= F(d) - F(a) = (F(b) - F(a)) + (F(d) - F(b))$$
$$= \mu_F\left(]a, b]\right) + \mu_F\left(]b, d]\right).$$

Finally, we prove that μ_F is σ-sub-additive. It suffices to consider $]a, b] \in \mathscr{A}$ and a sequence $(A_n)_{n \in \mathbb{N}}$ in $\mathscr{A}$, of the type $A_n =]a_n, b_n]$, such that $\bigcup_{n \in \mathbb{N}} A_n =]a, b]$ and prove that

$$\mu_F(A) \leq \sum_{n=1}^{\infty} \mu_F(A_n).$$

Fix $\varepsilon > 0$: due to the right-continuity of F, there exist $\delta > 0$ and a sequence of positive numbers $(\delta_n)_{n \in \mathbb{N}}$ such that

$$F(a + \delta) \leq F(a) + \varepsilon, \qquad F(b_n + \delta_n) \leq F(b_n) + \frac{\varepsilon}{2^n}. \qquad (1.5.15)$$

The family $(]a_n, b_n + \delta_n[)_{n \in \mathbb{N}}$ is an open cover[29] of the compact set $[a + \delta, b]$ and thus admits a finite sub-cover: to fix ideas, let us denote by $(n_k)_{k=1,\dots,N}$ the indices of such a sub-cover. Then, by the first inequality in (1.5.15), we have

$$F(b) - F(a) \leq \epsilon + F(b) - F(a + \delta)$$
$$\leq \epsilon + \mu_F\left(]a + \delta, b]\right) \leq$$

---

[28] It is not restrictive to assume $a \leq d$.

[29] Since, for each $n \in \mathbb{N}$, $]a_n, b_n + \delta_n[$ contains $]a_n, b_n]$.

(since μ_F is finitely additive and therefore also finitely sub-additive)

$$\leq \epsilon + \sum_{k=1}^{N} \mu_F \left( ]a_{n_k}, b_{n_k} + \delta_{n_k}] \right)$$

$$\leq \epsilon + \sum_{n=1}^{\infty} (F(b_n + \delta_n) - F(a_n)) \leq$$

(by the second inequality in (1.5.15))

$$\leq \epsilon + \sum_{n=1}^{\infty} \frac{\epsilon}{2^n} + \sum_{n=1}^{\infty} (F(b_n) - F(a_n))$$

$$= 2\epsilon + \sum_{n=1}^{\infty} (F(b_n) - F(a_n)),$$

and the thesis follows from the arbitrariness of $\epsilon > 0$.
[Part (ii)] Since

$$\mu_F(\mathbb{R}) = \lim_{x \to +\infty} F(x) - \lim_{x \to -\infty} F(x) = 1,$$

where the first equality is by construction and the second by hypothesis, then μ_F is a probability measure on $\mathbb{R}$, that is, a distribution.
[Part (iii)] If F is absolutely continuous, by Proposition 1.4.31, for each $a < b$ we have

$$\mu_F(]a, b]) = F(b) - F(a) = \int_a^b F'(x)dx.$$

Note that $F' \geq 0$ a.e. because it is the limit of the incremental ratio of a monotone increasing function: taking the limit as $a \to -\infty$ and $b \to +\infty$, by Beppo Levi's theorem, we get

$$1 = \mu_F(\mathbb{R}) = \int_{\mathbb{R}} F'(x)dx$$

and thus F' is a density. Let us consider the distribution defined by

$$\mu(H) := \int_H F'(x)dx, \qquad H \in \mathcal{B}.$$

Then μ_F coincides with μ on the semi-ring $\mathscr{A}$ of bounded intervals of the type $]a, b]$. Since $\mathscr{A}$ generates $\mathscr{B}$, by the uniqueness result of Carathéodory's theorem, we have $\mu_F = \mu$ on $\mathscr{B}$ and thus $\mu_F \in AC$ with density F'.

Conversely, if $\mu_F \in AC$ with density γ then

$$F(x) - F(a) = \int_a^x \gamma(t)dt, \qquad a < x,$$

and thus F is absolutely continuous and, by Proposition 1.4.31, $F' = \gamma$ almost everywhere. $\square$

Chapter 2
Random Variables

> *The theory of probability as a mathematical discipline can and should be developed from axioms in exactly the same way as geometry and algebra.*
>
> *Andrej N. Kolmogorov*

Random variables describe *quantities that depend on a random phenomenon or experiment*: for example, if the experiment is the *roll of two dice*, the quantity (random variable) that we might be interested in studying could be the *result of the sum of the two rolls*. The underlying random phenomenon is modeled with a probability space $(\Omega, \mathscr{F}, P)$ (in the example, the discrete space $\Omega = I_6 \times I_6$ with uniform probability) and the quantity of interest is described by the random variable X that associates the value $X(\omega)$ to each outcome $\omega \in \Omega$ (i.e., to each possible outcome of the random phenomenon): in the example, $\omega = (\omega_1, \omega_2) \in I_6 \times I_6$ and $X(\omega) = \omega_1 + \omega_2$.

2.1 Random Variables

Consider a probability space $(\Omega, \mathscr{F}, P)$ and fix $d \in \mathbb{N}$. Given $H \subseteq \mathbb{R}^d$ and a function

$$X : \Omega \longrightarrow \mathbb{R}^d,$$

we denote by

$$(X \in H) := X^{-1}(H) = \{\omega \in \Omega \mid X(\omega) \in H\}$$

the pre-image of H through X. Intuitively, $(X \in H)$ represents the set of outcomes ω (i.e., the states of the random phenomenon) such that $X(\omega) \in H$. Returning to the dice roll example, if $H = \{7\}$ then $(X \in H)$ represents the event "the result of the sum of the roll of two dice is 7" and consists of all pairs (ω_1, ω_2) such that $\omega_1 + \omega_2 = 7$. In the case $d = 1$, we will also use the following notations:

$$(X > c) := \{\omega \in \Omega \mid X(\omega) > c\}, \qquad (X = c) := \{\omega \in \Omega \mid X(\omega) = c\}, \qquad c \in \mathbb{R}.$$

Moreover, if X, Y are two functions from $(\Omega, \mathscr{F}, P)$ to $\mathbb{R}^d$, we write

$$(X = Y) := \{\omega \in \Omega \mid X(\omega) = Y(\omega)\}.$$

It is worth noting that it is not always the case that $(X \in H)$ is an event, i.e., in general $(X \in H) \notin \mathscr{F}$ (apart from the trivial case of discrete probability spaces, in which we assume that $\mathscr{F} = \mathscr{P}(\Omega)$ and therefore all subsets of Ω are events). In particular, without further assumptions, it does not make sense to write $P(X \in H)$. However, in practical applications, our focus lies in computing these probabilities. This rationale justifies the subsequent definition of a random variable.

Definition 2.1.1 (Random Variable) Let X be a function

$$X : \Omega \longrightarrow \mathbb{R}^d$$

defined on a probability space $(\Omega, \mathscr{F}, P)$, with values in $\mathbb{R}^d$. We say that X is a *random variable* (abbreviated as r.v.) if

$$(X \in H) \in \mathscr{F}, \qquad H \in \mathscr{B}_d$$

where $\mathscr{B}_d$ is the Borel σ-algebra. In that case, we write $X \in m\mathscr{F}$ and also say that X is *$\mathscr{F}$-measurable*. We denote by $m\mathscr{F}^+$ the class of non-negative, $\mathscr{F}$-measurable functions; moreover, $b\mathscr{F}$ is the class of bounded, $\mathscr{F}$-measurable functions.

Clearly, a r.v. X on $(\Omega, \mathscr{F}) = (\mathbb{R}^n, \mathscr{B}_n)$ is simply a Borel-measurable function.

Remark 2.1.2 [!] In this book, we mainly confine our discussion to random variables taking values in $\mathbb{R}^d$. However, it is good to know also the following general definition: given a measurable space $(E, \mathscr{E})$, a r.v. on $(\Omega, \mathscr{F}, P)$ with values in E is a function

$$X : \Omega \longrightarrow E$$

that is $(\mathscr{F}, \mathscr{E})$-measurable in the sense that $X^{-1}(\mathscr{E}) \subseteq \mathscr{F}$ i.e., $(X \in H) \in \mathscr{F}$ for every $H \in \mathscr{E}$. To consider a concrete example, if we are interested in "extended" random variables, taking values in $\overline{\mathbb{R}} = \mathbb{R} \cup \{-\infty, +\infty\}$, we can use the general definition of r.v. with $(E, \mathscr{E}) = (\overline{\mathbb{R}}, \overline{\mathscr{B}})$ where $\overline{\mathscr{B}}$ is the Borel σ-algebra associated with the metric $\varrho(x, y) := |\delta(x) - \delta(y)|$, where $\delta(x) = \frac{x}{1+|x|}$ maps $\overline{\mathbb{R}}$ into $[-1, 1]$ (cf. Sect. 1.4.2). It is not difficult to check that $\overline{\mathscr{B}} = \sigma\left(\mathscr{B} \cup \{\{-\infty\}, \{+\infty\}\}\right)$.

As we explained above, *in the case of discrete spaces* the measurability condition is automatically satisfied and *every function $X : \Omega \longrightarrow \mathbb{R}^d$ is a r.v.* In general, the condition $(X \in H) \in \mathscr{F}$ ensures the well-definedness of $P(X \in H)$, enabling us to consider the probability of X taking values in the Borel set H.

Remark 2.1.3 (σ-Algebra Generated by a r.v.) If

$$X : \Omega \longrightarrow \mathbb{R}^d$$

is any function, $H \subseteq \mathbb{R}^d$ and $(H_i)_{i \in I}$ is any family of subsets of $\mathbb{R}^d$, then we have

$$X^{-1}\left(H^c\right) = \left(X^{-1}\left(H\right)\right)^c, \qquad X^{-1}\left(\bigcup_{i \in I} H_i\right) = \bigcup_{i \in I} X^{-1}\left(H_i\right).$$

As a consequence, we have that

$$\sigma(X) := X^{-1}(\mathscr{B}_d) = \left\{X^{-1}(H) \mid H \in \mathscr{B}_d\right\}$$

is a σ-algebra, called the σ-algebra generated by X. Notice that $X \in m\mathscr{F}$ if and only if $\sigma(X) \subseteq \mathscr{F}$.

Example 2.1.4 Consider $X : I_6 \longrightarrow \mathbb{R}$ defined by

$$X(n) = \begin{cases} 1 & \text{if } n \text{ is even,} \\ 0 & \text{if } n \text{ is odd.} \end{cases}$$

We can interpret X as the r.v. that indicates whether the result of a die roll is an even or odd number. Then we have

$$\sigma(X) = \left\{\varnothing, \Omega, \{2, 4, 6\}, \{1, 3, 5\}\right\}$$

that is, $\sigma(X)$ contains precisely the "meaningful" events for the r.v. X. In probabilistic models for applications, $\sigma(X)$ is referred to as the *σ-algebra of information on X*, representing the collection of information regarding the random variable X. This is, at least partially, justified by the inclusion of events of the form $(X \in H)$ with $H \in \mathscr{B}$ within $\sigma(X)$: these events are deemed "relevant" for studying the random variable X, as knowing the probabilities of these events is tantamount to understanding the probabilities associated with X assuming its values.

Lemma 2.1.5 [!] *Let $\mathscr{H}$ be a family of subsets of $\mathbb{R}^d$ such that $\sigma(\mathscr{H}) = \mathscr{B}_d$. If $X^{-1}(\mathscr{H}) \subseteq \mathscr{F}$ then $X \in m\mathscr{F}$.*

Proof Let

$$\mathscr{E} = \{H \in \mathscr{B}_d \mid X^{-1}(H) \in \mathscr{F}\}.$$

Then $\mathscr{E}$ is a σ-algebra and since $\mathscr{E} \supseteq \mathscr{H}$ by assumption, then $\mathscr{E} \supseteq \sigma(\mathscr{H}) = \mathscr{B}_d$ which proves the claim.   □

Lemma 2.1.5 says that *to prove that a function X is a random variable, it is sufficient to verify the measurability condition on a family $\mathscr{H}$ of generators of $\mathscr{B}_d$.*

Corollary 2.1.6 *Let $X_k : \Omega \longrightarrow \mathbb{R}$ with $k = 1, \ldots, d$. The following properties are equivalent:*

(i) $X := (X_1, \ldots, X_d) \in m\mathscr{F}$;
(ii) $X_k \in m\mathscr{F}$ *for every* $k = 1, \ldots, d$;
(iii) $(X_k \leq x) \in \mathscr{F}$ *for every* $x \in \mathbb{R}$ *and* $k = 1, \ldots, d$.

Proof It is easy to prove that (i) implies (ii); the converse follows from Lemma 2.1.5, the fact that

$$((X_1, \ldots, X_d) \in H_1 \times \cdots \times H_d) = \bigcap_{k=1}^{d}(X_k \in H_k)$$

and $\mathscr{H} := \{H_1 \times \cdots \times H_d \mid H_k \in \mathscr{B}\}$ is a family of subsets of $\mathbb{R}^d$ such that $\sigma(\mathscr{H}) = \mathscr{B}_d$.

Finally, (ii) and (iii) are equivalent again by Lemma 2.1.5, since the family of intervals of the type $]-\infty, x]$ generates $\mathscr{B}$ (cf. Exercise 1.4.7-(iii)).   □

We now present the first simple examples of random variables, also explicitly writing the σ-algebra $\sigma(X)$ generated by X and the image $X(\Omega) = \{X(\omega) \mid \omega \in \Omega\}$ which is *the set of possible values of X.*

Example 2.1.7

(i) Given $c \in \mathbb{R}^d$, consider the constant function $X \equiv c$. We have

$$\sigma(X) = \{\emptyset, \Omega\}$$

and therefore X is a random variable. In this case $X(\Omega) = \{c\}$ and obviously c represents the only value that X can assume. Therefore, the variable X is "not really random".

(ii) Given an event $A \in \mathscr{F}$, the *indicator function of A* is defined by

$$X(\omega) = \mathbb{1}_A(\omega) = \begin{cases} 1 & \omega \in A, \\ 0 & \omega \in A^c. \end{cases}$$

X is a r.v. since

$$\sigma(X) = \{\emptyset, A, A^c, \Omega\},$$

and in this case $X(\Omega) = \{0, 1\}$.

(iii) Let $(C_h)_{h=1,\ldots,n}$ be a family of n repeated independent trials. Consider the r.v. S that counts the number of successes among the n tests: in other words

$$S(\omega) = \sum_{h=1}^{n} \mathbb{1}_{C_h}(\omega), \qquad \omega \in \Omega.$$

With reference to the canonical space of Proposition 1.3.30, we also have

$$S(\omega) = \sum_{h=1}^{n} \omega_h, \qquad \omega \in \Omega,$$

and, recalling formula (1.5.1), we have $(S = k) = \Omega_k$ with $k = 0, 1, \ldots, n$. Thus, $\sigma(X)$ contains $\emptyset$ and all unions of events $\Omega_0, \ldots, \Omega_n$. In this case $S(\Omega) = \{0, 1, \ldots, n\}$.

(iv) Let $(C_h)_{h=1,\ldots,n}$ be a family of n repeated independent trials. Consider the r.v. T that indicates the "first time" of success among the n tests: in other words

$$T(\omega) = \min\{h \mid \omega \in C_h\}, \qquad \omega \in \Omega,$$

and we set by convention $\min \emptyset = n+1$. In this case $T(\Omega) = \{1, \ldots, n, n+1\}$. With reference to the canonical space of Proposition 1.3.30, we also have

$$T(\omega) = \min\{h \mid \omega_h = 1\}, \qquad \omega \in \Omega.$$

$\sigma(X)$ contains $\emptyset$ and all unions of events $(T = 1), \ldots, (T = n + 1)$. Note that

$$(T = 1) = C_1, \qquad (T = n + 1) = C_1^c \cap \cdots \cap C_n^c$$

and, for $1 < k \leq n$,

$$(T = k) = C_1^c \cap \cdots \cap C_{k-1}^c \cap C_k.$$

Proposition 2.1.8 *The following properties of measurable functions hold true:*

(i) let

$$X : \Omega \longrightarrow \mathbb{R}^d, \qquad f : \mathbb{R}^d \longrightarrow \mathbb{R}^n,$$

with X being a r.v. and $f \in m\mathscr{B}_d$. Then we have

$$\sigma(f \circ X) \subseteq \sigma(X) \tag{2.1.1}$$

and consequently $f(X)$ is a r.v., that is $f(X) \in m\mathscr{F}$;

(ii) if $(X_n)_{n\in\mathbb{N}}$ is a sequence in $m\mathscr{F}$ then also

$$\inf_n X_n, \qquad \sup_n X_n, \qquad \liminf_{n\to\infty} X_n, \qquad \limsup_{n\to\infty} X_n,$$

belong[1] to $m\mathscr{F}$.

Proof Equation (2.1.1) follows from the assumption $f^{-1}(\mathscr{B}_n) \subseteq \mathscr{B}_d$ and the fact that $f(X) \in m\mathscr{F}$ is an immediate consequence.

(ii) follows from the fact that, for every $a \in \mathbb{R}$, we have

$$\left(\inf_n X_n < a\right) = \bigcup_n (X_n < a), \qquad \left(\sup_n X_n < a\right) = \bigcap_n (X_n < a),$$

and

$$\liminf_{n\to\infty} X_n = \sup_n \inf_{k\geq n} X_k, \qquad \limsup_{n\to\infty} X_n = \inf_n \sup_{k\geq n} X_k.$$

$\square$

Remark 2.1.9 From (i) of Proposition 2.1.8, it follows in particular that if $X, Y \in m\mathscr{F}$ and $\lambda \in \mathbb{R}$ then $X + Y, XY, \lambda X \in m\mathscr{F}$. In fact, it suffices to observe that $X + Y, XY$ and λX are continuous (and thus $\mathscr{B}$-measurable) functions of the pair (X, Y) which is a r.v. by Corollary 2.1.6.

Moreover, for every sequence $(X_n)_{n\in\mathbb{N}}$ of random variables, we have

$$A := \{\omega \in \Omega \mid \text{ there exists } \lim_{n\to\infty} X_n(\omega)\} = \{\omega \in \Omega \mid \limsup_{n\to\infty} X_n(\omega)$$

$$= \liminf_{n\to\infty} X_n(\omega)\} \in \mathscr{F}. \tag{2.1.2}$$

Definition 2.1.10 (Almost Sure Convergence) If A in (2.1.2) is almost sure, i.e., $P(A) = 1$, then we say that $(X_n)_{n\in\mathbb{N}}$ *converges almost surely.*

We recall from Remark 1.4.3 that a space $(\Omega, \mathscr{F}, P)$ is complete if $\mathscr{N} \subseteq \mathscr{F}$, i.e., negligible (and almost sure) sets are events, and it is always possible to complete $(\Omega, \mathscr{F}, P)$. The completeness assumption is often useful as the following remarks show.

Remark 2.1.11 (Almost Sure Properties and Completeness) Consider a "property" $\mathscr{P} = \mathscr{P}(\omega)$ whose validity depends on $\omega \in \Omega$: to fix ideas, in Remark 2.1.9 we have $\mathscr{P}(\omega)=$"there exists $\lim_{n\to\infty} X_n(\omega)$". We say that $\mathscr{P}$ is *almost sure* (or *holds a.s.*) if the set

$$A := \{\omega \in \Omega \mid \mathscr{P}(\omega) \text{ is true}\}$$

---

[1] In general, as extended random variables (cf. Remark 2.1.2).

is almost sure: this means that there exists $C \in \mathscr{F}$ such that $P(C) = 1$ and $C \subseteq A$ or, equivalently, there exists N negligible such that $\mathscr{P}(\omega)$ is true for every $\omega \in \Omega \setminus N$.

In a complete space, $\mathscr{P}$ holds a.s. if and only if $P(A) = 1$. If the space is not complete, it is not necessarily true that $A \in \mathscr{F}$ and therefore $P(A)$ is not defined. In the particular case of Remark 2.1.9, the fact that $A \in \mathscr{F}$ is a consequence of (2.1.2) and the fact that the X_n are random variables.

Definition 2.1.12 (Almost Sure Equality) Given two functions (not necessarily random variables)

$$X, Y : \Omega \longrightarrow \mathbb{R}^d,$$

we say that $X = Y$ almost surely, and write $X = Y$ a.s. (or $X \overset{a.s.}{=} Y$), if the set $(X = Y)$ is almost sure.

Remark 2.1.13 By Remark 1.1.18, in a complete space

$$X \overset{a.s.}{=} Y \quad \Longleftrightarrow \quad P(X = Y) = 1.$$

Without the completeness assumption, $(X = Y)$ is not necessarily an event (unless, for example, X and Y are both random variables). Consequently, $P(X = Y)$ is not well defined and it is not correct to state that $X = Y$ a.s. is equivalent to $P(X = Y) = 1$. We also note that, in a complete space, if $X = Y$ a.s. and Y is a r.v. then also X is a random variable: this is not necessarily true if the space is not complete.

Remark 2.1.14 [!] Let $(X_n)_{n \in \mathbb{N}}$ be a sequence of random variables that converges almost surely on the event A defined as in (2.1.2). Set

$$X(\omega) := \lim_{n \to \infty} X_n(\omega), \qquad \omega \in A,$$

and, by convention, $X(\omega) = 0$ for every $\omega \in \Omega \setminus A$. Then X is a random variable. If the space is incomplete, we can modify X on a negligible (non-measurable) set to get a function Y that is not a r.v. (i.e., Y is not measurable) and still $(X_n)_{n \in \mathbb{N}}$ converges to Y almost surely. In other words, *in an incomplete space, almost sure convergence does not preserve the measurability property.*

2.1.1   Random Variables and Distributions

Let $X : \Omega \longrightarrow \mathbb{R}^d$ be a r.v. on the probability space $(\Omega, \mathscr{F}, P)$. To X is naturally associated the distribution defined by

$$\mu_X(H) := P(X \in H), \qquad H \in \mathscr{B}_d. \tag{2.1.3}$$

It is easy to verify that μ_X in (2.1.3) is a distribution, i.e., a probability measure on $\mathscr{B}_d$: indeed, we have $\mu_X(\mathbb{R}^d) = P(X \in \mathbb{R}^d) = 1$ and also, for every disjoint sequence $(H_n)_{n\in\mathbb{N}}$ in $\mathscr{B}_d$, we have

$$\mu_X\left(\biguplus_{n=1}^{\infty} H_n\right) = P\left(X^{-1}\left(\biguplus_{n=1}^{\infty} H_n\right)\right) = P\left(\biguplus_{n=1}^{\infty} X^{-1}(H_n)\right) =$$

(by the σ-additivity of P)

$$= \sum_{n=1}^{\infty} P\left(X^{-1}(H_n)\right) = \sum_{n=1}^{\infty} \mu_X(H_n).$$

Definition 2.1.15 (Law, CDF, and Density of a r.v.) For a random variable $X : \Omega \longrightarrow \mathbb{R}^d$ defined on $(\Omega, \mathscr{F}, P)$, the distribution μ_X in (2.1.3) is called the *distribution* (or *law*) of X. To indicate that X has distribution μ_X, we write

$$X \sim \mu_X.$$

The function[2]

$$F_X(x) := P(X \le x), \qquad x \in \mathbb{R}^d,$$

is called the *cumulative distribution function or CDF* of X. Note that F_X is the CDF of μ_X. Finally, if $\mu_X \in AC$ with density γ_X, then we say that X is absolutely continuous and has density γ_X: in this case, we have

$$P(X \in H) = \int_H \gamma_X(x)dx, \qquad H \in \mathscr{B}_d.$$

To grasp the previous definition, we suggest examining in detail the following

Example 2.1.16 [!] On the probability space $(\Omega, \mathscr{F}, P) \equiv (\mathbb{R}, \mathscr{B}, \text{Exp}_\lambda)$, where λ is a positive parameter, consider the random variables

$$X(\omega) = \omega^2, \qquad Y(\omega) = \begin{cases} 0 & \text{if } \omega \le 2, \\ 1 & \text{if } \omega > 2, \end{cases} \qquad Z(\omega) = \omega, \qquad \omega \in \mathbb{R}.$$

---

[2] As usual, $(X \le x) = \bigcap_{k=1}^{d}(X_k \le x_k)$.

To determine the law of X, we calculate the corresponding CDF: for $x < 0$ we have $P(X \leq x) = 0$, while for $x \geq 0$ we have

$$F_X(x) = P(X \leq x) = \text{Exp}_\lambda(\{\omega \in \mathbb{R} \mid \omega^2 \leq x\}) = \int_0^{\sqrt{x}} \lambda e^{-\lambda t} dt = 1 - e^{-\lambda \sqrt{x}}.$$

Then X *is absolutely continuous* with density

$$\gamma_X(x) = \frac{dF_X(x)}{dx} = \frac{\lambda e^{-\lambda \sqrt{x}}}{2\sqrt{x}} \mathbb{1}_{\mathbb{R}_{\geq 0}}(x).$$

The r.v. Y takes only the values 0 and 1: therefore $Y \sim \text{Be}_p$ with

$$p = P(Y = 1) = \text{Exp}_\lambda(]2, +\infty]) = \int_2^{+\infty} \lambda e^{-\lambda t} dt = e^{-2\lambda}.$$

As an exercise, prove that $Z \sim \text{Exp}_\lambda$.

Remark 2.1.17 (Existence) [!] Given a distribution μ on $\mathbb{R}^d$, *there exists a r.v. X on a probability space $(\Omega, \mathscr{F}, P)$ such that $\mu = \mu_X$*: for example, it suffices to consider the identity r.v. $X(\omega) \equiv \omega$ defined on the probability space $(\mathbb{R}^d, \mathscr{B}_d, \mu)$. The choice of $(\Omega, \mathscr{F}, P)$ and X is not unique: in other words, different random variables, even defined on different probability spaces, may have the same distribution. For example, consider:

(i) rolling a die: $\Omega_1 = I_6 := \{1, 2, 3, 4, 5, 6\}$ with uniform probability and $X(\omega) = \omega$;

(ii) rolling two dice: $\Omega_2 = I_6 \times I_6$ with uniform probability and $Y(\omega_1, \omega_2) = \omega_1$.

Then X and Y have the same law (which is the discrete uniform distribution Unif_{I_6}) but *they are different random variables, defined on different probability spaces*.

Hence the law of a r.v. does not provide complete knowledge of the r.v. itself. Knowing the distribution of a r.v. X means knowing "how the probability is distributed among the various values that X can take". For many applications, this is more than enough; indeed, often *probabilistic models are defined starting from a distribution* (or, equivalently, assigning a CFD or a density, in the absolutely continuous case) rather than through the explicit definition of the probability space and the r.v. considered.

Definition 2.1.18 (Equivalence in Law) Let X, Y be random variables (not necessarily defined on the same probability space). We say that *X and Y are equal in law (or distribution)* if $\mu_X = \mu_Y$. In this case, we write

$$X \overset{d}{=} Y.$$

Exercise 2.1.19 Prove the following statements:

(i) if $X \overset{a.s.}{=} Y$ then $X \overset{d}{=} Y$;
(ii) there exist X, Y random variables defined on the same space $(\Omega, \mathscr{F}, P)$ such that $X \overset{d}{=} Y$ but $P(X = Y) < 1$;
(iii) if $X \overset{d}{=} Y$ and $f \in m\mathscr{B}$ then $f \circ X \overset{d}{=} f \circ Y$.

Solution

(i) We use the fact that $P(X = Y) = 1$ and, recalling Example 1.1.29, for every z we have

$$P(X \in H) = P((X \in H) \cap (X = Y))$$
$$= P((Y \in H) \cap (X = Y)) = P(Y \in H).$$

(ii) In a space $(\Omega, \mathscr{F}, P)$ let $A, B \in \mathscr{F}$ be such that $P(A) = P(B)$. Then the indicator random variables $X = \mathbb{1}_A$ and $Y = \mathbb{1}_B$ both have Bernoulli distribution equal to

$$P(A)\delta_1 + (1 - P(A))\,\delta_0,$$

since they assume only the values 1 and 0 with probability $P(A)$ and $1 - P(A)$, respectively. Regarding the CDF, we have

$$F_Y(x) = F_X(x) = P(X \leq x) = \begin{cases} 0 & \text{if } x < 0, \\ P(A^c) & \text{if } 0 \leq x < 1, \\ 1 & \text{if } x \geq 1. \end{cases}$$

(iii) For every $H \in \mathscr{B}$ we have

$$P\left((f \circ X)^{-1}(H)\right) = P\left(X^{-1}\left(f^{-1}(H)\right)\right) =$$

(since by assumption $X \overset{d}{=} Y$)

$$= P\left(Y^{-1}\left(f^{-1}(H)\right)\right) = P((f \circ Y)^{-1}(H)).$$

We now examine some examples of *absolutely continuous* and *discrete* random variables. Recall that X is *absolutely continuous* if

$$P(X \in H) = \int_H \gamma_X(x)dx, \qquad H \in \mathscr{B}_d,$$

where the density γ_X is a $\mathscr{B}_d$-measurable, non-negative function (i.e., $\gamma_X \in m\mathscr{B}_d^+$) such that $\int_{\mathbb{R}^d} \gamma_X(x)dx = 1$.

We say that a r.v. X is *discrete* if its law is a discrete distribution (cf. Definition 1.4.15), that is, it is a finite or countable combination of Dirac deltas

$$\mu_X = \sum_{k \geq 1} p_k \delta_{x_k}, \qquad (2.1.4)$$

where (x_k) is a sequence of distinct points in $\mathbb{R}^d$ and (p_k) is a sequence of non-negative numbers with sum equal to one. If $\bar{\mu}_X$ denotes the distribution function of μ_X, then we have

$$P(X = x_k) = \bar{\mu}_X(x_k) = p_k, \qquad k \in \mathbb{N}.$$

Remark 2.1.20 The plots of the density γ_X (in the case of absolutely continuous distributions) and the distribution function $\bar{\mu}_X$ (in the case of discrete distributions) provide a simple and clear representation of how the probability is distributed among the possible values of X: we illustrate this fact in the following section by means of some examples.

2.1.2 Discrete Random Variables

Example 2.1.21 (Binomial) For a r.v. S with binomial distribution, $S \sim \text{Bin}_{n,p}$ (see Example 1.4.17-(iii)), we have

$$P(S = k) = \binom{n}{k} p^k (1 - p)^{n-k}, \qquad k = 0, 1, \ldots, n. \qquad (2.1.5)$$

S represents the "number of successes in n repeated independent trials with probability p" (cf. Example 2.1.7-(iii)). Examples of binomial random variables are:

(i) with reference to Example 1.2.17, in which we consider drawing with replacement from an urn containing b white balls and r red balls, the r.v. S representing the "number of white balls drawn in n draws" has distribution $\text{Bin}_{n,\frac{b}{b+r}}$;

(ii) with reference to Example 1.3.43, in which we suppose to randomly arrange n objects in r boxes, the r.v. S representing the "number of objects in the first box" has distribution $\text{Bin}_{n,\frac{1}{r}}$.

In Fig. 2.1, we show the plot of the distribution function $k \mapsto P(X = k)$ of $X \sim \text{Bin}_{n,p}$ with $n = 40$ and $p = 10\%$: this plot depicts *the possible values of X, i.e., $X(\Omega)$, on the abscissa and the corresponding probabilities on the ordinate.*

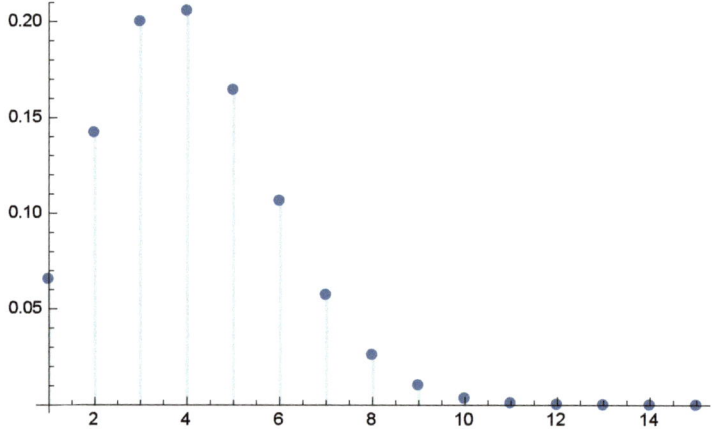

Fig. 2.1 Plot of the distribution function of a binomial random variable

Example 2.1.22 (Overbooking) Suppose that the probability that a traveler does not show up for boarding at the airport is 10%, independently of other travelers. How many reservations for a flight with 100 passengers can be accepted, wanting the probability that all travelers present at boarding find a seat to be greater than 99%?

Solution Suppose we accept n reservations and consider the r.v. X equal to the "number of passengers present at boarding": then $X \sim \mathrm{Bin}_{n,p}$ where $p = \frac{9}{10}$ is the probability that a traveler shows up. We have to determine the maximum value of n such that

$$P(X > 100) = \sum_{k=101}^{n} P(X = k) < 1\%.$$

Since $P(X > 100) = 0.57\%$ if $n = 104$ and $P(X > 100) = 1.67\%$ if $n = 105$, then *we can accept* 104 *reservations*.

Example 2.1.23 (Poisson) Let $\lambda > 0$ be a fixed constant. For each $n \in \mathbb{N}$, $n \geq \lambda$, let $q_n = \frac{\lambda}{n}$ and consider $X_n \sim \mathrm{Bin}_{n,q_n}$. For each $k = 0, 1, \ldots, n$, let

$$p_{n,k} := P(X_n = k) = \binom{n}{k} q_n^k (1 - q_n)^{n-k} = \frac{n!}{k!(n-k)!} \left(\frac{\lambda}{n}\right)^k \left(1 - \frac{\lambda}{n}\right)^{n-k}$$

(2.1.6)

$$= \frac{\lambda^k}{k!} \cdot \frac{n(n-1)\cdots(n-k+1)}{n^k} \cdot \frac{\left(1 - \frac{\lambda}{n}\right)^n}{\left(1 - \frac{\lambda}{n}\right)^k}$$

and observe that

$$\lim_{n \to \infty} p_{n,k} = \frac{e^{-\lambda} \lambda^k}{k!} =: p_k, \qquad k \in \mathbb{N}_0.$$

We thus find the Poisson distribution

$$\text{Poisson}_\lambda = \sum_{k=0}^{\infty} p_k \delta_k$$

of Example 1.4.17-(iv).

Intuitively, $X \sim \text{Poisson}_\lambda$ can be thought of as the limit of a sequence of random variables $X_n \sim \text{Bin}_{n,q_n}$. In other words, the Poisson distribution with parameter np approximates the binomial distribution $\text{Bin}_{n,p}$ as $n \to +\infty$ (and $p \to 0^+$). Therefore we write

$$\text{Bin}_{n,p} \approx \text{Poisson}_{np} \qquad n \to +\infty, \; p \to 0^+.$$

This result will be formalized later in Example 3.3.12. We note that in practice, for large n, the value of $p_{n,k}$ in (2.1.6) is "difficult" to calculate due to the presence of factorials[3] in the binomial coefficient $\binom{n}{k}$. Hence, employing the Poisson distribution as an approximation for the binomial distribution proves to be advantageous.

In Fig. 2.2, we show the plot of the distribution function $k \mapsto P(X = k)$ of $X \sim \text{Poisson}_\lambda$ with $\lambda = 3$.

Example 2.1.24 A machine produces bolts and for each bolt produced there is a probability of 0.01% that it is defective (independently of the others). Calculate the probability of having fewer than 3 defective bolts in a box containing 1000 bolts.

Solution The r.v. X indicating the number of defective bolts in a box of 1000 bolts has a binomial distribution $\text{Bin}_{1000,p}$ where $p = 0.01\%$ is the probability that the single bolt is defective. Then

$$P(X < 3) = \sum_{k=0}^{2} P(X = k) = \sum_{k=0}^{2} \binom{1000}{k} p^k (1-p)^{1000-k} \approx 99.9846\%.$$

---

[3] For example, $70! > 10^{100}$. To calculate $n!$ for $n \gg 1$ one can use Stirling's approximation

$$n! \approx \sqrt{2\pi n} \left(\frac{n}{e}\right)^n.$$

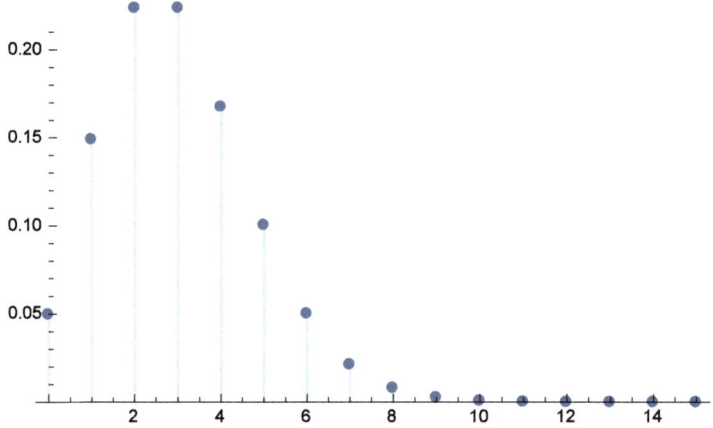

Fig. 2.2 Plot of the distribution function of a Poisson random variable

Using the approximation with a Poisson r.v., say $Y \sim$ Poisson$_\lambda$ where $\lambda = np = 0.1$, we obtain

$$P(Y < 3) = \sum_{k=0}^{2} P(Y = k) = e^{-\lambda} \sum_{k=0}^{2} \frac{\lambda^k}{k!} \approx 99.9845\%.$$

Example 2.1.25 (Geometric) For T with geometric distribution, $T \sim$ Geom$_p$ with $p \in \,]0, 1]$, we have[4]

$$P(T = k) = p(1 - p)^{k-1}, \qquad k \in \mathbb{N}.$$

The r.v. T represents the "first time of success" in a family of repeated independent trials with probability p: in this regard, recall Examples 2.1.7-(iv) and 1.3.31.

We now prove a fundamental property of the geometric distribution, known as *memoryless property*.

Theorem 2.1.26 *If $T \sim$ Geom$_p$ then*

$$P(T > n) = (1 - p)^n, \qquad n \in \mathbb{N}, \tag{2.1.7}$$

and we have the following memoryless property:

$$P(T > n + k \mid T > n) = P(T > k), \qquad k, n \in \mathbb{N}. \tag{2.1.8}$$

---

[4] By convention, we set $0^0 = 1$.

Conversely, if T is a r.v. with values in $\mathbb{N}$ and (2.1.8) holds, then $T \sim \mathrm{Geom}_p$ where $p = P(T = 1)$.

Proof If $T \sim \mathrm{Geom}_p$, then for each $n \in \mathbb{N}$ we have

$$P(T > n) = \sum_{k=n+1}^{\infty} P(T = k) = \sum_{k=n+1}^{\infty} p(1 - p)^{k-1} = \sum_{h=n}^{\infty} p(1 - p)^h$$

$$= p(1 - p)^n \sum_{h=0}^{\infty} (1 - p)^h = p(1 - p)^n \frac{1}{1 - (1 - p)} = (1 - p)^n,$$

and this proves (2.1.7). Then, since $(T > k + n) \subseteq (T > n)$, we have

$$P(T > n + k \mid T > n) = \frac{P(T > k + n)}{P(T > n)} = \frac{(1 - p)^{k+n}}{(1 - p)^n} = (1 - p)^k = P(T > k).$$

Conversely, suppose that T is a r.v. with values in $\mathbb{N}$ for which (2.1.8) holds. We note that (2.1.8) makes sense under the implicit assumption that $P(T > n) > 0$ for every $n \in \mathbb{N}$ and for $k = 1$ we have

$$P(T > 1) = P(T > n + 1 \mid T > n) = \frac{P(T > n + 1)}{P(T > n)}$$

from which

$$P(T > n + 1) = P(T > n)P(T > 1)$$

and therefore

$$P(T > n) = P(T > 1)^n.$$

Moreover, let $p = P(T = 1) = 1 - P(T > 1)$, then we have

$$P(T = k) = P(T > k - 1) - P(T > k) = P(T > 1)^{k-1} - P(T > 1)^k$$

$$= P(T > 1)^{k-1}(1 - P(T > 1)) = p(1 - p)^{k-1},$$

which proves the thesis. $\qquad\square$

Corollary 2.1.27 *Let $T \sim \mathrm{Geom}_p$ and $n \in \mathbb{N}$. We have*

$$P(T = n + k \mid T > n) = P(T = k), \qquad k \in \mathbb{N},$$

that is the law of the r.v. T with respect to probability P is equal to the law of the r.v. $(T - n)$ with respect to the conditional probability $P(\cdot \mid T > n)$.

Proof We have

$$P(T = n + k \mid T > n) = P(T > n + k - 1 \mid T > n)$$
$$- P(T > n + k \mid T > n) =$$

(by Theorem 2.1.26)

$$= P(T > k - 1) - P(T > k) = P(T = k).$$

$$\square$$

Exercise 2.1.28 In a lottery game, once a week 5 numbers are drawn from an urn containing 90 numbered balls. What is the probability that the number 13 is not drawn for 52 consecutive weeks? Knowing that 13 has not been drawn for 52 weeks, what is the probability that it is not drawn for the 53rd consecutive week?

Solution Let $p = \frac{|C_{89,4}|}{|C_{90,5}|} = \frac{5}{90}$ be the probability that in one draw the number 13 is drawn. If T indicates the first week in which 13 is drawn, then by (2.1.7) we have

$$P(T > 52) = (1 - p)^{52} \approx 5.11\%$$

Alternatively, we could have considered the binomial r.v. $X \sim \text{Bin}_{52,p}$ which indicates the number of times in which, among 52 draws, 13 is drawn and calculate

$$P(X = 0) = \binom{52}{0} p^0 (1 - p)^{52}$$

which gives the same result. For the second question, we need to calculate

$$P(T > 53 \mid T > 52) = P(T > 1) = \frac{85}{90},$$

where the first equality follows from (2.1.8).

Example 2.1.29 (Hypergeometric) A r.v. X with hypergeometric distribution represents the number of white balls drawn in n draws *without* replacement from an urn containing N balls of which b are white: in this regard, recall Example 1.2.19. In particular, let $n, b, N \in \mathbb{N}$ with $n, b \leq N$. Then $X \sim \text{Hyper}_{n,b,N}$ if[5]

$$P(X = k) = \frac{\binom{b}{k}\binom{N-b}{n-k}}{\binom{N}{n}} \qquad k = 0, 1, \ldots, n \wedge b. \qquad (2.1.9)$$

---

[5] By convention, we set $\binom{n}{k} = 0$ for $k > n$.

Exercise 2.1.30 Let $(b_N)_{N \in \mathbb{N}}$ be a sequence in $\mathbb{N}_0$ such that

$$\lim_{N \to \infty} \frac{b_N}{N} = p \in \,]0, 1[.$$

Then we have

$$\lim_{N \to \infty} \mathrm{Hyper}_{n,b_N,N}(\{k\}) = \mathrm{Bin}_{n,p}(\{k\})$$

for every $n \in \mathbb{N}$ and $k = 0, 1, \ldots, n$. Intuitively, if the number of white balls b and the total number of balls N are large, then the replacement or not of a ball after the draw modifies the composition of the urn in a negligible way.

Solution It is a direct calculation: for more details, see, for example, Remark 1.40 in [8].

2.1.3 Absolutely Continuous Random Variables

Example 2.1.31 (Exponential) A r.v. with exponential distribution $X \sim \mathrm{Exp}_\lambda$ has a *memoryless* property similar to that seen in Theorem 2.1.26 for the geometric distribution:

$$P\,(X > t + s \mid X > s) = P\,(X > t)\,, \qquad t, s \geq 0. \tag{2.1.10}$$

In fact, since $(X > t + s) \subseteq (X > s)$, we have

$$P\,(X > t + s \mid X > s) = \frac{P(X > t + s)}{P(X > s)} =$$

(by (1.4.8))

$$= \frac{e^{-\lambda(t+s)}}{e^{-\lambda s}} = e^{-\lambda t} = P\,(X > t)\,.$$

The exponential distribution belongs to a wide family of distributions that we will examine in Example 2.1.35.

Here is a simple yet useful result.

Proposition 2.1.32 (Linear Transformations and Densities) *Let X be a r.v. in $\mathbb{R}^d$, which is absolutely continuous with density γ_X. Then for every invertible matrix A of dimension $d \times d$, and $b \in \mathbb{R}^d$, the r.v. $Z := AX + b$ is absolutely continuous with density*

$$\gamma_Z(z) = \frac{1}{|\det A|} \gamma_X \left(A^{-1}(z - b) \right).$$

Proof For every $H \in \mathscr{B}_d$ we have

$$P(Z \in H) = P\left(X \in A^{-1}(H - b) \right) = \int_{A^{-1}(H-b)} \gamma_X(x)dx =$$

(by the change of variables $z = Ax + b$)

$$= \frac{1}{|\det A|} \int_H \gamma_X \left(A^{-1}(z - b) \right) dz$$

and this proves the thesis.   □

Example 2.1.33 (Uniform)   As in Example 1.4.22-(i), we consider a r.v. with uniform distribution on $K \in \mathscr{B}_d$, where K has positive and finite Lebesgue measure. In particular, let K be the triangle in $\mathbb{R}^2$ with vertices $(0, 0)$, $(1, 0)$, and $(0, 1)$. Let $(X, Y) \sim \text{Unif}_K$, with density $\gamma_{(X,Y)}(x, y) = 2\mathbb{1}_K(x, y)$: by Proposition 2.1.32 we can easily compute the density of $(X + Y, X - Y)$. In fact, being

$$\begin{pmatrix} X + Y \\ X - Y \end{pmatrix} = A \begin{pmatrix} X \\ Y \end{pmatrix}, \qquad A = \begin{pmatrix} 1 & 1 \\ 1 & -1 \end{pmatrix},$$

we have $\det A = -2$ and

$$\gamma_{(X+Y,X-Y)}(z, w) = \frac{2}{|\det A|} \mathbb{1}_K \left(A^{-1} \begin{pmatrix} z \\ w \end{pmatrix} \right) = \mathbb{1}_{AK}(z, w)$$

where AK is the triangle with vertices[6] $(0, 0)$, $(1, 1) = A \cdot (1, 0)$, and $(1, -1) = A \cdot (0, 1)$.

Example 2.1.34 (Normal) A r.v. X has a normal distribution with parameters $\mu \in \mathbb{R}$ and $\sigma > 0$, i.e. $X \sim \mathscr{N}_{\mu,\sigma^2}$, if

$$P(X \in H) = \int_H \frac{1}{\sqrt{2\pi\sigma^2}} e^{-\frac{1}{2}\left(\frac{x-\mu}{\sigma}\right)^2} dx, \qquad H \in \mathscr{B}.$$

---

[6] Here $A \cdot (1, 0) \equiv A \begin{pmatrix} 1 \\ 0 \end{pmatrix}$.

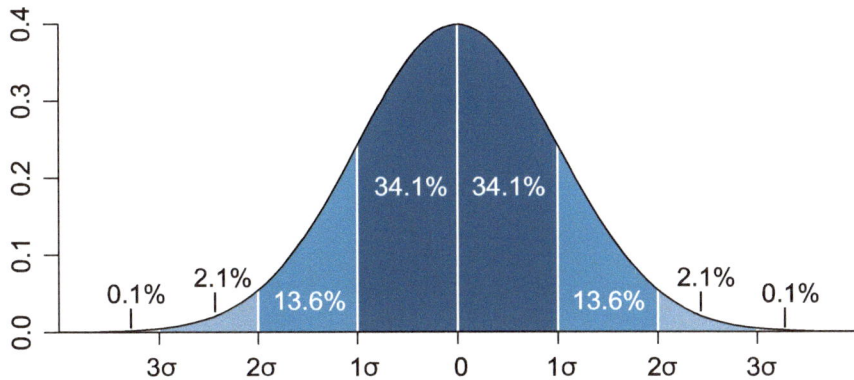

Fig. 2.3 Probability in the normal distribution

Note that $P(X \in H) > 0$ if and only if $\text{Leb}(H) > 0$, since the density is strictly positive. Obviously, $P(X = x) = 0$ for every $x \in \mathbb{R}$ because X is absolutely continuous.

Although X can take any real value, it is good to know that the probability is essentially concentrated around the value μ. In fact, we have

$$P(|X - \mu| \le \sigma) \approx 68.27\%$$

$$P(|X - \mu| \le 2\sigma) \approx 95.45\% \qquad (2.1.11)$$

$$P(|X - \mu| \le 3\sigma) \approx 99.73\%$$

and this means that extreme values (not even too far from μ) are very unlikely (see[7] Fig. 2.3). For this reason, we say that the Gaussian density has "thin tails".

At first glance, the fact that the values in (2.1.11) are *independent of μ and σ* may seem a bit strange. On the other hand, $P(|X - \mu| \le \lambda\sigma) = P(|Z| \le \lambda)$ where $Z = \frac{X - \mu}{\sigma}$ and by Proposition 2.1.32 we have

$$X \sim \mathcal{N}_{\mu,\sigma^2} \quad \Longleftrightarrow \quad Z \sim \mathcal{N}_{0,1}.$$

In other words, one can always *standardize* a normal r.v. through a simple linear transformation.

Note that the Gaussian density of $Z \sim \mathcal{N}_{0,1}$ is an even function and therefore, for every $\lambda > 0$ we have

$$P(Z \ge -\lambda) = P(-Z \le \lambda) = P(Z \le \lambda)$$

---

[7] Figure 2.3 is taken from https://commons.wikimedia.org/wiki/File:Standard_deviation_diagram.svg.

and consequently

$$P(|Z| \le \lambda) = P(Z \le \lambda) - P(Z \le -\lambda)$$
$$= P(Z \le \lambda) - (1 - P(Z \ge -\lambda))$$
$$= 2F_Z(\lambda) - 1, \tag{2.1.12}$$

where F_Z indicates the CDF of Z.

Example 2.1.35 (Gamma)   Let us recall the definition of Euler's Gamma function:

$$\Gamma(\alpha) := \int_0^{+\infty} x^{\alpha-1} e^{-x} dx, \qquad \alpha > 0. \tag{2.1.13}$$

We observe that Γ takes positive values, $\Gamma(1) = 1$ and $\Gamma(\alpha + 1) = \alpha\Gamma(\alpha)$ since, integrating by parts, we have

$$\Gamma(\alpha + 1) = \int_0^{+\infty} x^{\alpha} e^{-x} dx = \int_0^{+\infty} \alpha x^{\alpha-1} e^{-x} dx = \alpha\Gamma(\alpha).$$

It follows in particular that $\Gamma(n + 1) = n!$ for every $n \in \mathbb{N}$. Another significant value is obtained for $\alpha = \frac{1}{2}$:

$$\Gamma\left(\tfrac{1}{2}\right) = \int_0^{+\infty} \frac{e^{-x}}{\sqrt{x}} dx =$$

(by the change of variable $x = y^2$)

$$= 2 \int_0^{+\infty} e^{-y^2} dy = \sqrt{\pi}.$$

We also note that, fixing $\lambda > 0$, by the change of variable $x = \lambda t$ in (2.1.13) we obtain

$$\Gamma(\alpha) := \lambda^{\alpha} \int_0^{+\infty} t^{\alpha-1} e^{-\lambda t} dt, \qquad \alpha > 0.$$

It follows that the function

$$\gamma_{\alpha,\lambda}(t) := \frac{\lambda^{\alpha}}{\Gamma(\alpha)} t^{\alpha-1} e^{-\lambda t} \mathbb{1}_{\mathbb{R}>0}(t), \qquad t \in \mathbb{R}, \tag{2.1.14}$$

is a density for every $\alpha > 0$ and $\lambda > 0$. Figure 2.4 depicts the graph of the density in (2.1.14).

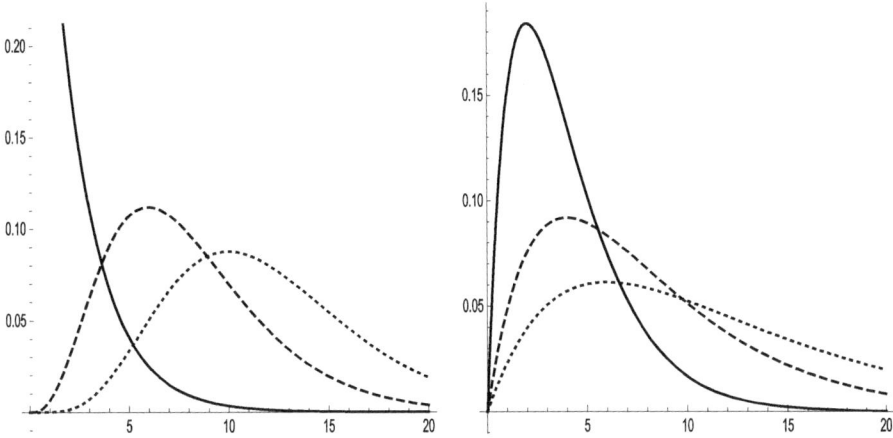

Fig. 2.4 On the left: plot of the density $\gamma_{\alpha,2}$ for $\alpha = 1$ (solid line), $\alpha = 4$ (dashed line) and $\alpha = 6$ (dotted line). **On the right:** plot of the density $\gamma_{2,\lambda}$ for $\lambda = \frac{1}{2}$ (solid line), $\lambda = \frac{1}{4}$ (dashed line) and $\lambda = \frac{1}{6}$ (dotted line)

Definition 2.1.36 The distribution with density $\gamma_{\alpha,\lambda}$ in (2.1.14) is called *Gamma distribution with parameters $\alpha, \lambda > 0$:*

$$\text{Gamma}_{\alpha,\lambda}(H) := \frac{\lambda^\alpha}{\Gamma(\alpha)} \int_{H \cap \mathbb{R}_{>0}} t^{\alpha-1} e^{-\lambda t} dt, \qquad H \in \mathscr{B}.$$

Notice that the exponential distribution corresponds to the particular case $\alpha = 1$:

$$\text{Gamma}_{1,\lambda} = \text{Exp}_\lambda.$$

The Gamma distribution has the following scale invariance property:

Lemma 2.1.37 *If $X \sim \text{Gamma}_{\alpha,\lambda}$ and $c > 0$ then $cX \sim \text{Gamma}_{\alpha,\frac{\lambda}{c}}$. In particular, $\lambda X \sim \text{Gamma}_{\alpha,1}$.*

Proof We use the cumulative distribution function to determine the distribution of cX:

$$P(cX \le y) = P(X \le y/c) = \int_0^{\frac{y}{c}} \frac{\lambda^\alpha e^{-\lambda t}}{\Gamma(\alpha) t^{1-\alpha}} dt =$$

(by the change of variable $x = ct$)

$$= \int_0^y \frac{\lambda^\alpha e^{-\frac{\lambda}{c}x}}{c^\alpha \Gamma(\alpha) x^{1-\alpha}} dx = \text{Gamma}_{\alpha,\frac{\lambda}{c}}(\,]-\infty, y]).$$

$\square$

2.1.4   Other Examples

Example 2.1.38 (χ^2 Distribution)   Let $X \sim \mathcal{N}_{0,1}$. We determine the distribution of the r.v. $Z = X^2$ by studying its CDF F_Z. Since $Z \geq 0$, we have $F_Z(x) = 0$ for $x \leq 0$, while for $x > 0$ we have

$$F_Z(x) = P(X^2 \leq x) = P\left(-\sqrt{x} \leq X \leq \sqrt{x}\right) =$$

(by symmetry)

$$= 2 \int_0^{\sqrt{x}} \frac{1}{\sqrt{2\pi}} e^{-\frac{y^2}{2}} dy = 2 \left(F_X(\sqrt{x}) - F_X(0)\right)$$

where F_X is the CDF of X. It follows that F_Z is absolutely continuous and therefore, by Theorem 1.4.33, the density of Z is given by

$$\frac{d}{dx} F_Z(x) = 2 \frac{d}{dx} F_X(\sqrt{x}) = F_X'(\sqrt{x}) \frac{1}{\sqrt{x}} = \frac{1}{\sqrt{2\pi x}} e^{-\frac{x}{2}}, \qquad x > 0.$$

We then recognize that

$$Z \sim \text{Gamma}_{\frac{1}{2}, \frac{1}{2}}.$$

The distribution $\text{Gamma}_{\frac{1}{2}, \frac{1}{2}}$ is called *chi-square distribution*. Sometimes it is indicated by the symbol χ^2.

Proposition 2.1.39   *Let*

$$X : \Omega \longrightarrow I \quad \text{and} \quad f : I \longrightarrow J$$

be a r.v. on the space $(\Omega, \mathcal{F}, P)$ *with values in the real interval* I *and a continuous and strictly increasing (thus invertible) function with values in the real interval* J. *Then the CDF of the r.v.* $Y := f(X)$ *is*

$$F_Y = F_X \circ f^{-1} \tag{2.1.15}$$

where F_X *denotes the CDF of* X.

Proof   Equation (2.1.15) simply follows from

$$P(Y \leq y) = P(f(X) \leq y) = P\left(X \leq f^{-1}(y)\right) = F_X(f^{-1}(y)), \qquad y \in J,$$

where in the second equality we used the fact that f is strictly increasing.   □

Exercise 2.1.40   Determine the density of $Y := e^X$ where $X \sim \text{Unif}_{[0,1]}$.

Corollary 2.1.41 *[!] If X is a r.v. with values within an interval I and with CDF F_X continuous and strictly increasing on I, then*

$$F_X(X) \sim \text{Unif}_{[0,1]}. \qquad (2.1.16)$$

Proof Let $Y := F_X(X)$. Clearly, we have $F_Y(y) = 0$ if $y \le 0$ and $F_Y(y) = 1$ if $y \ge 1$ since F_X takes values in $[0, 1]$ by definition and is continuous. Moreover, by Proposition 2.1.39, we have $F_Y(y) = y$ if $0 < y < 1$, which proves the thesis. $\qquad \square$

The previous corollary applies, for example, to $X \sim \mathcal{N}_{\mu,\sigma^2}$ with $I = \mathbb{R}$ and to $X \sim \text{Gamma}_{\alpha,\lambda}$ with $I = \mathbb{R}_{>0}$.

Exercise 2.1.42 Let $X \sim \frac{1}{2}(\delta_0 + \text{Unif}_{[0,1]})$. Prove that $F_X(X) \sim \frac{1}{2}\left(\delta_{\frac{1}{2}} + \text{Unif}_{\left[\frac{1}{2},1\right]}\right)$ and therefore the assumption of continuity of F_X in Corollary 2.1.41 cannot be removed.

Example 2.1.43 Proposition 2.1.39 is frequently employed to generate or simulate a random variable with a predetermined CDF using a uniform random variable as a basis. In fact, if $Y \sim \text{Unif}_{[0,1]}$ and F is a strictly increasing CDF, then the r.v.

$$X := F^{-1}(Y)$$

has the CDF equal to F.

For instance, suppose we aim to generate an exponential random variable using a uniform random variable: recalling that

$$F(x) = 1 - e^{-\lambda x}, \qquad x \in \mathbb{R},$$

is the CDF of the distribution Exp_λ, we have

$$F^{-1}(y) = -\frac{1}{\lambda} \log(1 - y), \qquad y \in]0, 1[.$$

Then, by Proposition 2.1.39, if $Y \sim \text{Unif}_{]0,1[}$ we have

$$-\frac{1}{\lambda} \log(1 - Y) \sim \text{Exp}_\lambda.$$

Corollary 2.1.41, and in particular (2.1.16), provides *a method for generating random numbers with a given CDF or density on a computer, starting from random numbers with distribution* $\text{Unif}_{[0,1]}$.

The following result extends Proposition 2.1.32.

Proposition 2.1.44 *If $X \in AC$ is a real r.v. with density γ_X and $f \in C^1$ with $f' \neq 0$ then $Y := f(X) \in AC$ and has density*

$$\gamma_Y = \frac{\gamma_X(f^{-1})}{|f'(f^{-1})|}. \tag{2.1.17}$$

Proof First, recall that the assumptions on f imply that f is invertible and there exists

$$(f^{-1})' = \frac{1}{f'(f^{-1})}. \tag{2.1.18}$$

Furthermore, for every $H \in \mathcal{B}$ we have

$$P(Y \in H) = P\left(X \in f^{-1}(H)\right) = \int_{f^{-1}(H)} \gamma_X(x)dx =$$

(by the change of variables $y = f(x)$)

$$= \int_H \gamma_X(f^{-1}(y))|(f^{-1})'(y)|dy =$$

(by (2.1.18) and with γ_Y defined as in (2.1.17))

$$= \int_H \gamma_Y(y)dy,$$

and this proves that $Y \in AC$ with density γ_Y in (2.1.17). Note that if f is strictly increasing, then $f' > 0$ and the absolute value in (2.1.17) is unnecessary. However, the result is also valid for f strictly *decreasing* and in that case the absolute value is required. $\qquad\qquad\square$

Example 2.1.45 (Log-Normal Distribution) Let $X \sim \mathcal{N}_{0,1}$ and $f(x) = e^x$. Then, by (2.1.17), the density of the r.v. $Y = e^X$ is

$$\gamma_Y(y) = \frac{1}{y\sqrt{2\pi}}e^{-\frac{(\log y)^2}{2}}, \qquad y \in \mathbb{R}_{>0}. \tag{2.1.19}$$

The function γ_Y in (2.1.19) is called the *density of the log-normal distribution:* note that if Y has a log-normal distribution, then $\log Y$ has a normal distribution.

Example 2.1.46 (Bivariate Normal Distribution) Let X and Y be random variables representing the temperature variation in Bologna from the beginning to the end of the months of September and October, respectively. We assume that (X, Y) has a bivariate normal density

$$\gamma(x, y) = \frac{1}{2\pi \sqrt{\det C}} e^{-\frac{1}{2} \langle C^{-1}(x,y),(x,y) \rangle}, \qquad (x, y) \in \mathbb{R}^2$$

where

$$C = \begin{pmatrix} 2 & 1 \\ 1 & 3 \end{pmatrix}.$$

Let us determine:

(i) $P(Y < -1)$;
(ii) $P(Y < -1 \mid X < 0)$.

We have $\gamma(x, y) = \frac{1}{2\sqrt{5}\pi} e^{-\frac{3x^2 - 2xy + 2y^2}{10}}$ and

$$P(Y < -1) = \int_{\mathbb{R}} \int_{-\infty}^{-1} \gamma(x, y) dy dx \approx 28\%,$$

$$P(Y < -1 \mid X < 0) = \frac{P((Y < -1) \cap (X < 0))}{P(X < 0)} \approx 39\%,$$

being

$$P((Y < -1) \cap (X < 0)) = \int_{-\infty}^{0} \int_{-\infty}^{-1} \gamma(x, y) dy dx \approx 19.7\%,$$

$$P(X < 0) = \int_{-\infty}^{0} \int_{\mathbb{R}} \gamma(x, y) dy dx = \frac{1}{2}.$$

Example 2.1.47 [!] Let $X \sim \mathrm{Exp}_\lambda$ and f be a Borel-measurable function from $\mathbb{R}$ to $\mathbb{R}$. Then the two-dimensional random variable $(X, f(X))$ is not absolutely continuous. In fact, the Lebesgue measure of the set $H := \{(x, y) \in \mathbb{R}^2 \mid y = f(x)\}$ is null, but $P((X, Y) \in H) = 1$. (cf. Remark 1.4.21).

2.2 Expected Value

In this section, we introduce the concept of *expected value* (or *expectation* or *mean*) of a random variable. If X is a real r.v. with a finite discrete distribution

$$X \sim \sum_{k=1}^{m} p_k \delta_{x_k},$$

that is, $P(X = x_k) = p_k$ for $k = 1, \ldots, m$, then the expected value of X is simply defined as

$$E[X] := \sum_{k=1}^{m} x_k P(X = x_k) = \sum_{k=1}^{m} x_k p_k. \qquad (2.2.1)$$

In other words, $E[X]$ is a weighted average of the values of X based on the probability that such values are assumed. If $m = \infty$ then the sum in (2.2.1) becomes a series and conditions for convergence must be imposed. For example, on the space $(\Omega, \mathscr{F}, P) \equiv (\mathbb{N} \cup \{0\}, \mathscr{B}, \text{Poisson}_\lambda)$, where λ is a positive parameter, consider the r.v. $X(k) := k!$ for $k \in \mathbb{N} \cup \{0\}$. Then, by definition (2.2.1) with $x_k = k!$ we have

$$E[X] = \sum_{k=0}^{\infty} k! \, P(X = k!) = \sum_{k=0}^{\infty} k! \, \text{Poisson}_\lambda(\{k\}) = e^{-\lambda} \sum_{k=0}^{\infty} \lambda^k$$

$$= \begin{cases} \frac{e^{-\lambda}}{1-\lambda} & \text{if } 0 < \lambda < 1, \\ +\infty & \text{if } \lambda \geq 1. \end{cases}$$

When X assumes an *uncountable* infinity of values, it is no longer possible to define $E[X]$ as a series: in the general case, the expected value $E[X]$ will be defined as the integral of X with respect to the probability measure P and indicated indifferently with

$$\int_\Omega X dP \quad \text{or} \quad \int_\Omega X(\omega) P(d\omega) \quad \text{or} \quad \int_\Omega P(d\omega) X(\omega).$$

To give the precise definition of expected value, we recall some elements of the so-called *abstract integration theory* on a probability space $(\Omega, \mathscr{F}, P)$, remembering that a r.v. is nothing more than a measurable function. The proofs that follow can be readily adapted to encompass σ-finite measure spaces, including $\mathbb{R}^d$ with Lebesgue measure.

We intend to present the following:

- the *theoretical* definition of abstract integral in Sects. 2.2.1, 2.2.2, and 2.2.3;
- an *operational* characterization of the abstract integral along with an explicit calculation method in Sects. 2.2.4 and 2.2.5.

2.2.1   Integral of Simple Random Variables

To introduce the abstract integral, we proceed step by step, starting from the case of "simple" real-valued random variables up to the general case. We say that a r.v. X on a probability space $(\Omega, \mathscr{F}, P)$ is *simple* if it assumes only a finite number of

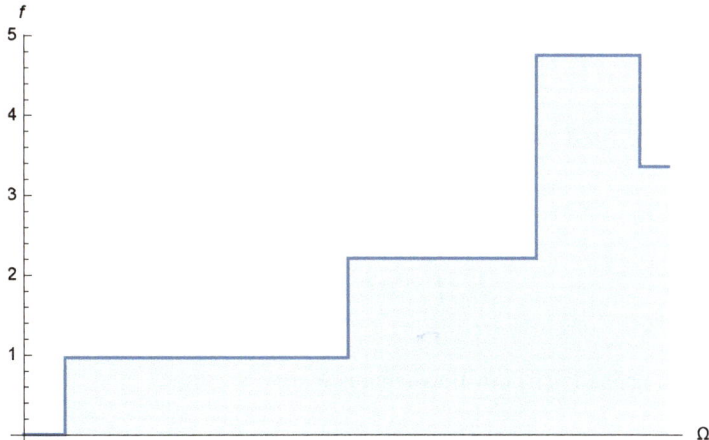

Fig. 2.5 Interpretation of the abstract integral as a Riemann sum

distinct values $x_1, \ldots, x_m \in \mathbb{R}$: then, we can write

$$X = \sum_{k=1}^{m} x_k \mathbb{1}_{(X=x_k)},$$

where $(X = x_1), \ldots, (X = x_m)$ are disjoint events. Note that the probability distribution of X is discrete, as discussed in the previous introductory section. Therefore, in accordance with Eq. (2.2.1), we define the abstract integral of X as

$$\int_{\Omega} X\,dP := \sum_{k=1}^{m} x_k P(X = x_k). \tag{2.2.2}$$

We have provided the "probabilistic" interpretation of this integral as a weighted average of the values of X; the "mathematical" interpretation of (2.2.2) is instead of a Riemann sum in which each term $x_k P(X = x_k)$ represents the area of a rectangle calculated as "base" × "height" where the measure of the base is $P(X = x_k)$ and the height x_k is the value of X on $(X = x_k)$: see Fig. 2.5.

Even though it may seem obvious, it is important to note and remember that *by definition* we have

$$\int_{\Omega} \mathbb{1}_A\,dP = P(A) \tag{2.2.3}$$

for every $A \in \mathscr{F}$. For every simple X and $A \in \mathscr{F}$, we will also use the notation

$$\int_A X\,dP := \int_{\Omega} X\mathbb{1}_A\,dP.$$

It is clear that the following properties hold:

(i) *linearity:* for every simple X, Y and $\alpha, \beta \in \mathbb{R}$ we have

$$\int_\Omega (\alpha X + \beta Y)\, dP = \alpha \int_\Omega X dP + \beta \int_\Omega Y dP; \tag{2.2.4}$$

(ii) *monotonicity:* for every simple X, Y such that $X \leq Y$ a.s. we have

$$\int_\Omega X dP \leq \int_\Omega Y dP. \tag{2.2.5}$$

Note that from property (ii) it follows that if $X = Y$ a.s. then

$$\int_\Omega X dP = \int_\Omega Y dP.$$

Before giving the general definition of integral, we prove some preliminary results.

Lemma 2.2.1 *Let $(X_n)_{n \in \mathbb{N}}$ be a sequence of simple random variables such that $0 \leq X_n \nearrow X$ a.s. If X is simple then*

$$\lim_{n \to \infty} \int_\Omega X_n dP = \int_\Omega X dP. \tag{2.2.6}$$

Proof By assumption, there exists $A \in \mathscr{F}$ with $P(A) = 1$, such that $0 \leq X_n(\omega) \nearrow X(\omega)$ for every $\omega \in A$. For any fixed $\varepsilon > 0$

$$A_{n,\varepsilon} := (X - X_n \geq \varepsilon) \cap A, \qquad n \in \mathbb{N},$$

is a decreasing sequence in $\mathscr{F}$ with empty intersection, i.e., $A_{n,\varepsilon} \searrow \emptyset$ as $n \to \infty$. Then by the continuity from above of P we have $\lim_{n \to \infty} P(A_{n,\varepsilon}) = 0$ and consequently

$$0 \leq \int_A (X - X_n) dP = \int_\Omega (X - X_n) dP = \int_{\Omega \setminus A_{n,\varepsilon}} (X - X_n) dP$$

$$+ \int_{A_{n,\varepsilon}} (X - X_n) dP \leq \varepsilon P(\Omega) + P(A_{n,\varepsilon}) \max_\Omega X$$

which yields (2.2.6). We explicitly note that $\max_\Omega X < \infty$ since X is simple by assumption. $\qquad\square$

Lemma 2.2.2 *Let* $(X_n)_{n\in\mathbb{N}}$ *and* $(Y_n)_{n\in\mathbb{N}}$ *be sequences of simple random variables such that* $0 \leq X_n \nearrow X$ *and* $0 \leq Y_n \nearrow Y$ *a.s. If* $X \leq Y$ *a.s. then*

$$\lim_{n\to\infty} \int_\Omega X_n dP \leq \lim_{n\to\infty} \int_\Omega Y_n dP.$$

Proof Fixed $k \in \mathbb{N}$, the sequence of simple random variables $(X_k \wedge Y_n)_{n\in\mathbb{N}}$ is such that $0 \leq X_k \wedge Y_n \nearrow X_k$ a.s. as n tends to infinity. Therefore we have

$$\int_\Omega X_k dP = \lim_{n\to\infty} \int_\Omega X_k \wedge Y_n dP \leq \lim_{n\to\infty} \int_\Omega Y_n dP$$

where the first equality follows from Lemma 2.2.1, while the inequality is due to the fact that $X_k \wedge Y_n \leq Y_n$ a.s. This concludes the proof.   □

2.2.2 Integral of Non-negative Random Variables

To extend the definition of the integral to $m\mathscr{F}^+$ we use the following

Lemma 2.2.3 *For every* $X \in m\mathscr{F}^+$ *there exists an* increasing *sequence* $(X_n)_{n\in\mathbb{N}}$ *in* $m\mathscr{F}^+$ *of* simple *random variables such that* $X_n \nearrow X$, *i.e.*

$$\lim_{n\to\infty} X_n(\omega) = X(\omega), \qquad \omega \in \Omega.$$

Proof We define a sequence of "step" functions on $[0, +\infty]$ in the following way: for each $n \in \mathbb{N}$ we consider the partition of $[0, n]$ formed by the points

$$\frac{0}{2^n}, \frac{1}{2^n}, \frac{2}{2^n}, \dots, \frac{n2^n}{2^n}$$

and set

$$\varphi_n(x) = \begin{cases} \frac{k-1}{2^n} & \text{if } \frac{k-1}{2^n} < x \leq \frac{k}{2^n} \text{ for } 1 \leq k \leq n2^n, \\ n & \text{if } x > n. \end{cases} \qquad (2.2.7)$$

Note that $0 \leq \varphi_n \leq \varphi_{n+1}$ for every $n \in \mathbb{N}$ and

$$x - \frac{1}{2^n} \leq \varphi_n(x) < x, \qquad x \in [0, n],$$

so that

$$\lim_{n\to\infty} \varphi_n(x) = x, \qquad x \geq 0. \qquad (2.2.8)$$

Then it suffices to consider the sequence $X_n := \varphi_n(X)$.

Notice that, since φ_n is left continuous by definition, we have that $Y_k \nearrow Y$ as $k \to \infty$ implies $\varphi_n(Y_k) \nearrow \varphi_n(Y)$.   □

Thanks to Lemmas 2.2.2 and 2.2.3, the following definition is well-posed, i.e., independent of the approximating sequence $(X_n)_{n \in \mathbb{N}}$.

Definition 2.2.4 (Integral of Non-negative Random Variables) For every $X \in m\mathscr{F}^+$ we define

$$* \int_\Omega X dP := \lim_{n \to \infty} \int_\Omega X_n dP \leq +\infty$$

where $(X_n)_{n \in \mathbb{N}}$ is any sequence in $m\mathscr{F}^+$ of *simple* random variables, such that $X_n \nearrow X$ a.s.

Remark 2.2.5 By the monotonicity property (2.2.5) for simple random variables, the above limit exists and is equal to

$$\sup_{n \in \mathbb{N}} \int_\Omega X_n dP.$$

Based on Definition 2.2.4, the properties of linearity (2.2.4) and monotonicity (2.2.5) easily extend to the integral of $X \in m\mathscr{F}^+$.

The definition of abstract integral is entirely analogous to that of the Lebesgue integral. Also in this case, the central result on which the entire development of the integration theory is based is the fundamental result on monotone convergence. Let us first prove a preliminary

Lemma 2.2.6 *Let $(x_n^k)_{n,k \in \mathbb{N}}$ be a sequence in $[0, +\infty]$ that is "doubly" increasing in the sense that*

$$x_n^k \leq x_{n+1}^k, \qquad x_n^k \leq x_n^{k+1}, \qquad n, k \in \mathbb{N},$$

so that there exist the limits

$$x^k := \lim_{n \to \infty} x_n^k, \qquad x_n := \lim_{k \to \infty} x_n^k.$$

Then for

$$x_\infty := \lim_{n \to \infty} x_n, \qquad x^\infty := \lim_{k \to \infty} x^k.$$

we have $x_\infty = x^\infty$.

Proof Fixed $\varepsilon > 0$, let $n_\varepsilon \in \mathbb{N}$ be such that $x_{n_\varepsilon} \geq x_\infty - \varepsilon$ and let $k_\varepsilon \in \mathbb{N}$ be such that $x_{n_\varepsilon}^{k_\varepsilon} \geq x_{n_\varepsilon} - \varepsilon$. Then we have

$$x^\infty \geq x^{k_\varepsilon} \geq x_{n_\varepsilon}^{k_\varepsilon} \geq x_\infty - 2\varepsilon.$$

Given the arbitrariness of ε, we have $x^\infty \geq x_\infty$ and similarly we prove that $x^\infty \leq x_\infty$. □

Theorem 2.2.7 (Beppo Levi's Theorem) *[!!!] If $(X_n)_{n \in \mathbb{N}}$ is a sequence in $m\mathscr{F}$ such that $0 \leq X_n \nearrow X$ a.s. then*

$$\lim_{n \to \infty} \int_\Omega X_n dP = \int_\Omega X dP.$$

Proof By (2.2.5) it is not restrictive to assume $0 \leq X_n(\omega) \nearrow X(\omega)$ for any $\omega \in \Omega$. We set $X_n^k := \varphi_k(X_n)$ with φ_k as in (2.2.7) and notice that for any $k \to \infty$

$$0 \leq \varphi_k(X_n) \nearrow X_n \qquad (2.2.9)$$

thanks to (2.2.8); moreover

$$0 \leq \varphi_k(X_n) \nearrow \varphi_k(X) \qquad (2.2.10)$$

as $n \to \infty$, since φ_k is left continuous. Thus by (2.2.9) and Lemma 2.2.1 we have

$$\lim_{n \to \infty} \int_\Omega X_n dP = \lim_{n \to \infty} \lim_{k \to \infty} \int_\Omega \varphi_k(X_n) dP =$$

(by Lemma 2.2.6)

$$= \lim_{k \to \infty} \lim_{n \to \infty} \int_\Omega \varphi_k(X_n) dP =$$

(by (2.2.10) and Lemma 2.2.1)

$$= \lim_{k \to \infty} \int_\Omega \varphi_k(X) dP =$$

(by Definition 2.2.4 and since $0 \leq \varphi_k(X) \nearrow X$ as $k \to \infty$)

$$= \int_\Omega X dP.$$

□

Lemma 2.2.8 (Fatou's Lemma) *[!] For any sequence* $(X_n)_{n\in\mathbb{N}}$ *in* $m\mathscr{F}^+$ *we have*

$$\int_\Omega \liminf_{n\to\infty} X_n dP \le \liminf_{n\to\infty} \int_\Omega X_n dP.$$

Proof By definition

$$\liminf_{n\to\infty} X_n := \sup_{n\in\mathbb{N}} Y_n, \qquad Y_n := \inf_{k\ge n} X_k,$$

and therefore $Y_n \nearrow X := \liminf_{n\to\infty} X_n$. Then we have

$$\int_\Omega \liminf_{n\to\infty} X_n dP = \int_\Omega \lim_{n\to\infty} Y_n dP =$$

(by Beppo Levi's theorem)

$$= \lim_{n\to\infty} \int_\Omega Y_n dP \le$$

(by monotonicity)

$$\le \lim_{n\to\infty} \inf_{k\ge n} \int_\Omega X_k dP = \liminf_{n\to\infty} \int_\Omega X_n dP.$$

$$\square$$

2.2.3   Integral of Random Vectors

Let x^+ and x^- denote the positive and negative parts of $x \in \mathbb{R}$, respectively.

Definition 2.2.9 (Integral of $\mathbb{R}^d$-Valued Random Variables)   Let $X \in m\mathscr{F}$ be a real r.v.: if at least one of $\int_\Omega X^+ dP$ and $\int_\Omega X^- dP$ is finite, then we say that X is *integrable* and set

$$\int_\Omega X dP := \int_\Omega X^+ dP - \int_\Omega X^- dP \in [-\infty, +\infty].$$

If both $\int_\Omega X^+ dP$ and $\int_\Omega X^- dP$ are finite, then we say that X *is absolutely integrable (or summable)* and write

$$X \in L^1(\Omega, P).$$

If $X = (X_1, \ldots, X_d)$ is a $\mathbb{R}^d$-valued r.v., we say that X is (absolutely) integrable if each component X_i is (absolutely) integrable: in this case we set

$$\int_\Omega X \, dP = \left(\int_\Omega X_1 \, dP, \ldots, \int_\Omega X_d \, dP \right) \in [-\infty, +\infty]^d.$$

For a real r.v. $X \in L^1(\Omega, P)$ we have

$$\int_\Omega |X| \, dP = \int_\Omega X^+ \, dP + \int_\Omega X^- \, dP \in \mathbb{R}.$$

For every integrable r.v. X the *triangle inequality* holds

$$\left| \int_\Omega X \, dP \right| = \left| \int_\Omega X^+ \, dP - \int_\Omega X^- \, dP \right| \leq \int_\Omega X^+ \, dP + \int_\Omega X^- \, dP = \int_\Omega |X| \, dP.$$

Notation 2.2.10 The following notations will be used interchangeably, especially when indicating the integration variable:

$$\int_\Omega X \, dP \equiv \int_\Omega X(\omega) P(d\omega) \equiv \int_\Omega P(d\omega) X(\omega).$$

For the Lebesgue integral, we simply write

$$\int_{\mathbb{R}^d} f(x) \, dx \quad \text{instead of} \quad \int_{\mathbb{R}^d} f \, d\text{Leb}.$$

Proposition 2.2.11 *The integral enjoys the following properties:*

(i) Linearity: *for every* $X, Y \in L^1(\Omega, P)$ *and* $\alpha, \beta \in \mathbb{R}$ *we have*

$$\int_\Omega (\alpha X + \beta Y) \, dP = \alpha \int_\Omega X \, dP + \beta \int_\Omega Y \, dP.$$

(ii) Monotonicity: *for every* $X, Y \in L^1(\Omega, P)$ *such that* $X \leq Y$ *a.s. we have*

$$\int_\Omega X \, dP \leq \int_\Omega Y \, dP.$$

In particular, if $X = Y$ a.s. then $\int_\Omega X \, dP = \int_\Omega Y \, dP$.
(iii) σ-additivity: *let* $A = \biguplus_{n \in \mathbb{N}} A_n$ *where* $(A_n)_{n \in \mathbb{N}}$ *is a disjoint sequence in* $\mathscr{F}$. *If* $X \in m\mathscr{F}^+$ *or* $X \in L^1(\Omega, P)$ *then we have*

$$\int_A X \, dP = \sum_{n \in \mathbb{N}} \int_{A_n} X \, dP.$$

Proof The proofs of the three properties are similar, therefore, we focus on providing a detailed proof for property (i) in the scalar case. Considering separately[8] the positive and negative parts of the random variables, it is sufficient to consider the case $X, Y \in m\mathscr{F}^+$ and $\alpha, \beta \in \mathbb{R}_{\geq 0}$. Consider the approximating sequences (X_n) and (Y_n) constructed as in Lemma 2.2.3: exploiting the linearity of the expected value in the case of simple r.v., we obtain by Beppo Levi's theorem

$$\int_\Omega (\alpha X + \beta Y) dP = \lim_{n\to\infty} \int_\Omega (\alpha X_n + \beta Y_n) dP$$

$$= \lim_{n\to\infty} \left(\alpha \int_\Omega X_n dP + \beta \int_\Omega Y_n dP \right)$$

$$= \alpha \int_\Omega X dP + \beta \int_\Omega Y dP.$$

$\square$

We conclude the section with the classical

Theorem 2.2.12 (Dominated Convergence Theorem) *[!!] Let $(X_n)_{n\in\mathbb{N}}$ be a sequence of random variables on $(\Omega, \mathscr{F}, P)$, such that $X_n \to X$ a.s. and $|X_n| \leq Y$ a.s. for every n, with $Y \in L^1(\Omega, P)$. Then we have*

$$\lim_{n\to\infty} \int_\Omega X_n dP = \int_\Omega X dP.$$

Proof Passing to the limit in $|X_n| \leq Y$ a.s. we also have $|X| \leq Y$ a.s. Then we have

$$0 \leq \limsup_{n\to\infty} \left| \int_\Omega X_n dP - \int_\Omega X dP \right| \leq$$

(by the triangle inequality)

$$\leq \limsup_{n\to\infty} \int_\Omega |X_n - X| dP =$$

$$= \int_\Omega 2Y dP - \liminf_{n\to\infty} \int_\Omega (2Y - |X_n - X|) dP \leq$$

---

[8] Notice that the linearity property for positive and negative parts is inherent to the definition of integral (cf. Definition 2.2.9).

(by Fatou's lemma)

$$\leq \int_{\Omega} 2Y dP - \int_{\Omega} \liminf_{n \to \infty} (2Y - |X_n - X|) \, dP =$$

$$= \int_{\Omega} 2Y dP - \int_{\Omega} 2Y dP = 0.$$

□

A generalization of the dominated convergence theorem is the Vitali's Theorem C.0.2. The following corollary of Theorem 2.2.12 is easily proved by contradiction.

Corollary 2.2.13 (Absolute Continuity of the Integral) *Let $X \in L^1(\Omega, P)$. For every $\varepsilon > 0$ there exists $\delta > 0$ such that $\int_A |X| dP < \varepsilon$ for every $A \in \mathscr{F}$ such that $P(A) < \delta$.*

We now give a simple but useful result.

Proposition 2.2.14 [!] *Let $X \in m\mathscr{F}$. Then*

$$\int_{(X>0)} X dP = 0 \implies X \leq 0 \text{ a.s.}$$

Proof Consider the increasing sequence defined by $A_n = \left(X \geq \frac{1}{n}\right)$ for $n \in \mathbb{N}$. By the monotonicity property of the integral, we have

$$0 = \int_{(X>0)} X dP \geq \int_{(X>0)} X \mathbb{1}_{A_n} dP \geq \frac{1}{n} \int_{(X>0)} \mathbb{1}_{A_n} dP = \frac{P(A_n)}{n},$$

and so $P(A_n) = 0$ for every $n \in \mathbb{N}$. Due to the continuity from below of P (cf. Proposition 1.1.32-(ii)) and being

$$(X > 0) = \bigcup_{n \in \mathbb{N}} A_n,$$

it follows that $P(X > 0) = 0$. □

Corollary 2.2.15 *If $X \in m\mathscr{F}^+$ is such that $\int_{\Omega} X dP = 0$ then $X = 0$ a.s.*

2.2.4 Integration with Distributions

In this section, we examine the abstract integral with respect to a distribution, with particular attention to the case of discrete and absolutely continuous distributions (or combinations of them). We begin with a simple

Example 2.2.16 [!] Consider the Dirac delta distribution δ_{x_0} on $(\mathbb{R}^d, \mathcal{B}_d)$. For every function $f \in m\mathcal{B}_d$, we have

$$\int_{\mathbb{R}^d} f(x)\delta_{x_0}(dx) = f(x_0).$$

In fact, f is equal to δ_{x_0}-almost everywhere to the simple function

$$\hat{f}(x) = \begin{cases} f(x_0) & \text{if } x = x_0, \\ 0 & \text{otherwise.} \end{cases}$$

Thus, by Proposition 2.2.11-(ii) we have

$$\int_{\mathbb{R}^d} f(x)\delta_{x_0}(dx) = \int_{\mathbb{R}^d} \hat{f}(x)\delta_{x_0}(dx) =$$

(by definition of the integral of a simple function)

$$= \hat{f}(x_0)\delta_{x_0}(\{x_0\}) = f(x_0).$$

Proposition 2.2.17 *Let*

$$\mu = \sum_{n=1}^{\infty} p_n\delta_{x_n}$$

be a discrete distribution on $(\mathbb{R}^d, \mathcal{B}_d)$ (cf. Definition 1.4.15). If $f \in m\mathcal{B}_d^+$ or $f \in L^1(\mathbb{R}^d, \mu)$ then

$$\int_{\mathbb{R}^d} f d\mu = \sum_{n=1}^{\infty} f(x_n)p_n.$$

Proof It follows directly by applying Proposition 2.2.11-(iii) with $A_n = \{x_n\}$. $\quad\square$

Example 2.2.18 For the Bernoulli distribution, $\text{Be}_p = p\delta_1 + (1-p)\delta_0$ with $0 \le p \le 1$, (cf. Example 1.4.17-(i)) we simply have

$$\int_{\mathbb{R}} f(x)\text{Be}_p(dx) = pf(1) + (1-p)f(0).$$

For the Poisson$_\lambda$ distribution, with $\lambda > 0$, we have

$$\int_{\mathbb{R}} f(x)\text{Poisson}_\lambda(dx) = e^{-\lambda} \sum_{k=0}^{\infty} \frac{\lambda^k}{k!} f(k),$$

assuming that f is integrable.

Exercise 2.2.19 Prove that if $\alpha, \beta > 0$, μ_1, μ_2 are distributions on $\mathbb{R}^d$ and $f \in L^1(\mathbb{R}^d, \mu_1) \cap L^1(\mathbb{R}^d, \mu_2)$ then $f \in L^1(\mathbb{R}^d, \alpha\mu_1 + \beta\mu_2)$ and

$$\int_{\mathbb{R}^d} f d(\alpha\mu_1 + \beta\mu_2) = \alpha \int_{\mathbb{R}^d} f d\mu_1 + \beta \int_{\mathbb{R}^d} f d\mu_2.$$

Now we see that in the case of an absolutely continuous distribution, *an abstract integral is reduced to a Lebesgue integral, weighted with the density of the distribution.*

Proposition 2.2.20 *[!] Let μ be an absolutely continuous distribution on $\mathbb{R}^d$ with density γ. Then f is μ-integrable (resp. $f \in L^1(\mathbb{R}^d, \mu)$) if and only if f is Lebesgue-integrable (resp.[9] $f\gamma \in L^1(\mathbb{R}^d)$) and in this case we have*

$$\int_{\mathbb{R}^d} f(x)\mu(dx) = \int_{\mathbb{R}^d} f(x)\gamma(x)dx.$$

Proof First consider the case where f is simple, i.e., $f(\mathbb{R}^d) = \{\alpha_1, \ldots, \alpha_m\}$ so that

$$f = \sum_{k=1}^{m} \alpha_k \mathbb{1}_{H_k}, \qquad H_k := \{x \in \mathbb{R}^d \mid f(x) = \alpha_k\}, \ k = 1, \ldots, m.$$

By linearity

$$\int_{\mathbb{R}^d} f d\mu = \sum_{k=1}^{m} \alpha_k \int_{\mathbb{R}^d} \mathbb{1}_{H_k} d\mu =$$

(by (2.2.3))

$$= \sum_{k=1}^{m} \alpha_k \mu(H_k) =$$

(being $\mu \in$ AC with density γ)

$$= \sum_{k=1}^{m} \alpha_k \int_{H_k} \gamma(x)dx = \sum_{k=1}^{m} \alpha_k \int_{\mathbb{R}^d} \mathbb{1}_{H_k}(x)\gamma(x)dx =$$

---

[9] Here $L^1(\mathbb{R}^d)$ is the usual space of absolutely integrable functions on $\mathbb{R}^d$ with respect to the Lebesgue measure.

(by linearity of the Lebesgue integral)

$$= \int_{\mathbb{R}^d} f(x)\gamma(x)dx,$$

which proves the thesis.

Now assume $f \geq 0$ and consider $f_n := \varphi_n(f)$ with φ_n as in (2.2.7). By Beppo
Levi's theorem, we have

$$\int_{\mathbb{R}^d} f d\mu = \lim_{n\to\infty} \int_{\mathbb{R}^d} f_n d\mu =$$

(for what has just been proven, since f_n is simple for each $n \in \mathbb{N}$)

$$= \lim_{n\to\infty} \int_{\mathbb{R}^d} f_n(x)\gamma(x)dx =$$

(reapplying Beppo Levi's theorem to the Lebesgue integral and using the fact that
$\gamma \geq 0$ by assumption and consequently $(f_n\gamma)$ is a monotone increasing sequence
of non-negative functions)

$$= \int_{\mathbb{R}^d} f(x)\gamma(x)dx.$$

Finally, if f is a generic function in $L^1(\mathbb{R}^d, \mu)$, then it is sufficient to consider
its positive and negative parts to which the previous result applies. Then the thesis
follows from the linearity of the integral.   □

Example 2.2.21 Consider the standard normal distribution $\mathcal{N}_{0,1}$ and the functions
$f(x) = x$ and $g(x) = x^2$. Then $f, g \in L^1(\mathbb{R}, \mathcal{N}_{0,1})$ and we have

$$\int_{\mathbb{R}} f(x)\mathcal{N}_{0,1}(dx) = \frac{1}{\sqrt{2\pi}} \int_{\mathbb{R}} x e^{-\frac{x^2}{2}} dx = 0,$$

$$\int_{\mathbb{R}} g(x)\mathcal{N}_{0,1}(dx) = \frac{1}{\sqrt{2\pi}} \int_{\mathbb{R}} x^2 e^{-\frac{x^2}{2}} dx = 1.$$

Remark 2.2.22 The proof of Proposition 2.2.20 is exemplary of a procedure
often used in the context of integration theory and probability. This procedure,
sometimes called *standard procedure*, consists in verifying the validity of the
thesis in 4 steps:

(continued)

Remark 2.2.22 (continued)
(1) the case of indicator functions: usually it is a direct check based on the definition of integral;
(2) the case of simple functions: it follows from the previous case and the linearity of the integral;
(3) the case of non-negative functions: it suffices to use an approximation argument based on Lemma 2.2.3 and Beppo Levi's theorem;
(4) the case of absolutely integrable functions: this case is related to the previous one due to linearity, considering the positive and negative parts.

A more general formulation of this procedure is given by the second Dynkin's theorem (cf. Theorem A.0.8).

We conclude the section with a useful result that we will prove later (cf. Corollary 2.5.8).

Corollary 2.2.23 *[!] If μ, ν are distributions such that*

$$\int_{\mathbb{R}} f\, d\mu = \int_{\mathbb{R}} f\, d\nu$$

for every $f \in bC(\mathbb{R})$ then $\mu \equiv \nu$. Here $bC(\mathbb{R})$ indicates the space of continuous and bounded functions.

2.2.5 Expected Value and Distributions

The expected value of a r.v. is nothing more than its integral under the probability measure. We give the precise definition.

Definition 2.2.24 (Expectation) Let X be an integrable r.v. on a probability space $(\Omega, \mathcal{F}, P)$. The expected value of X is defined as

$$E[X] := \int_{\Omega} X\, dP.$$

Example 2.2.25 [!] Starting from the definition (2.2.2) of abstract integral, it is easy to calculate the expected value in two particular cases: constant and indicator random variables. We have, in fact,

$$E[c] = c, \qquad c \in \mathbb{R}^d,$$

$$E[\mathbb{1}_A] = P(A), \qquad A \in \mathcal{F}.$$

Moreover, if X is a simple r.v. of the form

$$X = \sum_{k=1}^{m} x_k \mathbb{1}_{(X=x_k)}$$

by linearity, we have

$$E[X] = \sum_{k=1}^{m} x_k P(X = x_k),$$

that is, $E[X]$ *is a weighted average of the values of* X *based on the probability that such values are assumed.*

In general, computing an expected value defined as an abstract integral over a generic sample space Ω might not be easy: the following result shows that it is possible to express the expected value of a r.v. X as an integral on the Euclidean space with respect to its law μ_X.

Theorem 2.2.26 *[!] Let*

$$X : \Omega \longrightarrow \mathbb{R}^d \qquad and \qquad f : \mathbb{R}^d \longrightarrow \mathbb{R}^N$$

be respectively a r.v. on $(\Omega, \mathscr{F}, P)$ *with law* μ_X *and a* $\mathscr{B}_d$-*measurable function,* $f \in m\mathscr{B}_d$. *Then* $f \circ X$ *is integrable on* Ω *with respect to* P *if and only if* f *is integrable on* $\mathbb{R}^d$ *with respect to* μ_X *and in such case*

$$E[f(X)] = \int_{\mathbb{R}^d} f d\mu_X. \tag{2.2.11}$$

In particular, if $\mu_X = \sum_{k=1}^{\infty} p_k \delta_{x_k}$ *is a discrete distribution then*

$$E[f(X)] = \sum_{k=1}^{\infty} f(x_k) p_k, \tag{2.2.12}$$

while if μ_X *is absolutely continuous with density* γ_X *then we have*

$$E[f(X)] = \int_{\mathbb{R}^d} f(x) \gamma_X(x) dx. \tag{2.2.13}$$

Proof We prove (2.2.11) in the case $f = \mathbb{1}_H$ with $H \in \mathscr{B}_d$: we have

$$E[f(X)] = E[\mathbb{1}_H(X)] = P(X \in H) = \mu_X(H) = \int_{\mathbb{R}^d} \mathbb{1}_H d\mu_X.$$

The general case follows by applying the standard procedure of Remark 2.2.22. Finally, based on (2.2.11) and (2.2.12) follows from Proposition 2.2.17 and (2.2.13) follows from Proposition 2.2.20. $\square$

Remark 2.2.27 By applying Theorem 2.2.26 in the particular case of the identity function $f(x) = x$, we have that if X is integrable then

$$E[X] = \int_{\mathbb{R}^d} x \mu_X(dx).$$

Definition 2.2.28 (Variance) The variance of an integrable real r.v. X is defined as

$$\text{var}(X) := E\left[(X - E[X])^2\right].$$

The square root of the variance $\sqrt{\text{var}(X)}$ is called *standard deviation*.

Notice that $0 \le \text{var}(X) \le +\infty$ and if $X \in L^1(\Omega, P)$ then we have

$$\text{var}(X) = E[X^2] - E[X]^2.$$

The standard deviation is an average distance of X from its expected value. We will see in Example 2.2.31 that the standard deviation of $X \in \mathcal{N}_{\mu,\sigma^2}$ is equal to σ: we used σ in Fig. 2.3 to define the confidence intervals of X.

By linearity we have

$$\text{var}(aX + b) = a^2 \text{var}(X), \qquad a, b \in \mathbb{R}.$$

Furthermore, by Proposition 2.2.14, we have

$$\text{var}(X) = 0 \quad \text{if and only if} \quad X \overset{a.s.}{=} E[X].$$

Example 2.2.29 [!] We compute the mean and variance of some discrete random variables:

(i) if $X \sim \delta_{x_0}$, with $x_0 \in \mathbb{R}^d$, then by (2.2.11)–(2.2.12) we have

$$E[X] = \int_{\mathbb{R}^d} y \delta_{x_0}(dy) = x_0,$$

$$\text{var}(X) = \int_{\mathbb{R}^d} (y - x_0)^2 \delta_{x_0}(dy) = 0.$$

(ii) $X \sim \text{Unif}_n$ has distribution function $\gamma(k) = \frac{1}{n}$ for $k \in I_n = \{1, \ldots, n\}$ and we have

$$E[X] = \sum_{k=1}^{n} k\gamma(k) = \frac{1}{n} \sum_{k=1}^{n} k = \frac{1}{n} \cdot \frac{n(n+1)}{2} = \frac{n+1}{2},$$

$$\text{var}(X) = E\left[X^2\right] - E\left[X\right]^2 = \sum_{k=1}^{n} k^2 \gamma(k) - \left(\frac{n+1}{2}\right)^2$$

$$= \frac{1}{n}\sum_{k=1}^{n} k^2 - \left(\frac{n+1}{2}\right)^2$$

$$= \frac{1}{n} \cdot \frac{n(n+1)(2n+1)}{6} - \left(\frac{n+1}{2}\right)^2 = \frac{n^2-1}{12}.$$

(iii) $X \sim \text{Be}_p$ has a distribution function γ defined by $\gamma(1) = p$, $\gamma(0) = 1 - p$ and we have

$$E[X] = \sum_{k \in \{0,1\}} k\gamma(k) = 0 \cdot (1 - p) + p = p,$$

$$\text{var}(X) = E\left[X^2\right] - E[X]^2 = \sum_{k \in \{0,1\}} k^2 \gamma(k) - p^2 = p(1 - p).$$

(iv) If $X \sim \text{Bin}_{n,p}$ a direct calculation (see also Proposition 2.6.3) shows

$$E[X] = np, \qquad \text{var}(X) = np(1 - p). \tag{2.2.14}$$

(v) $X \sim \text{Poisson}_\lambda$ has a distribution function $\gamma(k) = e^{-\lambda}\frac{\lambda^k}{k!}$ for $k \in \mathbb{N}_0$ and we have

$$E[X] = \sum_{k=0}^{\infty} k\gamma(k) = \sum_{k=1}^{\infty} ke^{-\lambda}\frac{\lambda^k}{k!} = \lambda e^{-\lambda}\sum_{k=1}^{\infty}\frac{\lambda^{k-1}}{(k-1)!} = \lambda.$$

Similarly, we have $\text{var}(X) = \lambda$.

(vi) $X \sim \text{Geom}_p$ has a distribution function $\gamma(k) = p(1-p)^{k-1}$ for $k \in \mathbb{N}$ and therefore

$$E[X] = \sum_{k=1}^{\infty} k\gamma(k) = p\sum_{k=1}^{\infty} k(1-p)^{k-1} = p\sum_{k=1}^{\infty}\left(-\frac{d}{dp}(1-p)^k\right)$$

$$= -p\frac{d}{dp}\sum_{k=1}^{\infty}(1-p)^k = -p\frac{d}{dp}\left(\frac{1}{1-(1-p)}\right) = \frac{1}{p}.$$

Similarly, we have $\text{var}(X) = \frac{1-p}{p^2}$.

Example 2.2.30 [!] Consider a gambling game in which a non-rigged coin is tossed: if it comes up heads, you win one euro and if it comes up tails, you lose one euro. If X is the r.v. representing the result of the game, we have

$$E[X] = 1 \cdot \frac{1}{2} + (-1) \cdot \frac{1}{2} = 0$$

and so we say that the game is *fair*. The game is also fair if the winnings and losses were equal to 1000 euros, but intuitively we would be less inclined to play because we perceive a greater risk (of losing a lot of money). Mathematically, this is explained by the fact that

$$\text{var}(X) = E\left[X^2\right] = 1^2 \cdot \frac{1}{2} + (-1)^2 \cdot \frac{1}{2} = 1$$

while if Y represents the r.v. when the stake is 1000 euros, we have

$$\text{var}(Y) = E\left[Y^2\right] = 1000^2 \cdot \frac{1}{2} + (-1000)^2 \cdot \frac{1}{2} = 1000^2.$$

In practice, even if two bets have the same expected value, the one with the lower variance limits the extent of potential losses.

Example 2.2.31 [!] We consider some examples of absolutely continuous random variables:

(i) if $X \sim \text{Unif}_{[a,b]}$ we have

$$E[X] = \int_{\mathbb{R}} y \text{Unif}_{[a,b]}(dy) = \frac{1}{b-a} \int_a^b y \, dy = \frac{a+b}{2},$$

$$\text{var}(X) = \int_{\mathbb{R}} \left(y - \frac{a+b}{2}\right)^2 \text{Unif}_{[a,b]}(dy)$$

$$= \frac{1}{b-a} \int_a^b \left(y - \frac{a+b}{2}\right)^2 dy = \frac{(b-a)^2}{12}.$$

Compare this with the analogous discrete result in Example 2.2.29-(i).

(ii) If $X \sim \mathcal{N}_{\mu,\sigma^2}$ with $\sigma > 0$ then

$$E[X] = \int_{\mathbb{R}} y \mathcal{N}_{\mu,\sigma^2}(dy) = \frac{1}{\sqrt{2\pi\sigma^2}} \int_{\mathbb{R}} y e^{-\frac{(y-\mu)^2}{2\sigma^2}} dy =$$

(by the change of variables $z = \frac{y-\mu}{\sigma\sqrt{2}}$)

$$= \frac{1}{\sqrt{\pi}} \int_{\mathbb{R}} \left(\mu + z\sigma\sqrt{2}\right) e^{-z^2} dz = \frac{\mu}{\sqrt{\pi}} \int_{\mathbb{R}} e^{-z^2} dz = \mu.$$

Similarly, we see that

$$\mathrm{var}(X) = \int_{\mathbb{R}} (y - \mu)^2 \, \mathcal{N}_{\mu,\sigma^2}(dy) = \sigma^2.$$

(iii) If $X \sim \mathrm{Gamma}_{\alpha,1}$ we have

$$E\left[X\right] = \int_0^\infty t\gamma_{\alpha,1}(t)dt = \frac{1}{\Gamma(\alpha)} \int_0^\infty t^\alpha e^{-\lambda t} dt = \frac{\Gamma(\alpha + 1)}{\Gamma(\alpha)} = \alpha,$$

$$E\left[X^2\right] = \int_0^\infty t^2 \gamma_{\alpha,1}(t)dt = \frac{1}{\Gamma(\alpha)} \int_0^\infty t^{1+\alpha} e^{-\lambda t} dt$$

$$= \frac{\Gamma(\alpha + 2)}{\Gamma(\alpha)} = \alpha(\alpha + 1)$$

so that

$$\mathrm{var}(X) = E\left[X^2\right] - E\left[X\right]^2 = \alpha.$$

In general, by Lemma 2.1.37, if $X \sim \mathrm{Gamma}_{\alpha,\lambda}$ we have

$$E\left[X\right] = \frac{\alpha}{\lambda}, \qquad \mathrm{var}(X) = \frac{\alpha}{\lambda^2}.$$

In particular, if $X \sim \mathrm{Exp}_\lambda = \mathrm{Gamma}_{1,\lambda}$ then

$$E\left[X\right] = \int_{\mathbb{R}} y\mathrm{Exp}_\lambda(dy) = \lambda \int_0^{+\infty} ye^{-\lambda y} dy = \frac{1}{\lambda},$$

$$\mathrm{var}(X) = \int_{\mathbb{R}} \left(y - \frac{1}{\lambda}\right)^2 \mathrm{Exp}_\lambda(dy) = \lambda \int_0^{+\infty} \left(y - \frac{1}{\lambda}\right)^2 e^{-\lambda y} dy = \frac{1}{\lambda^2}.$$

2.2.6   Jensen's Inequality

We prove an important extension to convex functions of the triangular inequality for the expected value. Typical examples of convex functions that we will use later are:

(i) $f(x) = |x|^p$ with $p \in [1, +\infty[$;
(ii) $f(x) = e^{\lambda x}$ with $\lambda \in \mathbb{R}$;
(iii) $f(x) = -\log x$ for $x \in \mathbb{R}_{>0}$.

Theorem 2.2.32 (Jensen's Inequality) *[!!] Let* $-\infty \le a < b \le +\infty$ *and*

$$X : \Omega \longrightarrow]a, b[\qquad \text{and} \qquad f :]a, b[\longrightarrow \mathbb{R}$$

be respectively a r.v. on the space $(\Omega, \mathscr{F}, P)$ *and a convex function. If* $X, f(X) \in L^1(\Omega, P)$ *then we have*

$$f(E[X]) \le E[f(X)].$$

Proof Recall that if f is convex then for every $z \in]a, b[$ there exists $m \in \mathbb{R}$ such that

$$f(w) \ge f(z) + m(w - z), \qquad \forall w \in]a, b[. \tag{2.2.15}$$

We conclude the proof of Jensen's inequality first and prove (2.2.15) later. Setting $z = E[X]$ (note that $E[X] \in]a, b[$ since $X(\Omega) \subseteq]a, b[$ by assumption) we have

$$f(X(\omega)) \ge f(E[X]) + m(X(\omega) - E[X]), \qquad \omega \in \Omega,$$

from which, taking the expected value and using the monotonicity property,

$$E[f(X)] \ge E[f(E[X]) + m(X - E[X])] =$$

(by linearity and the fact that $E[c] = c$ for every constant c)

$$= f(E[X]) + mE[X - E[X]] = f(E[X]).$$

Now let us prove (2.2.15). Recall that f is convex if we have

$$f((1 - \lambda)x + \lambda y) \le (1 - \lambda)f(x) + \lambda f(y), \qquad \forall x, y \in]a, b[, \ \lambda \in [0, 1],$$

or equivalently, setting $z = (1 - \lambda)x + \lambda y$,

$$(y - x)f(z) \le (y - z)f(x) + (z - x)f(y), \qquad x < z < y. \tag{2.2.16}$$

Introduce the notation

$$\Delta_{y,x} = \frac{f(y) - f(x)}{y - x}, \qquad a < x < y < b.$$

It is not difficult to verify[10] that (2.2.16) is equivalent to

$$\Delta_{z,x} \le \Delta_{y,x} \le \Delta_{y,z}, \qquad x < z < y. \tag{2.2.17}$$

(2.2.17) implies[11] that f is a continuous function on $]a, b[$ and also that the functions

$$z \mapsto \Delta_{z,x}, \text{ for } z > x, \quad \text{and} \quad z \mapsto \Delta_{y,z}, \text{ for } z < y,$$

are monotone increasing. Consequently, the limits[12] exist

$$D^- f(z) := \lim_{x \to z^-} \Delta_{z,x} \le \lim_{y \to z^+} \Delta_{y,z} =: D^+ f(z), \qquad z \in]a, b[. \tag{2.2.18}$$

Now if $m \in [D^- f(z), D^+ f(z)]$ we have

$$\Delta_{z,x} \le m \le \Delta_{y,z}, \qquad x < z < y,$$

which implies (2.2.15).   □

Remark 2.2.33 The proof of Jensen's inequality is based, in addition to the convexity property, on the properties of monotonicity, linearity, and $E[1] = 1$ of the expectation. In particular, the fact that $E[1] = 1$ is fundamental: unlike the triangular inequality, Jensen's inequality does not hold for all integrals or sums.

---

[10] Let us prove for example the first inequality:

$$\Delta_{z,x} \le \Delta_{y,x} \iff \frac{f(z) - f(x)}{z - x} \le \frac{f(y) - f(x)}{y - x} \iff (f(z) - f(x))(y - x)$$
$$\le (f(y) - f(x))(z - x)$$

which is equivalent to (2.2.16).

[11] Indeed, from (2.2.17), in particular from $\Delta_{z,x} \le \Delta_{y,x}$, it follows

$$f(z) \le f(x) + (z - x)\frac{f(y) - f(x)}{y - x} \longrightarrow f(y) \quad \text{for } z \to y^-.$$

Moreover, fixing $y_0 \in]y, b[$, still from (2.2.17), in particular from $\Delta_{y,z} \le \Delta_{y_0,y}$, it follows

$$f(z) \ge f(y) - (y - z)\Delta_{y_0,y} \longrightarrow f(y) \quad \text{for } z \to y^-.$$

Combining the two inequalities, the left continuity of f is proved. For the right continuity, proceed analogously.

[12] To fix ideas, think of $f(x) = |x|$ for which we have $-1 = D^- f(0) < D^+ f(0) = 1$. By (2.2.18) the set of points z where $D^- f(z) < D^+ f(z)$, i.e. where f is not differentiable, is at most countable.

2.2.7 L^p Spaces and Inequalities

Definition 2.2.34 Let $(\Omega, \mathcal{F}, P)$ be a probability space and $p \in [1, +\infty[$. The *p-norm* of a r.v. X is defined by

$$\|X\|_p := \left(E\left[|X|^p \right] \right)^{\frac{1}{p}}.$$

We denote by

$$L^p(\Omega, P) = \{ X \in m\mathcal{F} \mid \|X\|_p < \infty \}$$

the space of random variables for which the p-th power of the absolute value is integrable.

Actually, $\| \cdot \|_p$ is not a norm because $\|X\|_p = 0$ only implies $X \overset{a.s.}{=} 0$ but not $X \equiv 0$: we will see in Theorem 2.2.40 that $\| \cdot \|_p$ is a *semi-norm* on the space $L^p(\Omega, P)$. In functional analysis, it is common practice to form a quotient space by considering equivalence classes of functions that are equal almost everywhere. With this identification $\| \cdot \|_p$ becomes a genuine norm. However, in certain areas of probability theory, especially in more advanced topics, such identification is not always possible. We will delve deeper into this issue later on (see, for instance, Remark 4.2.6).

Example 2.2.35 If $X \sim \mathcal{N}_{\mu,\sigma^2}$ then $X \in L^p(\Omega, P)$ for every $p \geq 1$ since

$$E\left[|X|^p \right] = \int_{\mathbb{R}} |x|^p \frac{1}{\sqrt{2\pi\sigma^2}} e^{-\frac{1}{2}\left(\frac{x-\mu}{\sigma}\right)^2} dx < \infty.$$

It is easy to give an example of $X, Y \in L^1(\Omega, P)$ such that $XY \notin L^1(\Omega, P)$: it is sufficient to consider $X(\omega) = Y(\omega) = \frac{1}{\sqrt{\omega}}$ in the space $([0, 1], \mathcal{B}, \text{Leb})$. We also give an example in a discrete space.

Example 2.2.36 Consider the probability space $\Omega = \mathbb{N}$ with the probability measure defined by

$$P(\{n\}) = \frac{c}{n^3}, \qquad n \in \mathbb{N},$$

where c is the positive constant[13] that normalizes the sum of the $P(\{n\})$ to 1 so that P is a probability measure. The r.v. $X(n) = n$ is integrable under P since

$$E[X] = \sum_{n=1}^{\infty} X(n)P(\{n\}) = \sum_{n=1}^{\infty} n \cdot \frac{c}{n^3} < +\infty.$$

On the other hand, $X \notin L^2(\Omega, P)$ since

$$E\left[X^2\right] = \sum_{n=1}^{\infty} n^2 \cdot \frac{c}{n^3} = +\infty.$$

Proposition 2.2.37 *If $1 \leq p_1 \leq p_2$ then*

$$\|X\|_{p_1} \leq \|X\|_{p_2}$$

and therefore

$$L^{p_2}(\Omega, P) \subseteq L^{p_1}(\Omega, P).$$

In general the inclusion is strict, as Example 2.2.36 shows.

Proof The thesis is a direct consequence of Jensen's inequality with $f(x) = |x|^q$ and $q = \frac{p_2}{p_1} \geq 1$:

$$E\left[|X|^{p_1}\right]^{\frac{p_2}{p_1}} \leq E\left[|X|^{p_2}\right].$$

$\square$

Theorem 2.2.38 (Hölder) *[!] Let $p, q > 1$ be conjugate exponents, i.e., such that $\frac{1}{p} + \frac{1}{q} = 1$. If $X \in L^p(\Omega, P)$ and $Y \in L^q(\Omega, P)$ are d-dimensional random variables then $X \cdot Y \in L^1(\Omega, P)$ and*

$$\|X \cdot Y\|_1 \leq \|X\|_p \|Y\|_q. \tag{2.2.19}$$

Proof We prove the thesis for $\|X\|_p > 0$ otherwise it is trivial. Then, (2.2.19) follows from

$$E\left[\widetilde{X}|Y|\right] \leq \|Y\|_q, \quad \text{where } \widetilde{X} = \frac{|X|}{\|X\|_p}.$$

---

[13] For precision, $c = \text{Zeta}(3) \approx 1.20206$ where Zeta indicates the Riemann zeta function.

Note that $\widetilde{X}^p \geq 0$ and $E\left[\widetilde{X}^p\right] = 1$: therefore, we consider the probability Q with density $\widetilde{X}^p$ with respect to P, defined by

$$Q(A) = E\left[\widetilde{X}^p \mathbb{1}_A\right], \qquad A \in \mathscr{F}.$$

Then we have

$$E^P\left[\widetilde{X}|Y|\right]^q = E^P\left[\widetilde{X}^p \frac{|Y|}{\widetilde{X}^{p-1}}\mathbb{1}_{(\widetilde{X}>0)}\right]^q = E^Q\left[\frac{|Y|}{\widetilde{X}^{p-1}}\mathbb{1}_{(\widetilde{X}>0)}\right]^q \leq$$

(by Jensen's inequality)

$$\leq E^Q\left[\frac{|Y|^q}{\widetilde{X}^{q(p-1)}}\mathbb{1}_{(\widetilde{X}>0)}\right] =$$

(since $q(p-1) = p$, being p, q conjugate)

$$= E^Q\left[\frac{|Y|^q}{\widetilde{X}^p}\mathbb{1}_{(\widetilde{X}>0)}\right] = E^P\left[|Y|^q \mathbb{1}_{(\widetilde{X}>0)}\right] \leq \|Y\|_q^q$$

which proves the thesis.   □

Corollary 2.2.39 (Cauchy-Schwarz) *[!] If $X, Y \in L^2(\Omega, P)$ are d-dimensional random variables then*

$$|E\left[X \cdot Y\right]| \leq \|X\|_2 \|Y\|_2. \tag{2.2.20}$$

Equality holds in (2.2.20) if and only if $aX + bY \overset{a.s.}{=} 0$ for some $(a, b) \in \mathbb{R}^2 \backslash \{(0, 0)\}$.

Proof Equation (2.2.20) follows from $|E\left[X \cdot Y\right]| \leq E\left[|X||Y|\right]$ and Hölder's inequality. If $aX + bY \overset{a.s.}{=} 0$ for some $(a, b) \in \mathbb{R}^2 \setminus \{(0, 0)\}$ then it is easy to verify that equality holds in (2.2.20). Conversely, it is not restrictive to assume $E\left[X \cdot Y\right] \geq 0$ (otherwise just consider $-X$ instead of X) and $\|X\|_2, \|Y\|_2 > 0$ (otherwise the thesis is obvious): in this case we set

$$\widetilde{X} = \frac{X}{\|X\|_2}, \qquad \widetilde{Y} = \frac{Y}{\|Y\|_2}.$$

We have $\|\widetilde{X}\|_2 = \|\widetilde{Y}\|_2 = 1$ and also, by assumption, $E\left[\widetilde{X} \cdot \widetilde{Y}\right] = 1$. Then

$$E\left[(\widetilde{X} - \widetilde{Y})^2\right] = E\left[\widetilde{X}^2\right] + E\left[\widetilde{Y}^2\right] - 2E\left[\widetilde{X} \cdot \widetilde{Y}\right] = 0$$

from which $\widetilde{X} \overset{a.s.}{=} \widetilde{Y}$.   □

Theorem 2.2.40 *For every* $p \geq 1$, $L^p(\Omega, P)$ *is a vector space on which* $\| \cdot \|_p$ *is a semi-norm, that is*

(i) $\|X\|_p = 0$ *if and only if* $X \stackrel{a.s.}{=} 0$;
(ii) $\|\lambda X\|_p = |\lambda| \|X\|_p$ *for every* $\lambda \in \mathbb{R}$ *and* $X \in L^p(\Omega, P)$;
(iii) *the* Minkowski *inequality*

$$\|X + Y\|_p \leq \|X\|_p + \|Y\|_p,$$

holds for every $X, Y \in L^p(\Omega, P)$.

Proof It is clear that, if $X \in L^p(\Omega, P)$ and $\lambda \in \mathbb{R}$, then $\lambda X \in L^p(\Omega, P)$. Moreover, since

$$(a + b)^p \leq 2^p (a \vee b)^p \leq 2^p \left(a^p + b^p \right), \qquad a, b \geq 0, \ p \geq 1,$$

then the fact that $X, Y \in L^p(\Omega, P)$ implies that $(X + Y) \in L^p(\Omega, P)$. Hence $L^p(\Omega, P)$ is a vector space. Properties (i) and (ii) follow easily from the general properties of expectation. As for (iii), it is sufficient to consider the case $p > 1$: by the triangular inequality we have

$$E \left[|X + Y|^p \right] \leq E \left[|X| |X + Y|^{p-1} \right] + E \left[|Y| |X + Y|^{p-1} \right] \leq$$

(by the Hölder inequality, indicating with q the conjugate exponent of $p > 1$)

$$\leq \left(\|X\|_p + \|Y\|_p \right) E \left[|X + Y|^{(p-1)q} \right]^{\frac{1}{q}} =$$

(since $(p - 1)q = p$)

$$\leq \left(\|X\|_p + \|Y\|_p \right) E \left[|X + Y|^p \right]^{1 - \frac{1}{p}},$$

which implies the Minkowski inequality. $\square$

2.2.8   Covariance and Correlation

Definition 2.2.41 (Covariance) The *covariance* of two real random variables $X, Y \in L^2(\Omega, P)$ is the real number

$$\mathrm{cov}(X, Y) := E \left[(X - E[X])(Y - E[Y]) \right] = E[XY] - E[X] E[Y].$$

Example 2.2.42 Assume that (X, Y) has density

$$\gamma_{(X,Y)}(x, y) = ye^{-xy} \mathbb{1}_{\mathbb{R}_{\geq 0} \times [1,2]}(x, y).$$

Then we have

$$E[X] = \iint_{\mathbb{R}^2} x\gamma_{(X,Y)}(x, y)dxdy = \log 2,$$

$$E[Y] = \iint_{\mathbb{R}^2} y\gamma_{(X,Y)}(x, y)dxdy = \frac{3}{2}$$

and

$$\text{cov}(X, Y) = \iint_{\mathbb{R}^2} (x - \log 2)\left(y - \frac{3}{2}\right)\gamma_{(X,Y)}(x, y)dxdy = 1 - \frac{3}{2}\log 2.$$

In this section we use the following notations:

- $e_X := E[X]$ for the expectation of X;
- $\sigma_{XY} := \text{cov}(X, Y) := e_{(X-e_X)(Y-e_Y)} = e_{XY} - e_X e_Y$ for the covariance of X, Y;
- $\sigma_X = \sqrt{\text{var}(X)}$ for the standard deviation of X, where

$$\text{var}(X) = \text{cov}(X, X) = e_{(X-e_X)^2} = e_{X^2} - e_X^2.$$

We observe that:

(i) for every $c \in \mathbb{R}$ we have

$$\text{var}(X) = E\left[(X - E[X])^2\right] \leq E\left[(X - c)^2\right]$$

and the equality holds if and only if $c = E[X]$. In fact,

$$E\left[(X - c)^2\right] = E\left[(X - e_X + e_X - c)^2\right]$$

$$= \sigma_X^2 + 2\underbrace{E[X - e_X]}_{=0}(e_X - c) + (e_X - c)^2$$

$$= \sigma_X^2 + (e_X - c)^2 \geq \sigma_X^2.$$

(ii) If $\sigma_X > 0$ we can always "normalize" the r.v. X by setting

$$Z = \frac{X - e_X}{\sigma_X},$$

so that $E[Z] = 0$ and $\text{var}(Z) = 1$.

(iii) We have

$$\text{var}(X + Y) = \text{var}(X) + \text{var}(Y) + 2\text{cov}(X, Y). \qquad (2.2.21)$$

If $\text{cov}(X, Y) = 0$ we say that X, Y are *uncorrelated*.

(iv) The covariance $\text{cov}(\cdot, \cdot)$ is a *bilinear and symmetric operator on* $L^2(\Omega, P) \times L^2(\Omega, P)$, that is, for every $X, Y, Z \in L^2(\Omega, P)$ and $\alpha, \beta \in \mathbb{R}$ we have

$$\text{cov}(X, Y) = \text{cov}(Y, X) \quad \text{and}$$

$$\text{cov}(\alpha X + \beta Y, Z) = \alpha\text{cov}(X, Z) + \beta\text{cov}(Y, Z).$$

(v) By the Cauchy-Schwarz inequality (2.2.20) we have $|\text{cov}(X, Y)| \le \sqrt{\text{var}(X)\text{var}(Y)}$ that is

$$|\sigma_{XY}| \le \sigma_X \sigma_Y \qquad (2.2.22)$$

and we have equality in (2.2.22) if and only if $aX + bY \overset{a.s.}{=} c$ for some $(a, b) \in \mathbb{R}^2 \setminus \{(0, 0)\}$ and $c \in \mathbb{R}$: in particular, if $\sigma_X > 0$ then Y is a linear function of X in the sense that $Y \overset{a.s.}{=} \bar{a}X + \bar{b}$, where the constants $\bar{a}$ and $\bar{b}$ are given by

$$\bar{a} = \frac{\sigma_{XY}}{\sigma_X^2}, \qquad \bar{b} = e_Y - e_X \frac{\sigma_{XY}}{\sigma_X^2}. \qquad (2.2.23)$$

As we will see in Sect. 2.2.9, the line of equation $y = \bar{a}x + \bar{b}$ is called *regression line*, and intuitively provides a representation of the *linear dependence* between two samples of data.

Definition 2.2.43 (Correlation) Let $X, Y \in L^2(\Omega, P)$ be such that $\sigma_X, \sigma_Y > 0$. The *correlation coefficient* of X, Y is defined by

$$\varrho_{XY} := \frac{\sigma_{XY}}{\sigma_X \sigma_Y}.$$

By (2.2.22) we have $\varrho_{XY} \in [-1, 1]$; in particular, $\varrho_{XY} = \pm 1$ if and only if $Y \overset{a.s.}{=} \bar{a}X + \bar{b}$, where $\bar{a}$ shares the same sign as ρ_{XY}. Thus, the correlation coefficient ρ_{XY} quantifies the *degree of linear dependence* between X and Y.

Let now $X = (X_1, \ldots, X_d) \in L^2(\Omega, P)$ be a random vector in $\mathbb{R}^d$. The *covariance matrix of* X is the $d \times d$ symmetric matrix

$$\text{cov}(X) = \left(\sigma_{X_i X_j}\right)_{i, j=1,\ldots,d} = E\Big[\underbrace{(X - E[X])}_{d \times 1} \underbrace{(X - E[X])^*}_{1 \times d}\Big],$$

where M^* denotes the transpose of the matrix M. Since

$$\langle \operatorname{cov}(X)y, y\rangle = E\left[|(X - E[X])^* y|^2\right] \geq 0, \qquad y \in \mathbb{R}^d,$$

the covariance matrix is *positive semi-definite*. Note that the elements of the diagonal are the variances $\sigma_{X_i}^2$ for $i = 1, \ldots, d$. If $\sigma_{X_i} > 0$ for each $i = 1, \ldots, d$, we define the *correlation matrix* analogously:

$$\varrho(X) = \left(\varrho_{X_i X_j}\right)_{i,j=1,\ldots,d}.$$

The matrix $\varrho(X)$ is symmetric, positive semi-definite and the elements of the diagonal are equal to one: for example, in the case $d = 2$ we have

$$\varrho(X) = \begin{pmatrix} 1 & \varrho_{X_1 X_2} \\ \varrho_{X_1 X_2} & 1 \end{pmatrix} \qquad \operatorname{cov}(X) = \begin{pmatrix} \sigma_{X_1}^2 & \sigma_{X_1}\sigma_{X_2}\varrho_{X_1 X_2} \\ \sigma_{X_1}\sigma_{X_2}\varrho_{X_1 X_2} & \sigma_{X_2}^2 \end{pmatrix}.$$

If A is a constant $N \times d$ matrix and $b \in \mathbb{R}^N$, then $Z := AX + b$ is $\mathbb{R}^N$-valued r.v. with expectation

$$E[Z] = AE[X] + b,$$

and covariance matrix

$$\operatorname{cov}(Z) = E\left[(AX + b - E[AX + b])(AX + b - E[AX + b])^*\right] = A\operatorname{cov}(X)A^*.$$

Remark 2.2.44 (Cholesky Decomposition) A symmetric and positive semi-definite matrix C can be factorized in the form $C = AA^*$: this follows from the fact that, by the spectral theorem, $C = UDU^*$ with U orthogonal (i.e., such $U^{-1} = U^*$) and D diagonal matrix; hence it suffices to set $A = U\sqrt{D}U^*$ where $\sqrt{D}$ denotes the diagonal matrix whose elements are the square roots of the elements of D (which are non-negative real numbers, since C is symmetric and positive semi-definite).

The factorization $C = AA^*$ is not unique: the Cholesky algorithm allows to determine a lower triangular matrix A such that $C = AA^*$. For example, given the correlation matrix in dimension two

$$C = \begin{pmatrix} 1 & \varrho \\ \varrho & 1 \end{pmatrix}$$

we have the Cholesky factorization $C = AA^*$ with

$$A = \begin{pmatrix} 1 & 0 \\ \varrho & \sqrt{1 - \varrho^2} \end{pmatrix}.$$

2.2.9   *Linear Regression*

In Statistics, we often deal with *time series* (or *samples*) of data that provide the dynamics of a certain phenomenon over time (for example, a temperature, the price of a financial asset, the number of employees of a company, etc.). In the case of one-dimensional data, a time series is a vector $x = (x_1, \ldots, x_M)$ of $\mathbb{R}^M$. We can consider of the vector x as a "realization" of a discrete r.v. X, which is defined as follows

$$X : I_M \longrightarrow \mathbb{R}, \qquad X(i) := x_i, \quad i \in I_M = \{1, \ldots, M\}.$$

Endowing the sample space I_M with uniform probability, mean and variance of X are given by

$$E[X] = \frac{1}{M} \sum_{i=1}^{M} x_i, \qquad \mathrm{var}(X) = \frac{1}{M} \sum_{i=1}^{M} (x_i - E[X])^2.$$

In Statistics, $E[X]$ and $\mathrm{var}(X)$ are called the *sample mean* and the *sample variance* of the time series x and are often denoted by $E[x]$ and $\mathrm{var}(x)$ respectively.

Now, let $x = (x_1, \ldots, x_M)$ and $y = (y_1, \ldots, y_M)$ be two time series. A simple tool to visualize the degree of "dependence" between x and y is the so-called *scatter plot*: in it, the points with coordinates $(x_i, y_i)_{i \in I_M}$ are represented on the Cartesian plane. An example is given in Fig. 2.6.

The *regression line*, drawn in the scatter plot in Fig. 2.6, is the line with equation $y = ax + b$ where a, b minimize the differences between $ax_i + b$ and y_i in the sense that they minimize the mean squared error

$$Q(a, b) = \sum_{i=1}^{M} (ax_i + b - y_i)^2.$$

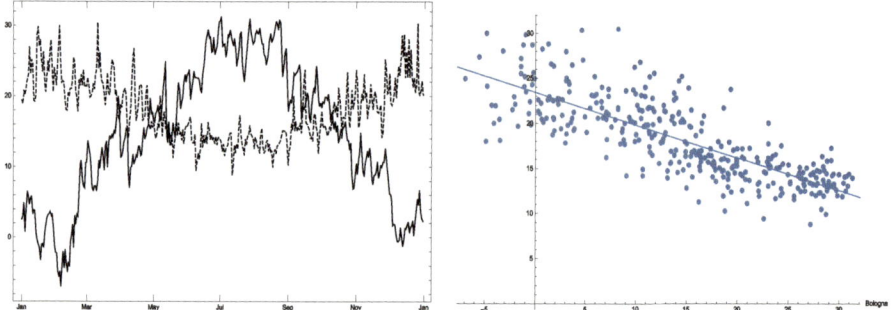

Fig. 2.6 *On the left:* temperatures in 2012 for Bologna (solid line) and Cape Town (dashed line). *On the right:* scatter plot of temperatures in 2012 for Bologna (on the x-axis) and Cape Town (on the y-axis)

By setting the gradient of Q to zero

$$(\partial_a Q(a, b), \partial_b Q(a, b)) = \left(2 \sum_{i=1}^{M} (ax_i + b - y_i) x_i \, , \, 2 \sum_{i=1}^{M} (ax_i + b - y_i) \right)$$

a, b are determined: precisely a simple calculation shows that

$$a = \frac{\sigma_{xy}}{\sigma_x^2}, \qquad b = E[y] - \frac{\sigma_{xy}}{\sigma_x^2} E[x], \tag{2.2.24}$$

if $\sigma_x^2 = \text{var}(x) > 0$ and

$$\sigma_{xy} = \text{cov}(x, y) = \frac{1}{M} \sum_{i=1}^{M} (x_i - E[x]) (y_i - E[y])$$

is the *sample (or empirical) covariance* of x and y. Note the analogy with formulas (2.2.23).

The covariance σ_{xy} is proportional and has the same sign as the slope of the regression line. σ_{xy} *is an indicator of the linear dependence between x and y:* if $\sigma_{xy} = 0$, i.e., x and y are *uncorrelated* samples, there is no linear dependence (but there could be dependence of another type); if $\sigma_{xy} > 0$ the samples depend linearly in a positive way, the regression line is increasing and this indicates that y tends to increase as x increases.

The quantity

$$\varrho_{xy} = \frac{\sigma_{xy}}{\sigma_x \sigma_y}$$

is called *sample (or empirical) correlation* between x and y. The correlation has the advantage of being invariant for scale changes: for every $\alpha, \beta > 0$ the correlation between αx and βy is equal to the correlation between x and y. By the Cauchy-Schwarz inequality, we have $\varrho_{xy} \in [-1, 1]$. Moreover, $\varrho_{xy} = \pm 1$ if and only if $Q(a, b) = 0$ with a, b as in (2.2.24).

2.2.10 Random Vectors: Marginal and Joint Distributions

A random vector is a multi-variate r.v. $X = (X_1, \ldots, X_n)$ on a probability space $(\Omega, \mathscr{F}, P)$. In this section, we examine the relationship between X and its components

$$X_i : \Omega \longrightarrow \mathbb{R}^{d_i}, \qquad i = 1, \ldots, n,$$

where $d_i \in \mathbb{N}$ and we set $d = d_1 + \cdots + d_n$.

Notation 2.2.45 As usual, we denote by μ_X and F_X the distribution and the cumulative distribution function (CDF) of X respectively. We will examine in particular the cases in which:

(i) X is absolutely continuous: in this case, we denote by γ_X its density (which is uniquely defined up to Lebesgue-negligible sets);
(ii) X is discrete: in this case, we denote by $\bar{\mu}_X$ its distribution function defined by $\bar{\mu}_X(x) = P(X = x)$.

Later, we will always use vector notation: in particular, if $x, y \in \mathbb{R}^d$ then $x \leq y$ means $x_i \leq y_i$ for each $i = 1, \ldots, d$, and

$$] - \infty, x] :=] - \infty, x_1] \times \cdots \times] - \infty, x_d].$$

Definition 2.2.46 We say that μ_X and F_X are the *joint distribution* and the *joint CDF* of the random variables $X_1, \ldots, X_n$, respectively. Similarly, if they exist, γ_X and $\bar{\mu}_X$ are the *joint density* and the *joint distribution function* of $X_1, \ldots, X_n$.

Conversely, the distributions μ_{X_i}, $i = 1, \ldots, n$, of the random variables $X_1, \ldots, X_n$ are called *marginal distributions* of X. Similarly, we speak of *marginal CDFs*, *marginal densities*, and *marginal distribution functions of X*.

The following proposition shows that the marginals can be easily derived from the joint distribution. In the statement, to simplify the notations, we consider only the marginals for the first component X_1 but an analogous result is valid for each component.

Proposition 2.2.47 *[!] Let $X = (X_1, \ldots, X_n)$ be a random vector. We have:*

$$\mu_{X_1}(H) = \mu_X(H \times \mathbb{R}^{d-d_1}), \qquad\qquad H \in \mathscr{B}_{d_1}, \qquad\qquad (2.2.25)$$

$$F_{X_1}(x_1) = F_X(x_1, +\infty, \ldots, +\infty), \qquad x_1 \in \mathbb{R}^{d_1}.$$

Moreover, if $X \in AC$ then $X_1 \in AC$ and

$$\gamma_{X_1}(x_1) := \int_{\mathbb{R}^{d-d_1}} \gamma_X(x_1, x_2, \ldots, x_n) dx_2 \cdots dx_n, \; x_1 \in \mathbb{R}^{d_1} \qquad (2.2.26)$$

is a density of X_1. If X is discrete then X_1 is discrete and we have

$$\bar{\mu}_{X_1}(x_1) = \sum_{(x_2,\ldots,x_n)\in\mathbb{R}^{d-d_1}} \bar{\mu}_X(x_1, x_2, \ldots, x_n), \qquad x_1 \in \mathbb{R}^{d_1}. \qquad (2.2.27)$$

Proof It suffices to observe that

$$\mu_{X_1}(H) = P(X_1 \in H) = P(X \in H \times \mathbb{R}^{d-d_1}) = \mu_X(H \times \mathbb{R}^{d-d_1}), \quad H \in \mathscr{B}_{d_1}.$$

Taking $H =]-\infty, x_1]$ the second equality is proved. Moreover, if $X \in AC$, by (2.2.25) we have

$$P(X_1 \in H) = P(X \in H \times \mathbb{R}^{d-d_1})$$

$$= \int_{H \times \mathbb{R}^{d-d_1}} \gamma_X(x)dx =$$

(by Fubini's theorem for Lebesgue integral, being γ_X non-negative)

$$= \int_H \left(\int_{\mathbb{R}^{d-d_1}} \gamma_X(x_1, \ldots, x_n)dx_2 \cdots dx_n \right) dx_1$$

which proves (2.2.26). Finally, we have

$$\bar{\mu}_{X_1}(x_1) = P(X_1 = x_1) = P(X \in \{x_1\} \times \mathbb{R}^{d-d_1}) =$$

(by (1.4.3))

$$= \sum_{x \in \{x_1\} \times \mathbb{R}^{d-d_1}} \bar{\mu}_X(x) = \sum_{(x_2,\ldots,x_n) \in \mathbb{R}^{d-d_1}} \bar{\mu}_X(x_1, x_2, \ldots, x_n).$$

$\square$

Remark 2.2.48 (Sylvester's Criterion) Recall that a $d \times d$ matrix $\mathbf{C}$ is said to be *positive definite* if

$$\langle \mathbf{C}x, x \rangle > 0, \qquad x \in \mathbb{R}^d \setminus \{0\}.$$

According to the useful *Sylvester's criterion*, a real symmetric matrix $\mathbf{C}$ is positive definite if and only if $d_k > 0$ for each $k = 1, \ldots, d$, where d_k denotes the determinant of the matrix obtained by deleting from $\mathbf{C}$ the last $d - k$ rows and the last $d - k$ columns.

Example 2.2.49 [!] By Sylvester's criterion, the symmetric matrix

$$\mathbf{C} = \begin{pmatrix} v_1 & c \\ c & v_2 \end{pmatrix}$$

is positive definite if and only if

$$v_1 > 0 \quad \text{and} \quad \det \mathbf{C} = v_1 v_2 - c^2 > 0.$$

Then $\mathbf{C}$ is invertible with

$$\mathbf{C}^{-1} = \frac{1}{v_1 v_2 - c^2} \begin{pmatrix} v_2 & -c \\ -c & v_1 \end{pmatrix}$$

and the two-dimensional Gaussian function

$$\Gamma(x) = \frac{1}{2\pi \sqrt{\det \mathbf{C}}} e^{-\frac{1}{2} \langle \mathbf{C}^{-1} x, x \rangle}, \qquad x \in \mathbb{R}^2,$$

is a density since it is a positive function and

$$\int_{\mathbb{R}^2} \Gamma(x) dx = 1.$$

The function Γ is called *density of the bivariate normal distribution*: if $X = (X_1, X_2)$ has density Γ then we say that X has a bivariate normal distribution and we write $X \sim \mathcal{N}_{0,\mathbf{C}}$.

According to Proposition 2.2.47, the marginal densities of X_1 and X_2 are respectively

$$\gamma_{X_1}(x_1) = \int_{\mathbb{R}} \Gamma(x_1, x_2) dx_2 = \frac{1}{\sqrt{2\pi v_1}} e^{-\frac{x_1^2}{2v_1}}, \qquad x_1 \in \mathbb{R},$$

$$\gamma_{X_2}(x_2) = \int_{\mathbb{R}} \Gamma(x_1, x_2) dx_1 = \frac{1}{\sqrt{2\pi v_2}} e^{-\frac{x_2^2}{2v_2}}, \qquad x_2 \in \mathbb{R},$$

that is, $X_1 \sim \mathcal{N}_{0,v_1}$ and $X_2 \sim \mathcal{N}_{0,v_2}$, *regardless the value of* $c \in \mathbb{R}$. On the other hand, we have

$$\text{cov}(X_1, X_2) = E[X_1 X_2] = \int_{\mathbb{R}^2} x_1 x_2 \Gamma(x_1, x_2) dx_1 dx_2 = c.$$

Thus, the joint distribution provides information not only on the individual marginal distributions but also *on the relationships between the different components of* X. Conversely, starting from the knowledge of the marginal distributions, $X_1 \sim \mathcal{N}_{0,v_1}$ and $X_2 \sim \mathcal{N}_{0,v_2}$, we cannot say anything about the covariance of X_1, X_2: *in conclusion, it is generally not possible to derive the joint distribution from the marginals*. In this regard, see also Example 2.3.24.

2.3 Independence

In probability theory, one of the main theoretical and practical issues revolves around the existence and degree of dependence between random quantities. For instance, correlation serves as an indicator of a specific form of dependence, namely, the *linear* relationship between random variables. In this paragraph, we provide a general treatment of the topic by introducing the concepts of *deterministic dependence* and *stochastic independence*.

2.3.1 Deterministic Dependence and Stochastic Independence

In this section, for the sake of simplicity, we confine our analysis to the case of two real random variables X, Y on the space $(\Omega, \mathscr{F}, P)$. Since we will systematically use the concept of σ-algebra generated by X, we recall its definition:

$$\sigma(X) = X^{-1}(\mathscr{B}) = \{(X \in H) \mid H \in \mathscr{B}\}.$$

Definition 2.3.1 We say that:

(i) X and Y are *stochastically independent under P* if the events $(X \in H)$ and $(Y \in K)$ are independent under P for every $H, K \in \mathscr{B}$. In other words, X and Y are independent under P if if the generated σ-algebras are mutually independent, in the sense that the elements of $\sigma(X)$ and $\sigma(Y)$ are pairwise independent in P;

(ii) X *depends deterministically on Y* if we have the following inclusion

$$\sigma(X) \subseteq \sigma(Y), \tag{2.3.1}$$

that is, if X *is $\sigma(Y)$-measurable* and in this case we write $X \in m\sigma(Y)$.

Remark 2.3.2 [!] Let Y be a r.v. and $f \in m\mathscr{B}$. As seen in (2.1.1), we have

$$\sigma(f(Y)) = (f \circ Y)^{-1}(\mathscr{B}) = Y^{-1}\left(f^{-1}(\mathscr{B})\right) \subseteq Y^{-1}(\mathscr{B}) = \sigma(Y)$$

and therefore

$$\sigma(f(Y)) \subseteq \sigma(Y). \tag{2.3.2}$$

Thus $X := f(Y)$ depends deterministically on Y. From the inclusion (2.1.9), we also deduce the following useful result: *if $f, g \in m\mathscr{B}$ and X, Y are independent random variables, then also $f(X), g(Y)$ are independent.*

The following theorem clarifies the meaning of the inclusion (2.3.1), characterizing it in terms of functional dependence of X on Y.

Theorem 2.3.3 (Doob's Theorem) *[!!] Let X, Y be real random variables on $(\Omega, \mathscr{F}, P)$. Then $X \in m\sigma(Y)$ if and only if there exists $f \in m\mathscr{B}$ such that $X = f(Y)$.*

Remark 2.3.4 Doob's theorem remains valid (with an almost identical proof) when X takes values in $\mathbb{R}^d$ and Y takes values in a generic measurable space $(E, \mathscr{E})$. The general statement is as follows: $X \in m\sigma(Y)$ if and only if there exists a measurable function[14] $f : E \longrightarrow \mathbb{R}^d$ such that $X = f(Y)$.

$$(\Omega, \mathscr{F}) \xrightarrow{\ X\ } \left(\mathbb{R}^d, \mathscr{B}_d\right)$$

$$Y \searrow \qquad \nearrow f$$

$$(E, \mathscr{E})$$

Proof of Theorem 2.3.3 If $X = f(Y)$ with $f \in m\mathscr{B}$ then $X \in m\sigma(Y)$: this follows directly from (2.3.2). Conversely, let $X \in m\sigma(Y)$. Using a transformation of the type

$$Z = \frac{1}{2} + \frac{1}{\pi} \arctan X$$

it is not restrictive to assume that X takes values in $]0, 1[$.

Consider first the case where X is simple, that is, X takes only the distinct values $x_1, \ldots, x_m \in]0, 1[$ and therefore can be written in the form

$$X = \sum_{k=1}^{m} x_k \mathbb{1}_{(X=x_k)}.$$

By assumption, we have $(X = x_k) = (Y \in H_k)$ with $H_k \in \mathscr{B}, k = 1, \ldots, m$. Then, setting

$$f(y) = \sum_{k=1}^{m} x_k \mathbb{1}_{H_k}(y), \qquad y \in \mathbb{R},$$

we get

$$f(Y) = \sum_{k=1}^{m} x_k \mathbb{1}_{H_k}(Y) = \sum_{k=1}^{m} x_k \mathbb{1}_{(Y \in H_k)} = \sum_{k=1}^{m} x_k \mathbb{1}_{(X=x_k)} = X.$$

---

[14] $f \in m\mathscr{E}$, that is, $f^{-1}(H) \in \mathscr{E}$ for every $H \in \mathscr{B}_d$.

Now, let us consider the general case in which X takes values in $]0, 1[$: by Lemma 2.2.3 there exists a sequence $(X_n)_{n \geq 1}$ of simple and $\sigma(Y)$-measurable random variables such that

$$0 \leq X_n(\omega) \nearrow X(\omega), \qquad \omega \in \Omega. \tag{2.3.3}$$

As shown earlier, we have $X_n = f_n(Y)$ with $f_n \in m\mathscr{B}$ taking values in $[0, 1[$. We define

$$f(y) := \limsup_{n \to \infty} f_n(y), \qquad y \in \mathbb{R}.$$

Then, $f \in m\mathscr{B}$ (cf. Proposition 2.1.8) is bounded and by (2.3.3) we get

$$X(\omega) = \lim_{n \to \infty} X_n(\omega) = \lim_{n \to \infty} f_n(Y(\omega)) = f(Y(\omega)), \qquad \omega \in \Omega.$$

$\square$

Corollary 2.3.5 *Let X, Y, Z be real random variables on $(\Omega, \mathscr{F}, P)$ with $X \geq Z$. If $X, Z \in m\sigma(Y)$, there exist $f, g \in m\mathscr{B}$ such that $X = f(Y)$, $Z = g(Y)$, and $f \geq g$.*

Proof In the case $Z \equiv 0$, the thesis is a consequence of the construction of f made in the proof of Theorem 2.3.3. In the general case, since $0 \leq X - Z \in m\sigma(Y)$, there exists $0 \leq h \in m\mathscr{B}$ such that $X - Z = h(Y)$. Moreover, there exists $f \in m\sigma(Y)$ such that $Z + h(Y) = X = f(Y)$ and therefore $Z = (f - h)(Y)$ with $f \geq f - h \in m\sigma(Y)$. $\square$

To grasp the concept of deterministic dependence, carefully consider the following

Exercise 2.3.6 [!] Let $\Omega = \{1, 2, 3\}$ and the Bernoulli random variables X, Y defined on Ω as follows

$$X(\omega) = \begin{cases} 1 & \text{if } \omega \in \{1, 2\}, \\ 0 & \text{if } \omega = 3, \end{cases} \qquad Y(\omega) = \begin{cases} 1 & \text{if } \omega = 1, \\ 0 & \text{if } \omega \in \{2, 3\}. \end{cases}$$

Note that

$$\sigma(X) = \{\emptyset, \Omega, \{1, 2\}, \{3\}\}, \qquad \sigma(Y) = \{\emptyset, \Omega, \{1\}, \{2, 3\}\}.$$

(i) Verify directly that *there is no function f such that $X = f(Y)$*.
(ii) Are random variables X and Y independent under the uniform probability?
(iii) Is there a probability measure on Ω under which X and Y are independent?

Solution

(i) If such a function f existed, then we would have

$$1 = X(2) = f(Y(2)) = f(0) = f(Y(3)) = X(3) = 0$$

which is absurd. Therefore, there is no deterministic dependence between X and Y. Note that, in accordance with Theorem 2.3.3, there are no inclusions between $\sigma(X)$ and $\sigma(Y)$.

(ii) X and Y are not independent under the uniform probability because the events $(X = 1) = \{1, 2\}$ and $(Y = 0) = \{2, 3\}$ are not independent since

$$P\left((X = 1) \cap (Y = 0)\right) = P(\{2\}) = \frac{1}{3}$$

but

$$P(X = 1)P(Y = 0) = \frac{4}{9}.$$

(iii) Yes, for example, the probability defined by $P(1) = P(3) = 0$ and $P(2) = 1$: more generally, X and Y are independent under a probability like the Dirac delta centered at 1 or 2 or 3 (see in this regard point i) of Exercise 2.3.8).

> **Remark 2.3.7** Exercise 2.3.6 allows us to reiterate that *the concept of stochastic independence is always relative to a particular fixed probability measure.* On the contrary, *deterministic dependence is a general property that does not depend on the considered probability measure.* In particular, the concepts of stochastic independence and deterministic dependence are not "one the opposite of the other". Moreover, deterministic dependence "goes in one direction": if X depends deterministically on Y, it is not necessarily true that Y depends deterministically on X.

Exercise 2.3.8 Let X, Y be discrete random variables on (Ω, P). Prove the following statements:

(i) if X is almost surely constant, $X \overset{a.s.}{=} c$, then X, Y are independent;

(ii) let

$$f : X(\Omega) \longrightarrow \mathbb{R}$$

be an injective function. Then X and $f(X)$ are independent under P if and only if X is almost surely constant.

Solution (i) Observing that $P(X \in H) \in \{0, 1\}$ for every $H \in \mathcal{B}$, it is not
difficult to prove the thesis.

(ii) It is sufficient to prove that if X and $f(X)$ are independent, then X is almost
surely constant. Let $y \in X(\Omega)$: since f is injective, we have $(X = y) = (f(X) = f(y))$ or more explicitly

$$\{\omega \in \Omega \mid X(\omega) = y\} = \{\omega \in \Omega \mid f(X(\omega)) = f(y)\}.$$

Then we get

$$P(X = y) = P\Big((X = y) \cap (f(X) = f(y))\Big)$$
$$= P(X = y)P(f(X) = f(y)) = P(X = y)^2$$

from which it follows $P(X = y) \in \{0, 1\}$ and therefore the thesis.

2.3.2 Product Measure and Fubini's Theorem

To explore the concept of stochastic independence among two or more random
variables, we introduce some basic results on the product of measures, which will be
crucial for the upcoming discussions. Given two finite measure spaces $(\Omega_1, \mathcal{F}_1, \mu_1)$
and $(\Omega_2, \mathcal{F}_2, \mu_2)$, we consider the Cartesian product

$$\Omega := \Omega_1 \times \Omega_2 = \{(x, y) \mid x \in \Omega_1, \ y \in \Omega_2\},$$

and the family of *rectangles* defined as follows

$$\mathcal{R} := \{A \times B \mid A \in \mathcal{F}_1, \ B \in \mathcal{F}_2\}.$$

We denote by

$$\mathcal{F}_1 \otimes \mathcal{F}_2 := \sigma(\mathcal{R})$$

the σ-algebra generated by the rectangles, also called *product σ-algebra of $\mathcal{F}_1$ and
$\mathcal{F}_2$*. We have the following generalization of Corollary 2.1.6 and Remark 2.1.9.

Corollary 2.3.9 *For $k = 1, 2$, let $X_k : \Omega_k \longrightarrow \mathbb{R}$ be functions on the measure
spaces $(\Omega_k, \mathcal{F}_k)$. The following properties are equivalent:*

(i) $(X_1, X_2) \in m(\mathcal{F}_1 \otimes \mathcal{F}_2)$;
(ii) $X_k \in m\mathcal{F}_k$ for $k = 1, 2$.

Moreover, if (i) or (ii) holds, then for every $f \in m\mathcal{B}_2$, we have that $f(X_1, X_2) \in m(\mathcal{F}_1 \otimes \mathcal{F}_2)$.

Remark 2.3.10 Every disk in $\mathbb{R}^2$ is a countable union of rectangles and therefore $\mathscr{B} \otimes \mathscr{B} = \mathscr{B}_2$. On the other hand, if $\mathscr{L}_d$ denotes the σ-algebra of Lebesgue measurable sets in $\mathbb{R}^d$, then $\mathscr{L}_1 \otimes \mathscr{L}_1$ is *strictly included* in $\mathscr{L}_2$. In fact, for example, if $H \subseteq \mathbb{R}$ is not Lebesgue measurable, then $H \times \{0\} \in \mathscr{L}_2 \setminus (\mathscr{L}_1 \otimes \mathscr{L}_1)$.

Lemma 2.3.11 *Let*

$$f : \Omega_1 \times \Omega_2 \longrightarrow \mathbb{R}$$

be a $\mathscr{F}_1 \otimes \mathscr{F}_2$-measurable function. Then we have:

(i) $f(\cdot, y) \in m\mathscr{F}_1$ *for every* $y \in \Omega_2$;
(ii) $f(x, \cdot) \in m\mathscr{F}_2$ *for every* $x \in \Omega_1$.

Proof It is not restrictive to assume that f is bounded. Let $\mathscr{H}$ be the family of $\mathscr{F}_1 \otimes \mathscr{F}_2$-measurable, bounded functions that verify properties i) and ii). Then $\mathscr{H}$ is a monotone family of functions (cf. Definition A.0.7). The family $\mathscr{R}$ is $\cap$-closed, generates $\mathscr{F}_1 \otimes \mathscr{F}_2$, and it is clear that $\mathbb{1}_{A \times B} \in \mathscr{H}$ for every $(A \times B) \in \mathscr{R}$. Then the thesis follows from Dynkin's Theorem A.0.8.   □

Remark 2.3.12 Fubini's theorem for the Lebesgue integral states that if $f = f(x, y) \in m\mathscr{L}_2$ (i.e., f is measurable with respect to the σ-algebra $\mathscr{L}_2$ of Lebesgue measurable sets in $\mathbb{R}^2$) then $f(x, \cdot) \in m\mathscr{L}_1$ *for almost every* $x \in \mathbb{R}$. Note the difference with Lemma 2.3.11 which claims that "$f(x, \cdot) \in m\mathscr{F}_2$ for every $x \in \Omega_1$". This is due to the fact that, as we have already observed, $\mathscr{L}_1 \otimes \mathscr{L}_1$ is *strictly included* in $\mathscr{L}_2$. For more details, we refer to the section "Completion of product measure", Cap.8 in [40].

Lemma 2.3.13 *Let f be $\mathscr{F}_1 \otimes \mathscr{F}_2$-measurable and bounded. Then, we have:*

(i) $x \mapsto \int\limits_{\Omega_2} f(x, y)\mu_2(dy) \in m\mathscr{F}_1$;
(ii) $y \mapsto \int\limits_{\Omega_1} f(x, y)\mu_1(dx) \in m\mathscr{F}_2$;
(iii)

$$\int_{\Omega_1} \left(\int_{\Omega_2} f(x, y)\mu_2(dy) \right) \mu_1(dx) = \int_{\Omega_2} \left(\int_{\Omega_1} f(x, y)\mu_1(dx) \right) \mu_2(dy).$$

Proof As in the previous lemma, the thesis follows from the second Dynkin's theorem applied to the family $\mathscr{H}$ of $\mathscr{F}_1 \otimes \mathscr{F}_2$-measurable and bounded functions that verify the properties (i), (ii) and (iii). In fact, $\mathscr{H}$ is a monotone family of functions and $\mathbb{1}_{A \times B} \in \mathscr{H}$ for every $(A \times B) \in \mathscr{R}$.   □

Proposition 2.3.14 (Product Measure) *The function defined by*

$$\mu(H) := \int_{\Omega_1} \left(\int_{\Omega_2} \mathbb{1}_H d\mu_2 \right) d\mu_1 = \int_{\Omega_2} \left(\int_{\Omega_1} \mathbb{1}_H d\mu_1 \right) d\mu_2, \qquad H \in \mathscr{F}_1 \otimes \mathscr{F}_2,$$

is the unique finite measure on $\mathscr{F}_1 \otimes \mathscr{F}_2$ *such that*

$$\mu(A \times B) = \mu_1(A)\mu_2(B), \qquad A \in \mathscr{F}_1, \ B \in \mathscr{F}_2.$$

We write $\mu = \mu_1 \otimes \mu_2$ *and say that* μ *is the product measure of the finite measures* μ_1 *and* μ_2.

Proof The fact that μ is a measure follows from the linearity of the integral and Beppo Levi's theorem. The uniqueness follows from Corollary A.0.5, since $\mathscr{R}$ is $\cap$-closed and generates $\mathscr{F}_1 \otimes \mathscr{F}_2$. $\qquad\square$

Theorem 2.3.15 (Fubini's Theorem) *[!!!] On the product space* $(\Omega_1 \times \Omega_2, \mathscr{F}_1 \otimes \mathscr{F}_2, \mu_1 \otimes \mu_2)$, *let* f *be a real-valued* $(\mathscr{F}_1 \otimes \mathscr{F}_2)$-*measurable function. If* f *is non-negative or absolutely integrable (i.e.,* $f \in L^1(\Omega_1 \times \Omega_2, \mu_1 \otimes \mu_2)$) *then we have:*

$$\int_{\Omega_1 \times \Omega_2} f\, d(\mu_1 \otimes \mu_2) = \int_{\Omega_1} \left(\int_{\Omega_2} f(x, y)\mu_2(dy) \right) \mu_1(dx)$$
$$= \int_{\Omega_2} \left(\int_{\Omega_1} f(x, y)\mu_1(dx) \right) \mu_2(dy). \tag{2.3.4}$$

Proof Equation (2.3.4) is true if $f = \mathbb{1}_{A \times B}$ and therefore, by second Dynkin's theorem, also for f measurable and bounded. Beppo Levi's theorem and the linearity of the integral ensure the validity of (2.3.4) for f non-negative and $f \in L^1$. $\qquad\square$

Remark 2.3.16 Theorem 2.3.15 remains valid for σ-finite spaces $(\Omega_1, \mathscr{F}_1, \mu_1)$ and $(\Omega_2, \mathscr{F}_2, \mu_2)$. Starting from Theorem 2.3.15, the product measure $\mu_1 \otimes \cdots \otimes \mu_n$ of more than two measures is defined by induction.

Example 2.3.17 [!] Let $\mu = \mathrm{Exp}_\lambda \otimes \mathrm{Be}_p$ be the product measure on $\mathbb{R}^2$ of the exponential distribution Exp_λ and the Bernoulli distribution Be_p. By Fubini's theorem, the calculation of the integral of $f \in L^1(\mathbb{R}^2, \mu)$ can be carried out as follows:

$$\iint_{\mathbb{R}^2} f(x, y)\mu(dx, dy) = \int_{\mathbb{R}} \left(\int_{\mathbb{R}} f(x, y)\mathrm{Be}_p(dy) \right) \mathrm{Exp}_\lambda(dx)$$
$$= \int_{\mathbb{R}} (pf(x, 1) + (1 - p)f(x, 0))\, \mathrm{Exp}_\lambda(dx)$$
$$= p\lambda \int_0^{+\infty} f(x, 1)e^{-\lambda x}\, dx$$
$$+ (1 - p)\lambda \int_0^{+\infty} f(x, 0)e^{-\lambda x}\, dx.$$

2.3.3 *Independence of σ-Algebras*

Since the general definition of independence of random variables is given in terms of independence of their generated σ-algebras, we first examine the concept of independence between σ-algebras. Afterwards, $(\Omega, \mathscr{F}, P)$ is a fixed probability space and I is any family of indices.

Definition 2.3.18 We say that the families of events $\mathscr{F}_i$, with $i \in I$, are *independent under P* if

$$P\left(\bigcap_{k=1}^{n} A_k\right) = \prod_{k=1}^{n} P(A_k),$$

for every choice of a *finite* number of indices $i_1, \ldots, i_n$ and $A_k \in \mathscr{F}_{i_k}$ for $k = 1, \ldots, n$.

Exercise 2.3.19 Let $\sigma(A) = \{\emptyset, \Omega, A, A^c\}$ be the σ-algebra generated by $A \in \mathscr{F}$. Prove that $A_1, \ldots, A_n \in \mathscr{F}$ are independent under P (cf. Definition 1.3.27) if and only if $\sigma(A_1), \ldots, \sigma(A_n)$ are independent under P.

Sometimes the following corollary of Dynkin's theorem may be useful.

Lemma 2.3.20 *[!] Let $\mathscr{A}_1, \ldots, \mathscr{A}_n$ be families of events in $(\Omega, \mathscr{F}, P)$, closed with respect to intersection. Then $\mathscr{A}_1, \ldots, \mathscr{A}_n$ are independent under P if and only if $\sigma(\mathscr{A}_1), \ldots, \sigma(\mathscr{A}_n)$ are independent under P.*

Proof We prove the case $n = 2$: the general proof is analogous. Fix $A \in \mathscr{A}_1$ and define the finite measures

$$\mu(B) = P(A \cap B), \qquad \nu(B) = P(A)P(B), \qquad B \in \sigma(\mathscr{A}_2).$$

By assumption, $\mu = \nu$ on $\mathscr{A}_2$ and also $\mu(\Omega) = P(A) = \nu(\Omega)$, so by Corollary A.0.5 $\mu = \nu$ on $\sigma(\mathscr{A}_2)$ or, in other words

$$P(A \cap B) = P(A)P(B), \qquad B \in \sigma(\mathscr{A}_2).$$

Now fix $B \in \sigma(\mathscr{A}_2)$ and define the finite measures

$$\mu(B) = P(A \cap B), \qquad \nu(B) = P(A)P(B), \qquad A \in \sigma(\mathscr{A}_1).$$

We have proved that $\mu = \nu$ on $\mathscr{A}_1$ and obviously $\mu(\Omega) = P(B) = \nu(\Omega)$, so again by Corollary A.0.5 we have $\mu = \nu$ on $\sigma(\mathscr{A}_1)$ which is equivalent to the thesis. $\square$

2.3.4 Independence of Random Vectors

We assume the hypotheses and notations of Sect. 2.2.10 and introduce the important concept of independence among random variables.

Definition 2.3.21 (Independence of Random Variables) We say that the random variables $X_1, \ldots, X_n$, defined on the space $(\Omega, \mathscr{F}, P)$, are independent under P if their respective σ-algebras generated $\sigma(X_1), \ldots, \sigma(X_n)$ are independent under P or, equivalently, if

$$P\left(\bigcap_{i=1}^{n}(X_i \in H_i)\right) = \prod_{i=1}^{n} P(X_i \in H_i), \qquad H_i \in \mathscr{B}_{d_i}, \ i = 1, \ldots, n.$$

A simple but useful application of Lemma 2.3.20 shows that two real random variables X, Y are independent under P if and only if

$$P((X \le x) \cap (Y \le y)) = P(X \le x)P(Y \le y), \qquad x, y \in \mathbb{R}.$$

In light of Definition 2.3.18, the notion of independence extends to a family $(X_i)_{i \in I}$ of random variables, where I is any set (not necessarily finite) of indices.

Remark 2.3.22 [!] As a consequence of (2.3.2), if $X_1, \ldots, X_n$ are independent random variables on $(\Omega, \mathscr{F}, P)$ and $f_1, \ldots, f_n \in m\mathscr{B}$ then also the random variables $f_1(X_1), \ldots, f_n(X_n)$ are independent under P: in other words, *the property of independence is invariant for deterministic transformations* (specifically, the operation of composition with measurable functions).

For example, suppose that $X_1, \ldots, X_n, Y_1, \ldots, Y_m$ are *real* random variables and $X := (X_1, \ldots, X_n)$ and $Y := (Y_1, \ldots, Y_m)$ are independent. Then, also the following pairs of random variables are independent[15]

(i) X_i and Y_j for every i and j;
(ii) $X_{i_1} + X_{i_2}$ and $Y_{j_1} + Y_{j_2}$ for every i_1, i_2, j_1, j_2;
(iii) X_i^2 and Y for every i.

The following result provides an important characterization of the property of independence. It also shows that, *in the case of independent random variables, the joint distribution can be obtained from the marginal distributions*. For clarity of exposition, we first state the result in the particular case of two random variables and then give the general result.

Theorem 2.3.23 [!!] *Let X_1, X_2 be random variables on $(\Omega, \mathscr{F}, P)$ with values respectively in $\mathbb{R}^{d_1}$ and $\mathbb{R}^{d_2}$. The following three properties are equivalent:*

(i) X_1, X_2 are independent under P;

---

[15] As an exercise, determine the measurable functions with which X and Y are composed.

(ii) $F_{(X_1,X_2)}(x_1, x_2) = F_{X_1}(x_1)F_{X_2}(x_2)$ *for every* $x_1 \in \mathbb{R}^{d_1}$ *and* $x_2 \in \mathbb{R}^{d_2}$;

(iii) $\mu_{(X_1,X_2)} = \mu_{X_1} \otimes \mu_{X_2}$;

(iv) *if* $(X_1, X_2) \in AC$ *then the previous properties are also equivalent to*

$$\gamma_{(X_1,X_2)}(x_1, x_2) = \gamma_{X_1}(x_1)\gamma_{X_2}(x_2) \tag{2.3.5}$$

for almost every $(x_1, x_2) \in \mathbb{R}^{d_1} \times \mathbb{R}^{d_2}$;

(v) *if* (X_1, X_2) *is discrete then properties (i), (ii), and (iii) are also equivalent to*

$$\bar{\mu}_{(X_1,X_2)}(x_1, x_2) = \bar{\mu}_{X_1}(x_1)\bar{\mu}_{X_2}(x_2) \tag{2.3.6}$$

for every $(x_1, x_2) \in \mathbb{R}^{d_1} \times \mathbb{R}^{d_2}$.

Proof [(i) $\Longrightarrow$ (ii)] We have

$$F_{(X_1,X_2)}(x_1, x_2) = P((X_1 \leq x_1) \cap (X_2 \leq x_2)) =$$

(by the independence hypothesis)

$$= P(X_1 \leq x_1)P(X_2 \leq x_2) = F_{X_1}(x_1)F_{X_2}(x_2).$$

[(ii) $\Longrightarrow$ (iii)] The hypothesis $F_{(X_1,X_2)} = F_{X_1}F_{X_2}$ implies that the distributions $\mu_{(X_1,X_2)}$ and $\mu_{X_1} \otimes \mu_{X_2}$ coincide on the family of multi-intervals $] - \infty, x_1] \times] - \infty, x_2]$: the thesis follows from the uniqueness of the measure extension of Carathéodory's Theorem 1.4.29 (or see Corollary A.0.5, since the family of multi-intervals is $\cap$-closed and generates $\mathscr{B}_{d_1+d_2}$).

[(iii) $\Longrightarrow$ (i)] For every $H \in \mathscr{B}_{d_1}$ and $K \in \mathscr{B}_{d_2}$ we have

$$P((X_1 \in H) \cap (X_2 \in K)) = \mu_{(X_1,X_2)}(H \times K) =$$

(since by assumption $\mu_{(X_1,X_2)} = \mu_{X_1} \otimes \mu_{X_2}$)

$$= \mu_{X_1}(H)\mu_{X_2}(K) = P(X_1 \in H)P(X_2 \in K)$$

from which the independence of X_1 and X_2.

Now assume that $(X_1, X_2) \in AC$ and therefore, by Proposition 2.2.47, also $X_1, X_2 \in AC$.

[(i) $\Longrightarrow$ (iv)] By the independence hypothesis, we have

$$P((X_1, X_2) \in H \times K) = P(X_1 \in H)P(X_2 \in K)$$

$$= \int_H \gamma_{X_1}(x_1)dx_1 \int_K \gamma_{X_2}(x_2)dx_2 =$$

(by Fubini's theorem and with the notation $x = (x_1, x_2)$ for the point of $\mathbb{R}^{d_1+d_2}$)

$$= \int_{H \times K} \gamma_{X_1}(x_1)\gamma_{X_2}(x_2)dx$$

and therefore $\gamma_{X_1}\gamma_{X_2}$ is the density of (X_1, X_2).
[(iv) $\Longrightarrow$ (i)] We have

$$P((X_1, X_2) \in H \times K) = \int_{H \times K} \gamma_{(X_1,X_2)}(x)dx =$$

(by assumption)

$$= \int_{H \times K} \gamma_{X_1}(x_1)\gamma_{X_2}(x_2)dx$$

(by Fubini's theorem)

$$= \int_H \gamma_{X_1}(x_1)dx_1 \int_K \gamma_{X_2}(x_2)dx_2 = P(X_1 \in H)P(X_2 \in K),$$

from which the independence of X_1 and X_2.

Finally, assume that the r.v. (X_1, X_2) is discrete and therefore, by Proposition 2.2.47, also X_1, X_2 are. The proof is completely analogous to the previous case.
[(i) $\Longrightarrow$ (v)] By the independence hypothesis, we have

$$\begin{aligned} \bar{\mu}_{(X_1,X_2)}(x_1, x_2) &= P((X_1 = x_1) \cap (X_2 = x_2)) \\ &= P(X_1 = x_1)P(X_2 = x_2) = \bar{\mu}_{X_1}(x_1)\bar{\mu}_{X_2}(x_2) \end{aligned}$$

from which (2.3.6).
[(v) $\Longrightarrow$ (i)] We have

$$P((X_1, X_2) \in H \times K) = \sum_{(x_1,x_2) \in H \times K} \bar{\mu}_{(X_1,X_2)}(x_1, x_2)$$

$$= \sum_{(x_1,x_2) \in H \times K} \bar{\mu}_{X_1}(x_1)\bar{\mu}_{X_2}(x_2) =$$

(since the terms of the sum are non-negative)

$$= \sum_{x_1 \in H} \bar{\mu}_{X_1}(x_1) \sum_{x_2 \in K} \bar{\mu}_{X_2}(x_2) = P(X_1 \in H)P(X_2 \in K),$$

which proves the independence of X_1 and X_2. $\qquad\square$

The following example shows two pairs of random variables with equal marginal distributions but different joint distributions.

Example 2.3.24 [!] Consider a urn containing n numbered balls. Let:

(i) X_1, X_2 be the results of two successive extractions *with replacement*;
(ii) Y_1, Y_2 be the results of two successive extractions *without replacement*.

It is natural to assume that the random variables X_1, X_2 have a uniform distribution Unif_n and are independent: by Theorem 2.3.23-(v) the joint distribution function is

$$\bar{\mu}_{(X_1,X_2)}(x_1, x_2) = \bar{\mu}_{X_1}(x_1)\bar{\mu}_{X_2}(x_2) = \frac{1}{n^2}, \qquad (x_1, x_2) \in I_n \times I_n,$$

where, as usual, $I_n = \{1, \ldots, n\}$.

The r.v. Y_1 has a uniform distribution Unif_n but is not independent of Y_2. To derive the joint distribution function, we use the knowledge of the conditional probability that the second draw is y_2, given that the first ball drawn is y_1:

$$P(Y_2 = y_2 \mid Y_1 = y_1) = \begin{cases} \frac{1}{n-1} & \text{if } y_2 \in I_n \setminus \{y_1\}, \\ 0 & \text{if } y_2 = y_1. \end{cases}$$

Then we have

$$P\big((Y_1, Y_2) = (y_1, y_2)\big) = P\big((Y_1 = y_1) \cap (Y_2 = y_2)\big)$$
$$= P(Y_2 = y_2 \mid Y_1 = y_1)\, P(Y_1 = y_1) \qquad (2.3.7)$$

so that

$$\bar{\mu}_{(Y_1,Y_2)}(y_1, y_2) = \begin{cases} \frac{1}{n(n-1)} & \text{if } y_1, y_2 \in I_n, \ y_1 \neq y_2, \\ 0 & \text{otherwise.} \end{cases}$$

We emphasize the importance of the passage (2.3.7) in which, due to the lack of independence, we employed the multiplication rule (1.3.5). Having $\bar{\mu}_{(Y_1,Y_2)}$, we can now calculate $\bar{\mu}_{Y_2}$ using (2.2.27) of Proposition 2.2.47: for each $y_2 \in I_n$ we have

$$\bar{\mu}_{Y_2}(y_2) = \sum_{y_1 \in I_n} \bar{\mu}_{(Y_1,Y_2)}(y_1, y_2) = \sum_{y_1 \in I_n \setminus \{y_2\}} \frac{1}{n(n-1)} = \frac{1}{n},$$

that is, $Y_2 \sim \text{Unif}_n$. In conclusion, Y_1, Y_2 have the same marginal uniform distributions as X_1, X_2, but different joint distribution.

Theorem 2.3.23 extends to the case of a finite number of random variables in the following way:

Theorem 2.3.25 [!!] *Let* $X_1, \ldots, X_n$ *be random variables on* $(\Omega, \mathscr{F}, P)$ *with values respectively in* $\mathbb{R}^{d_1}, \ldots, \mathbb{R}^{d_n}$. *Let* $X = (X_1, \ldots, X_n)$ *and* $d = d_1 + \cdots + d_n$, *the following three properties are equivalent:*

(i) $X_1, \ldots, X_n$ *are independent under P;*
(ii) *for every* $x = (x_1, \ldots, x_n) \in \mathbb{R}^d$

$$F_X(x_1, \ldots, x_n) = \prod_{i=1}^{n} F_{X_i}(x_i); \qquad (2.3.8)$$

(iii) $\mu_X = \mu_{X_1} \otimes \cdots \otimes \mu_{X_n}$.

Moreover, if $X \in AC$ *then the previous properties are also equivalent to:*

(iv) *for almost every* $x = (x_1, \ldots, x_n) \in \mathbb{R}^d$

$$\gamma_X(x) = \prod_{i=1}^{n} \gamma_{X_i}(x_i).$$

Finally, if X *is discrete then properties (i), (ii) and (iii) are also equivalent to:*

(v) *for every* $x \in \mathbb{R}^d$

$$\bar{\mu}_X(x) = \prod_{i=1}^{n} \bar{\mu}_{X_i}(x_i).$$

In Sect. 2.1.1 we proved that it is possible to construct a random vector with predetermined distribution (cf. Remark 2.1.17). As a simple consequence, we also have:

Corollary 2.3.26 (Existence of Independent Random Variables) [!] *Let* μ_k *be distributions on* $\mathbb{R}^{d_k}$, $k = 1, \ldots, n$. *There exists a probability space* $(\Omega, \mathscr{F}, P)$ *supporting random variables* $X_1, \ldots, X_n$ *that are* independent *under P and such that* $X_k \sim \mu_k$ *for* $k = 1, \ldots, n$.

Proof Consider the product distribution $\mu = \mu_1 \otimes \cdots \otimes \mu_n$ on $\mathbb{R}^d$ with $d = d_1 + \cdots + d_n$. By Remark 2.1.17, the identity function $X(\omega) = \omega$ is a r.v. on $(\mathbb{R}^d, \mathscr{B}_d, \mu)$ with $X \sim \mu$. By Theorem 2.3.25, the components of X satisfy the thesis. □

Remark 2.3.27 In the previous proof, we constructed a set of n independent random variables by using a sample space represented by the n-dimensional Euclidean space. This observation implies that constructing a sequence, or even more challenging, an uncountable collection, of independent random variables is inherently complex because it would require a sample space with infinite dimension.

2.3.5 Independence and Expected Value

An remarkable consequence of Theorem 2.3.23 is the following

Theorem 2.3.28 *[!!] Let X, Y be real independent random variables on the space $(\Omega, \mathscr{F}, P)$. Then, X, Y are integrable if and only if XY is integrable and in that case we have*

$$E[XY] = E[X] E[Y].$$

Proof We have

$$E[XY] = \int_{\mathbb{R}^2} xy \mu_{(X,Y)}(d(x, y))$$

(by (iii) of Theorem 2.3.23)

$$= \int_{\mathbb{R}^2} xy(\mu_X \otimes \mu_Y)(d(x, y))$$

(by Fubini's theorem)

$$= \int_{\mathbb{R}} x\mu_X(dx) \int_{\mathbb{R}} y\mu_Y(dy) = E[X] E[Y].$$

$\square$

Remark 2.3.29 In general, if $X, Y \in L^1(\Omega, P)$ it is not generally true that $XY \in L^1(\Omega, P)$ (cf. Exercise 2.2.36): however, this is the case if X, Y are independent, by Theorem 2.3.28.

Corollary 2.3.30 *If $X, Y \in L^2(\Omega, P)$ are independent then they are* uncorrelated, *that is, we have*

$$\mathrm{cov}(X, Y) = 0 \qquad \text{and} \qquad \mathrm{var}(X + Y) = \mathrm{var}(X) + \mathrm{var}(Y). \qquad (2.3.9)$$

Proof If X, Y are independent, then so are $\widetilde{X} := X - E[X]$ and $\widetilde{Y} := Y - E[Y]$, by Remark 2.3.22: therefore we have

$$\mathrm{cov}(X, Y) = E[\widetilde{X}\widetilde{Y}] = E[\widetilde{X}] E[\widetilde{Y}] = 0.$$

By (2.2.21), we also have that $\mathrm{var}(X + Y) = \mathrm{var}(X) + \mathrm{var}(Y)$.

$\square$

Example 2.3.31 An example of *uncorrelated but not independent random variables* is the following: let $\Omega = \{0, 1, 2\}$ with uniform probability P. Define

$$X(\omega) = \begin{cases} 1 & \omega = 0, \\ 0 & \omega = 1, \\ -1 & \omega = 2, \end{cases} \qquad Y(\omega) = \begin{cases} 0 & \omega = 0, \\ 1 & \omega = 1, \\ 0 & \omega = 2. \end{cases}$$

Then we have $E[X] = 0$ and $XY = 0$, so $\mathrm{cov}(X, Y) = E[XY] - E[X]E[Y] = 0$, that is, X, Y are uncorrelated. However,

$$P((X = 1) \cap (Y = 1)) = 0 \quad \text{and} \quad P(X = 1) = P(Y = 1) = \frac{1}{3}$$

and therefore X, Y are not independent under P.

Example 2.3.32 [!] The previous example shows that two uncorrelated random variables are not necessarily independent. However, in the case of the bivariate normal distribution (see Example 2.2.49), we have the following result: if $(X_1, X_2) \sim \mathcal{N}_{0,\mathbf{C}}$ and $\mathrm{cov}(X_1, X_2) = 0$ then X_1, X_2 are independent. This follows from Theorem 2.3.23-iv and the fact that if X_1, X_2 are uncorrelated then the joint density is equal to the product of the marginal densities. The assumption that X_1, X_2 have a joint normal distribution is crucial as underscored by Example 2.5.19.

Example 2.3.33 Consider two independent random variables $X \sim \mathcal{N}_{0,1}$ and $Y \sim \mathrm{Poisson}_\lambda$. By Theorem 2.3.25, the joint distribution of X, Y is

$$\mathcal{N}_{0,1} \otimes \mathrm{Poisson}_\lambda$$

and therefore, for every measurable and bounded function, we have

$$E[f(X, Y)] = \int_{\mathbb{R}^2} f(x, y) \left(\mathcal{N}_{0,1} \otimes \mathrm{Poisson}_\lambda\right)(dx, dy) =$$

(by Fubini's theorem)

$$= \int_{\mathbb{R}} \int_{\mathbb{R}} f(x, y) \mathcal{N}_{0,1}(dx) \mathrm{Poisson}_\lambda(dy)$$

$$= e^{-\lambda} \sum_{n=0}^{\infty} \frac{\lambda^n}{n!} \int_{\mathbb{R}} f(x, n) \frac{e^{-\frac{x^2}{2}}}{\sqrt{2\pi}} dx.$$

As an exercise, calculate $E\left[e^{X+Y}\right]$ and $E\left[e^{XY}\right]$.

Example 2.3.34 Consider the bivariate uniform distribution on the following three domains:

(i) the square $Q = [0, 1] \times [0, 1]$;
(ii) the circle $C = \{(x, y) \in \mathbb{R}^2 \mid x^2 + y^2 \leq 1\}$;
(iii) the triangle $T = \{(x, y) \in \mathbb{R}^2_{\geq 0} \mid x + y \leq 1\}$.

(i) The density function of $(X, Y) \sim \mathrm{Unif}_Q$ is

$$\gamma_{(X,Y)} = \mathbb{1}_{[0,1] \times [0,1]}.$$

Therefore,

$$E[X] = \int_{\mathbb{R}^2} x \mathbb{1}_{[0,1] \times [0,1]}(x, y) dx dy = \frac{1}{2},$$

$$\mathrm{var}(X) = \int_{\mathbb{R}^2} \left(x - \frac{1}{2}\right)^2 \mathbb{1}_{[0,1] \times [0,1]}(x, y) dx dy = \frac{1}{12},$$

$$\mathrm{cov}(X, Y) = \int_{\mathbb{R}^2} \left(x - \frac{1}{2}\right) \left(y - \frac{1}{2}\right) \mathbb{1}_{[0,1] \times [0,1]}(x, y) dx dy = 0,$$

and therefore X, Y are *uncorrelated*. Moreover, since by (2.2.26), the density of X is

$$\gamma_X(x) = \int_{\mathbb{R}} \mathbb{1}_{[0,1] \times [0,1]}(x, y) dy = \mathbb{1}_{[0,1]}(x), \qquad x \in \mathbb{R},$$

and similarly $\gamma_Y = \mathbb{1}_{[0,1]}$, it follows that X, Y are *independent* because of (2.3.5).

(ii) The density function of $(X, Y) \sim \mathrm{Unif}_C$ is

$$\gamma_{(X,Y)} = \frac{1}{\pi} \mathbb{1}_C.$$

Thus

$$E[X] = \frac{1}{\pi} \int_{\mathbb{R}^2} x \mathbb{1}_C(x, y) dx dy = 0 = E[Y],$$

$$\mathrm{var}(X) = \frac{1}{\pi} \int_{\mathbb{R}^2} x^2 \mathbb{1}_C(x, y) dx dy = \frac{1}{4},$$

$$\mathrm{cov}(X, Y) = \frac{1}{\pi} \int_{\mathbb{R}^2} xy \mathbb{1}_C(x, y) dx dy = 0,$$

and therefore X, Y are *uncorrelated*. However, X, Y *are not independent* because, by (2.2.26), the density of X is

$$\gamma_X(x) = \frac{1}{\pi} \int_{\mathbb{R}} \mathbb{1}_C(x, y) dy = \frac{2\sqrt{1 - x^2}}{\pi} \mathbb{1}_{[-1,1]}(x), \qquad x \in \mathbb{R},$$

and similarly $\gamma_Y(y) = \frac{2\sqrt{1-y^2}}{\pi} \mathbb{1}_{[-1,1]}(y)$: thus *the joint density is not the product of the marginals*. Alternatively, a direct check shows that

$$P\left(X \geq \frac{1}{2}\right) = \frac{1}{\pi} \int_{\mathbb{R}^2} \mathbb{1}_{\left[\frac{1}{2}, +\infty\right[}(x) \mathbb{1}_C(x, y) dx dy$$

$$= \frac{4\pi - 3\sqrt{3}}{12\pi} = P\left(Y \geq \frac{1}{2}\right),$$

$$P\left(\left(X \geq \frac{1}{2}\right) \cap \left(Y \geq \frac{1}{2}\right)\right) = \frac{3 - 3\sqrt{3} + \pi}{12\pi}$$

$$\neq P\left(X \geq \frac{1}{2}\right) P\left(Y \geq \frac{1}{2}\right).$$

This example, along with Example 2.3.31, shows that independence is a stronger property than uncorrelation.

(iii) The density function of $(X, Y) \sim \text{Unif}_T$ is

$$\gamma_{(X,Y)} = 2 \mathbb{1}_T.$$

Thus

$$E[X] = 2 \int_{\mathbb{R}^2} x \mathbb{1}_T(x, y) dx dy = \frac{1}{3} = E[Y],$$

$$\text{var}(X) = 2 \int_{\mathbb{R}^2} \left(x - \frac{1}{3}\right)^2 \mathbb{1}_T(x, y) dx dy = \frac{1}{18},$$

$$\text{cov}(X, Y) = 2 \int_{\mathbb{R}^2} \left(x - \frac{1}{3}\right)\left(y - \frac{1}{3}\right) \mathbb{1}_T(x, y) dx dy = -\frac{1}{36},$$

and thus X, Y are *negatively correlated* (and therefore *not independent*). By (2.2.26), the density of X is

$$\gamma_X(x) = 2 \int_{\mathbb{R}} \mathbb{1}_T(x, y) dy = 2(1 - x) \mathbb{1}_{[0,1]}(x), \qquad x \in \mathbb{R}.$$

□

2.4   Conditional Distribution and Expectation Given an Event

In a probability space $(\Omega, \mathscr{F}, P)$, let B be a non-negligible event, i.e. $B \in \mathscr{F}$ with $P(B) > 0$. Recall that $P(\cdot \mid B)$ denotes the *conditional probability given B*, which is the probability measure on $(\Omega, \mathscr{F})$ defined by

$$P(A \mid B) = \frac{P(A \cap B)}{P(B)}, \qquad A \in \mathscr{F}.$$

Definition 2.4.1 Let X be a $\mathbb{R}^d$-valued r.v. on $(\Omega, \mathscr{F}, P)$:

(i) *the conditional distribution of X given B* is the distribution of X under the conditional probability $P(\cdot \mid B)$, that is

$$\mu_{X \mid B}(H) := P(X \in H \mid B), \qquad H \in \mathscr{B}_d;$$

(ii) if X is integrable under P, *the conditional expectation of X given B* is the expected value of X under the conditional probability $P(\cdot \mid B)$, that is

$$E[X \mid B] := \int_{\Omega} X \, dP(\cdot \mid B).$$

Proposition 2.4.2 *[!] For every $f \in m\mathscr{B}_d$ such that $f(X)$ is integrable on $(\Omega, \mathscr{F}, P)$, we have*

$$E[f(X) \mid B] = \frac{1}{P(B)} \int_B f(X) dP \qquad\qquad (2.4.1)$$

$$= \int_{\mathbb{R}^d} f(x) \mu_{X \mid B}(dx). \qquad\qquad (2.4.2)$$

Proof It is sufficient to prove (2.4.1) for $f = \mathbb{1}_H$ with $H \in \mathscr{B}_d$: the general case follows from the standard procedure of Remark 2.2.22. Being $\mathbb{1}_H(X) = \mathbb{1}_{(X \in H)}$, we have

$$E\left[\mathbb{1}_{(X \in H)} \mid B\right] = P(X \in H \mid B) = \frac{P((X \in H) \cap B)}{P(B)} = \frac{1}{P(B)} \int_B \mathbb{1}_H(X) dP.$$

Since, by (2.4.1), $f(X)$ is integrable under $P(\cdot \mid B)$ then (2.4.2) follows from Theorem 2.2.26.   □

Exercise 2.4.3 Verify that if X and B are independent under P then

$$\mu_{X \mid B} = \mu_X \qquad \text{and} \qquad E[X \mid B] = E[X].$$

Remark 2.4.4 Similarly to the concept of conditional distribution of X given B, we define the *conditional density of X given B* which we will indicate by $\gamma_{X|B}$ and the *conditional CDF of X given B* which we will indicate by $F_{X|B}$.

The conditional distribution is the natural tool for studying problems of the following type.

Example 2.4.5 From an urn containing 90 numbered balls, two balls are drawn in sequence and without replacement. Let X_1 and X_2 be the random variables that indicate respectively the number of the first and second ball drawn. Clearly we have $\mu_{X_1} = \text{Unif}_{I_{90}}$ and we know that also $\mu_{X_2} = \text{Unif}_{I_{90}}$ (cf. Example 2.3.24).

Now we add the information that the first ball drawn has the number k, i.e., we condition on the event $B = (X_1 = k)$: we have

$$P(X_2 = h \mid X_1 = k) = \begin{cases} \frac{1}{89}, & \text{if } h, k \in I_{90}, \ h \neq k, \\ 0 & \text{otherwise,} \end{cases}$$

and therefore

$$\mu_{X_2|X_1=k} = \text{Unif}_{I_{90}\setminus\{k\}}.$$

So the additional information given by the event B modifies the distribution of X_2.

Using (2.4.2), as an exercise, calculate $\text{var}(X_2 \mid X_1 = k)$ to verify that $\text{var}(X_2 \mid X_1 = k) < \text{var}(X_2)$: intuitively this means that the uncertainty about the value of X_2 decreases by adding the information $(X_1 = k)$.

The remainder of the section presents additional examples.

Example 2.4.6 Let $T \sim \text{Exp}_\lambda$ and $B = (T > t_0)$ with $\lambda, t_0 \in \mathbb{R}_{>0}$. To determine the conditional distribution $\mu_{T|B}$, we calculate the conditional CDF of T to B or equivalently

$$P(T > t \mid T > t_0) = \begin{cases} 1 & \text{if } t \leq t_0, \\ P(T > t - t_0) & \text{if } t > t_0, \end{cases}$$

which follows from the memoryless property (2.1.10). It follows that $\mu_{T|B}$ is the *shifted exponential distribution* that has the density

$$\gamma_{T|B}(t) = \lambda e^{-\lambda(t-t_0)} \mathbb{1}_{[t_0,+\infty[}(t).$$

Example 2.4.7 Let $X \in \mathcal{N}_{0,1}$ and $B = (X \geq 0)$. Then $P(B) = \frac{1}{2}$ and, for $H \in \mathcal{B}$, we have

$$\mu_{X|B}(H) = P(X \in H \mid B) = \frac{P((X \in H) \cap B)}{P(B)} = 2P(X \in H \cap \mathbb{R}_{\geq 0})$$

$$= 2 \int_{H \cap \mathbb{R}_{\geq 0}} \frac{1}{\sqrt{2\pi}} e^{-\frac{x^2}{2}} dx.$$

In other words, $\mu_{X|B}$ is an absolutely continuous distribution and for each $H \in \mathcal{B}$ we have

$$\mu_{X|B}(H) = \int_H \gamma_{X|B}(x) dx, \qquad \gamma_{X|B}(x) := \sqrt{\frac{2}{\pi}} e^{-\frac{x^2}{2}} \mathbb{1}_{\mathbb{R}_{\geq 0}}(x).$$

Finally, by (2.4.2) we have

$$E[X \mid B] = \int_0^{+\infty} x \mu_{X|B}(dx)$$

$$= \int_0^{+\infty} x \gamma_{X|B}(x) dx$$

$$= \sqrt{\frac{2}{\pi}} \left[-e^{-\frac{x^2}{2}} \right]_{x=0}^{x=+\infty} = \sqrt{\frac{2}{\pi}}.$$

Example 2.4.8 Let $X, Y \sim \mathrm{Be}_p$, with $0 < p < 1$, independent and $B = (X + Y = 1)$. Determine:

(i) the conditional distribution $\mu_{X|B}$;
(ii) the conditional mean and variance, $E[X \mid B]$ and $\mathrm{var}(X \mid B)$.

First of all, we know that $X + Y \sim \mathrm{Bin}_{2,p}$ and therefore $P(B) = 2p(1-p) > 0$. Since X takes only the values 0 and 1, we have

$$\mu_{X|B}(\{0\}) = \frac{P((X = 0) \cap (X + Y = 1))}{2p(1-p)}$$

$$= \frac{P((X = 0) \cap (Y = 1))}{2p(1-p)}$$

$$= \frac{P(X = 0)P(Y = 1)}{2p(1-p)} = \frac{1}{2}.$$

In conclusion, $\mu_X = \mathrm{Be}_p$ but, regardless of the value of p, $\mu_{X|B} = \mathrm{Be}_{\frac{1}{2}}$ that is, *conditionally on the event* $(X + Y = 1)$, *X has a Bernoulli distribution with parameter* $\frac{1}{2}$. Then, by (2.4.2) and recalling the formulas (2.2.14) for mean and variance of a binomial variable, we get

$$E[X \mid B] = \frac{1}{2}, \qquad \mathrm{var}(X \mid B) = \frac{1}{4}.$$

A concrete interpretation is the following: how can one adjust a biased coin to make it fair (without knowing the probability $p \in]0, 1[$ of getting heads)? The result X of a toss of the rigged coin has distribution Be_p where $T := (X = 1)$ is the event "heads". Based on what has been seen above, to make the coin fair it is sufficient to toss it twice, considering the toss valid only if exactly one head is obtained: then the two events TC or CT have probability $1/2$, whatever $p \in]0, 1[$.

Example 2.4.9 Three draws are made without replacement from an urn containing 3 white balls, 2 black and 2 red balls. Let X and Y be respectively the number of white balls and black balls drawn. We determine the conditional distribution of X given $(Y = 0)$ and the conditional expectation $E[X \mid Y = 0]$:

$$P(X = 0 \mid Y = 0) = 0, \qquad P(X = 1 \mid Y = 0) = \frac{3}{10},$$

$$P(X = 2 \mid Y = 0) = \frac{6}{10}, \qquad P(X = 0 \mid Y = 0) = \frac{1}{10},$$

and

$$E[X \mid Y = 0] = \sum_{k=0}^{3} k P(X = k \mid Y = 0) = \frac{9}{5}.$$

Example 2.4.10 Let (X, Y) be an absolutely continuous random vector with density $\gamma_{(X,Y)}$ and $B = (Y \in K)$ with $K \in \mathscr{B}$ such that $P(B) > 0$. Then, for every $H \in \mathscr{B}$, we have

$$\mu_{X \mid Y \in K}(H) = \frac{P((X \in H) \cap (Y \in K))}{P(Y \in K)} \tag{2.4.3}$$

$$= \frac{\mu_{(X,Y)}(H \times K)}{\mu_Y(K)}$$

$$= \frac{1}{P(Y \in K)} \iint_{H \times K} \gamma_{(X,Y)}(x, y) dx dy =$$

(by Fubini's theorem)

$$= \int_H \left(\frac{1}{P(Y \in K)} \int_K \gamma_{(X,Y)}(x, y) dy \right) dx;$$

thus, we get the expression

$$\gamma_{X|Y\in K}(x) = \frac{1}{P(Y \in K)} \int_K \gamma_{(X,Y)}(x,y)dy \tag{2.4.4}$$

for the *conditional density of X given* $(Y \in K)$. Note that when $K = \mathbb{R}$ (and thus $(Y \in K) = \Omega$) (2.4.4) coincides with the formula (2.2.26) that expresses the marginal density from the joint one.

As a particular example, consider a two-dimensional normal random vector $(X, Y) \sim \mathcal{N}_{0,C}$ with covariance matrix

$$C = \begin{pmatrix} 1 & 1 \\ 1 & 2 \end{pmatrix}$$

and let $B = (Y > 0)$. By (2.5.18), the two-dimensional Gaussian density of (X, Y) is

$$\Gamma(x, y) = \frac{1}{2\pi} e^{-x^2 + xy - \frac{y^2}{2}}.$$

As in (2.4.3) we have

$$\mu_{X|Y>0}(H) = \int_H \left(\frac{1}{P(Y > 0)} \int_0^{+\infty} \Gamma(x, y)dy \right) dx, \qquad H \in \mathcal{B},$$

and therefore we get the expression of the conditional density of X given $(Y > 0)$:

$$\Gamma_{X|Y>0}(x) = \frac{1}{P(Y > 0)} \int_0^{+\infty} \Gamma(x, y)dy = \frac{e^{-\frac{x^2}{2}} \left(1 + \mathrm{erf}\frac{x}{\sqrt{2}}\right)}{\sqrt{2\pi}}, \qquad x \in \mathbb{R}.$$

Note that $E[X] = 0$ but

$$E[X \mid Y > 0] = \int_{\mathbb{R}} x \Gamma_{X|Y>0}(x)dx = \frac{1}{\sqrt{\pi}}.$$

2.5   Characteristic Function

Definition 2.5.1 (Characteristic Function) Let

$$X : \Omega \longrightarrow \mathbb{R}^d$$

be a r.v. on the probability space $(\Omega, \mathscr{F}, P)$. The function

$$\varphi_X : \mathbb{R}^d \longrightarrow \mathbb{C}$$

defined by

$$\varphi_X(\eta) = E\left[e^{i\langle\eta, X\rangle}\right] = E\left[\cos\langle\eta, X\rangle\right] + i E\left[\sin\langle\eta, X\rangle\right], \qquad \eta \in \mathbb{R}^d,$$

is called the *characteristic function of the r.v. X*. We also use the abbreviation CHF for φ_X.

Remark 2.5.2 By definition, if $X \sim \mu_X$ we have

$$\varphi_X(\eta) = \int_{\mathbb{R}^d} e^{i\eta \cdot x} \mu_X(dx)$$

where $x \cdot \eta \equiv \langle x, \eta \rangle$ denotes the scalar product in $\mathbb{R}^d$. If X has discrete distribution $\sum_{n=1}^{\infty} p_n \delta_{x_n}$ then φ_X is given by the Fourier series

$$\varphi_X(\eta) = \sum_{n=1}^{\infty} p_n e^{i\eta \cdot x_n}.$$

We also recall the definition[16] of *Fourier transform of* $f \in L^1(\mathbb{R}^d)$:

$$\hat{f}(\eta) = \int_{\mathbb{R}^d} e^{i\eta \cdot x} f(x) dx. \qquad (2.5.1)$$

---

[16] Depending on the fields of application, different conventions are used for the definition of the Fourier transform: in mathematical analysis it is usually defined as

$$\hat{f}(\eta) = \int_{\mathbb{R}^d} e^{-i\eta \cdot x} f(x) dx$$

while in engineering applications, sometimes the following definition is used

$$\hat{f}(\eta) = \frac{1}{(2\pi)^{\frac{d}{2}}} \int_{\mathbb{R}^d} e^{i\eta \cdot x} f(x) dx.$$

The latter is also the definition used in the Mathematica® scientific computing software. We will always use (2.5.1) which is the definition commonly adopted in probability theory. In particular, care must be taken with the inversion formula for the Fourier transform, which is different depending on the notation used.

Then, if $X \in AC$ with density γ_X, the CHF

$$\varphi_X(\eta) = \int_{\mathbb{R}^d} e^{i\eta \cdot x} \gamma_X(x) dx = \hat{\gamma}_X(\eta)$$

is the Fourier transform of the density of X.

Proposition 2.5.3 *The following properties hold:*

(i) $\varphi_X(0) = 1$;

(ii) $|\varphi_X(\eta)| \leq E\left[|e^{i\eta \cdot X}|\right] = 1$ *for every* $\eta \in \mathbb{R}^d$;

(iii) $|\varphi_X(\eta + h) - \varphi_X(\eta)| \leq E\left[|e^{ih \cdot X} - 1|\right]$ *and therefore, by the dominated convergence theorem,* φ_X *is uniformly continuous on* $\mathbb{R}^d$;

(iv) *denoting by* α^* *the transpose matrix of* α, *we have*

$$\varphi_{\alpha X+b}(\eta) = E\left[e^{i\langle \eta, \alpha X + b\rangle}\right] = e^{i\langle b, \eta\rangle} E\left[e^{i\langle \alpha^* \eta, X\rangle}\right] = e^{i\langle b, \eta\rangle} \varphi_X(\alpha^* \eta);$$

$$(2.5.2)$$

(v) *in the case* $d = 1$, $\varphi_X(-\eta) = \varphi_{-X}(\eta) = \overline{\varphi_X(\eta)}$ *where* $\bar{z}$ *denotes the conjugate of* $z \in \mathbb{C}$. *Consequently, if* X *has an even distribution,*[17] *that is* $\mu_X = \mu_{-X}$, *then* φ_X *takes real values and in this case we have*

$$\varphi_X(\eta) = \int_{\mathbb{R}} e^{i\eta x} \mu_X(dx) = \int_{\mathbb{R}} \cos(x\eta) \mu_X(dx).$$

Let us consider a few examples.

(i) If $X \sim \delta_{x_0}$, with $x_0 \in \mathbb{R}^d$, then

$$\varphi_X(\eta) = e^{i\eta \cdot x_0}.$$

Note that in this case $\varphi_X \notin L^1(\mathbb{R}^d)$ because $|\varphi_X(\eta)| = 1$ for every $\eta \in \mathbb{R}^d$. As a particular case, if $X \sim \delta_0$ then $\varphi_X \equiv 1$. Moreover, if $X \sim \frac{1}{2}(\delta_{-1} + \delta_1)$ then $\varphi_X(\eta) = \cos \eta$.

(ii) If $X \sim Be_p$, with $p \in [0, 1]$, then

$$\varphi_X(\eta) = 1 + p\left(e^{i\eta} - 1\right).$$

Moreover, since $X \sim Bin_{n,p}$ is equal in law to the sum $X_1 + \cdots + X_n$ of n independent Bernoulli random variables (cf. Proposition 2.6.3) then

$$\varphi_X(\eta) = E\left[e^{i\eta(X_1 + \cdots + X_n)}\right] = \left(E\left[e^{i\eta X_1}\right]\right)^n = \left(1 + p\left(e^{i\eta} - 1\right)\right)^n.$$

$$(2.5.3)$$

---

[17] This is true in particular if X has density γ_X which is an even function, i.e., $\gamma_X(x) = \gamma_X(-x)$, $x \in \mathbb{R}$.

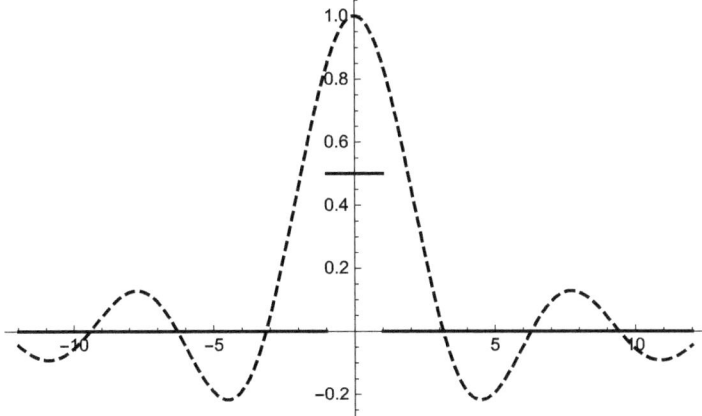

Fig. 2.7 Plot of the uniform density on $[-1, 1]$ (solid line) and the corresponding CHF (dashed line)

(iii) If $X \sim \text{Poisson}_\lambda$, with $\lambda > 0$, then

$$\varphi_X(\eta) = \sum_{k=0}^{\infty} e^{-\lambda} \frac{\lambda^k}{k!} e^{ik\eta} = e^{\lambda(e^{i\eta}-1)}.$$

(iv) If $X \sim \text{Unif}_{[-1,1]}$ then

$$\varphi_X(\eta) = \frac{\sin \eta}{\eta}, \qquad \eta \in \mathbb{R}. \tag{2.5.4}$$

See Fig. 2.7 for the plot of the uniform density and its Fourier transform. Also in this case[18] $\varphi_X \notin L^1(\mathbb{R})$.

(v) If X is a r.v. with Cauchy distribution, that is X has density

$$\gamma_X(x) = \frac{1}{\pi \left(1 + x^2\right)}, \qquad x \in \mathbb{R}, \tag{2.5.5}$$

then

$$\varphi_X(\eta) = e^{-|\eta|}, \qquad \eta \in \mathbb{R}. \tag{2.5.6}$$

See Fig. 2.8 for the plot of the Cauchy density and its Fourier transform. In this case the function φ_X is continuous *but not differentiable* at the origin.

---

[18] See, for example, [25] Ch.5 Sec.12.

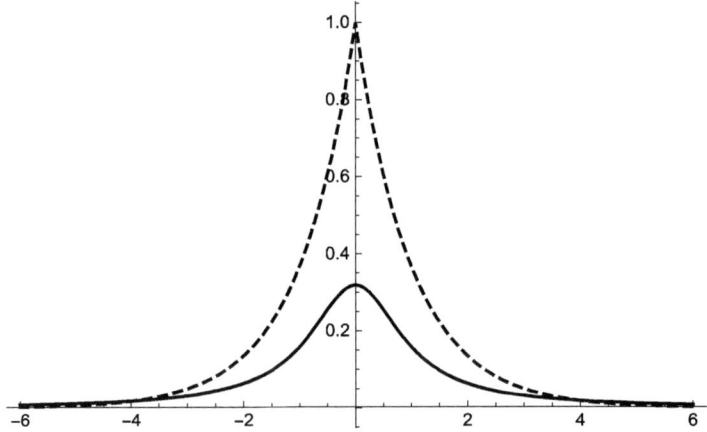

Fig. 2.8 Plot of the Cauchy density (2.5.5) (solid line) and the corresponding CHF (dashed line)

(vi) If $X \sim \mathcal{N}_{\mu,\sigma^2}$, with $\mu \in \mathbb{R}$ and $\sigma \geq 0$, then

$$\varphi_X(\eta) = e^{i\eta\mu - \frac{1}{2}\sigma^2\eta^2}, \qquad \eta \in \mathbb{R}. \qquad (2.5.7)$$

For $\sigma = 0$ we obtain the CHF of the Dirac delta centered at μ. We prove (2.5.7) in the standard case $\mu = 0$ and $\sigma = 1$. To avoid using more advanced tools of complex analysis, we employ a simple trick: first, note that by Proposition 2.5.3-(v) we have

$$\varphi_X(\eta) = \int_{\mathbb{R}} \cos(\eta x) \frac{e^{-\frac{x^2}{2}}}{\sqrt{2\pi}} dx.$$

Then, we compute the derivative of φ_X: by differentiating under the integral sign, we have

$$\frac{d}{d\eta} \varphi_X(\eta) = \int_{\mathbb{R}} \sin(\eta x)(-x) \frac{e^{-\frac{x^2}{2}}}{\sqrt{2\pi}} dx =$$

(since $-xe^{-\frac{x^2}{2}} = \frac{d}{dx} e^{-\frac{x^2}{2}}$)

$$= \int_{\mathbb{R}} \sin(\eta x) \frac{d}{dx} \frac{e^{-\frac{x^2}{2}}}{\sqrt{2\pi}} dx =$$

(integrating by parts)

$$= \frac{1}{\sqrt{2\pi}} \left[\sin(\eta x) e^{-\frac{x^2}{2}} \right]_{x=-\infty}^{x=+\infty} - \int_{\mathbb{R}} \eta \cos(\eta x) \frac{e^{-\frac{x^2}{2}}}{\sqrt{2\pi}} dx = -\eta \varphi_X(\eta).$$

In conclusion, φ_X is the solution of the Cauchy problem

$$\begin{cases} \frac{d}{d\eta} \varphi_X(\eta) = -\eta \varphi_X(\eta), \\ \varphi_X(0) = 1, \end{cases}$$

and therefore

$$\varphi_X(\eta) = e^{-\frac{\eta^2}{2}}. \tag{2.5.8}$$

For the general case where $Y \sim \mathcal{N}_{\mu,\sigma^2}$, it suffices to consider $X := \frac{Y-\mu}{\sigma} \sim \mathcal{N}_{0,1}$ and combine (2.5.8) with (2.5.2).

(vii) If $X \sim \text{Exp}_\lambda$, with $\lambda \in \mathbb{R}_{>0}$, then

$$\varphi_X(\eta) = \lambda \int_0^{+\infty} e^{i\eta x - \lambda x} dx = \frac{\lambda}{\lambda - i\eta}.$$

Example 2.5.4 [!] Let N and $Z_1, Z_2, \ldots$ be independent random variables with $N \sim \text{Poisson}_\lambda$ and Z_n identically distributed for $n \in \mathbb{N}$. We determine the CHF of

$$X := \begin{cases} 0 & \text{if } N = 0, \\ \sum_{k=1}^{N} Z_k & \text{if } N \geq 1. \end{cases}$$

We have

$$\varphi_X(\eta) = E\left[e^{i\eta X} \right] = \sum_{n=0}^{\infty} E\left[e^{i\eta \sum_{k=1}^{n} Z_k} \mathbb{1}_{(N=n)} \right] =$$

(by the independence of N and $Z_k, k \geq 1$)

$$= \sum_{n=0}^{\infty} E\left[e^{i\eta \sum_{k=1}^{n} Z_k} \right] P(N = n)$$

(because the Z_k are independent and identically distributed)

$$= e^{-\lambda} \sum_{n=0}^{\infty} E\left[e^{i\eta Z_1}\right]^n \frac{\lambda^n}{n!} = e^{\lambda(\varphi_{Z_1}(\eta)-1)}$$

where φ_{Z_1} denotes the CHF of Z_1.

2.5.1   The Inversion Theorem

The main result of this section is Theorem 2.5.6, which provides a significant inversion formula for the characteristic function. We start with a basic exercise to lay the groundwork.

Exercise 2.5.5 Let us prove the following formula for the improper integral of $\frac{\sin x}{x}$:

$$\int_0^{+\infty} \frac{\sin x}{x} dx := \lim_{a \to +\infty} \int_0^a \frac{\sin x}{x} dx = \frac{\pi}{2}. \tag{2.5.9}$$

We set

$$f(x, y) = e^{-xy} \sin x, \qquad x > 0, \ y > 0,$$

and notice that, for every $x, y, a > 0$, we have

$$\int_0^{+\infty} f(x, y) dy = \frac{\sin x}{x},$$

$$\int_0^a f(x, y) dx = \frac{1}{1+y^2} - \frac{e^{-ay}}{1+y^2} \cos a - \frac{y e^{-ay}}{1+y^2} \sin a.$$

Hence, by Fubini's theorem we have

$$\int_0^a \frac{\sin x}{x} dx = \frac{\pi}{2} - \cos a \int_0^{+\infty} \frac{e^{-ay}}{1+y^2} dy - \sin a \int_0^{+\infty} \frac{y e^{-ay}}{1+y^2} dy, \qquad a > 0,$$

and consequently, since $\frac{1}{1+y^2} \leq 1$,

$$\left| \int_0^a \frac{\sin x}{x} - \frac{\pi}{2} \right| \leq \int_0^{+\infty} (1+y) e^{-ay} dy = \frac{1+a}{a^2}, \qquad a > 0.$$

This proves (2.5.9). Note that $\frac{\sin x}{x}$ is integrable in the generalized sense but not absolutely integrable.

Theorem 2.5.6 (Inversion Theorem) *[!!] Let μ be a distribution on $(\mathbb{R}, \mathscr{B})$ and*

$$\varphi(\eta) := \int_{\mathbb{R}} e^{ix\eta} \mu(dx), \qquad \eta \in \mathbb{R}. \tag{2.5.10}$$

Then for every $a < b$ we have

$$\mu(]a, b[) + \frac{\mu(\{a\}) + \mu(\{b\})}{2} = \lim_{R \to +\infty} \frac{1}{2\pi} \int_{-R}^{R} \frac{e^{-ia\eta} - e^{-ib\eta}}{i\eta} \varphi(\eta) d\eta. \tag{2.5.11}$$

Moreover, if $\varphi \in L^1(\mathbb{R})$ then μ is absolutely continuous and has density function

$$\gamma(x) := \frac{1}{2\pi} \int_{\mathbb{R}} e^{-ix\eta} \varphi(\eta) d\eta, \qquad x \in \mathbb{R}. \tag{2.5.12}$$

Remark 2.5.7 *[!]* By Theorem 2.5.6, *the CHF of a r.v. identifies its law*:

$$\mu_X = \mu_Y \quad \Longleftrightarrow \quad \varphi_X = \varphi_Y.$$

In other words, X and Y have equal characteristic functions,

$$\varphi_X(\eta) = \varphi_Y(\eta), \qquad \eta \in \mathbb{R},$$

if and only if their respective laws μ_X and μ_Y coincide:

$$\mu_X(H) = \mu_Y(H), \qquad H \in \mathscr{B}.$$

Indeed, by (2.5.11) we have $\mu_X(]a, b[) = \mu_Y(]a, b[)$ for every $a, b \in \mathbb{R} \setminus A$ where

$$A := \{x \in \mathbb{R} \mid \mu_X(\{x\}) + \mu_Y(\{x\}) > 0\}.$$

On the other hand, by Remark 1.4.11, A is *finite or at most countable* and therefore $\mathbb{R} \setminus A$ is dense in $\mathbb{R}$: from Caratheodory's theorem it follows that $\mu_X \equiv \mu_Y$.

Corollary 2.5.8 *[!] If μ, ν are distributions such that*

$$\int_{\mathbb{R}} f d\mu = \int_{\mathbb{R}} f d\nu$$

for every $f \in bC(\mathbb{R})$ then $\mu \equiv \nu$. Similarly, if μ, ν are distributions on $\mathbb{R}^d$ such that

$$\int_{\mathbb{R}^d} \prod_{j=1}^{d} f_j(x_j) \mu(dx) = \int_{\mathbb{R}^d} \prod_{j=1}^{d} f_j(x_j) \nu(dx)$$

for every $f_1, \ldots, f_d \in bC(\mathbb{R})$ then $\mu \equiv \nu$. In particular, two random variables X, Y are equal in law if and only if $E[f(X)] = E[f(Y)]$ for every $f \in bC$.

Proof For $d = 1$, choosing f of the form $f(x) = \cos(x\eta)$ or $f(x) = \sin(x\eta)$, with $\eta \in \mathbb{R}$, we deduce from the assumption that the CHFs of μ and ν are equal: the thesis follows from Theorem 2.5.6. If $d \geq 2$, we proceed in a similar way using the multi-dimensional version of Theorem 2.5.6 and the fact that $e^{i\eta x} = \prod_{j=1}^{d} e^{i\eta_j x_j}$. $\quad\square$

Remark 2.5.9 [!] Let μ be a distribution with density g such that $\hat{g} \in L^1(\mathbb{R})$: by Theorem 2.5.6 also γ defined by (2.5.10)–(2.5.12) is a density of μ and therefore by Remark 1.4.19 we have $g = \gamma$ a.e. that is

$$g(x) = \frac{1}{2\pi} \int_{\mathbb{R}} e^{-ix\eta} \hat{g}(\eta) d\eta \quad \text{for almost every } x \in \mathbb{R}. \tag{2.5.13}$$

Equation (2.5.13) is the classical inversion formula of the Fourier transform. The integral on the right-hand side of (2.5.13) is a bounded and uniformly continuous function of $x \in \mathbb{R}$ (cf. Proposition 2.5.3). Clearly, a density function g is not necessarily bounded nor continuous (indeed, it can be modified on every Lebesgue-negligible Borel set): however, if $\hat{g} \in L^1(\mathbb{R})$ then g *is necessarily equal a.e.* to a bounded and continuous function.

Remark 2.5.10 [!] Based on Theorem 2.5.6, if $\varphi_X \in L^1(\mathbb{R})$ then $X \in AC$ and a density of X is given by the inversion formula

$$\gamma_X(x) = \frac{1}{2\pi} \int_{\mathbb{R}} e^{-ix\eta} \varphi_X(\eta) d\eta, \qquad x \in \mathbb{R}.$$

Condition $\varphi_X \in L^1(\mathbb{R})$ is only *sufficient but not necessary* for the absolute continuity of μ. In fact, by Remark 2.5.9, if $\varphi_X \in L^1(\mathbb{R})$ then necessarily the density of X is equal a.e. to a continuous function: however, for example, the uniform distribution on $[-1, 1]$ is absolutely continuous but has density $\gamma(x) = \frac{1}{2} \mathbb{1}_{[-1,1]}(x)$ which is not equal a.e. to a continuous function; in fact, its CHF in (2.5.4) is not absolutely integrable.

Proof of Theorem 2.5.6 Fix $a, b \in \mathbb{R}$, with $a < b$, and let

$$g_{a,b}(\eta) := \int_a^b e^{-ix\eta} dx = \frac{e^{-ia\eta} - e^{-ib\eta}}{i\eta}, \qquad \eta \in \mathbb{R}. \tag{2.5.14}$$

Observe that, by the triangle inequality, $|g_{a,b}(\eta)| \leq b - a$. Therefore, by Fubini's theorem, for every $R > 0$ we have

$$\int_{-R}^{R} g_{a,b}(\eta) \varphi(\eta) d\eta = \int_{\mathbb{R}} \left(\int_{-R}^{R} g_{a,b}(\eta) e^{ix\eta} d\eta \right) \mu(dx). \tag{2.5.15}$$

Since cosine and sine are even[19] and odd functions respectively, we have

$$\int_{-R}^{R} g_{a,b}(\eta)e^{ix\eta}d\eta = 2\int_0^R \left(\frac{\sin((x-a)\eta)}{\eta} - \frac{\sin((x-b)\eta)}{\eta}\right)d\eta \longrightarrow G_{a,b}(x)$$

$$:= \begin{cases} \pi & \text{if } x = a \text{ or } x = b, \\ 2\pi & \text{if } a < x < b, \\ 0 & \text{if } x < a \text{ or } x > b, \end{cases} \qquad (2.5.16)$$

as $R \to +\infty$: this follows from the fact that by (2.5.9), we have[20]

$$\int_0^R \frac{\sin \lambda\eta}{\eta}d\eta = \int_0^{\lambda R} \frac{\sin \eta}{\eta}d\eta = \mathrm{sgn}(\lambda)\int_0^{|\lambda|R} \frac{\sin \eta}{\eta}d\eta \longrightarrow \begin{cases} \frac{\pi}{2} & \text{if } \lambda > 0, \\ 0 & \text{if } \lambda = 0, \\ -\frac{\pi}{2} & \text{if } \lambda < 0. \end{cases}$$

Now we use the dominated convergence[21] Theorem 2.2.12 to pass to the limit in (2.5.15) as $R \to +\infty$: we have

$$\lim_{R\to+\infty} \frac{1}{2\pi}\int_{-R}^{R} g_{a,b}(\eta)\varphi(\eta)d\eta = \frac{1}{2\pi}\int_{\mathbb{R}} G_{a,b}(x)\mu(dx) = \frac{1}{2}\int_{\{a\}} \mu(dx)$$

$$+ \int_{]a,b[} \mu(dx) + \frac{1}{2}\int_{\{b\}} \mu(dx)$$

and this proves (2.5.11).

Let us prove the second part of the thesis: if $\varphi \in L^1(\mathbb{R})$ then, recalling that $|g_{a,b}(\eta)\varphi(\eta)| \leq (b-a)|\varphi(\eta)|$ and applying the dominated convergence theorem to take the limit in (2.5.11), we obtain

$$\frac{1}{2\pi}\int_{\mathbb{R}} g_{a,b}(\eta)\varphi(\eta)d\eta = \mu(]a,b[) + \frac{1}{2}\mu(\{a,b\}) \geq \mu(\{b\}). \qquad (2.5.17)$$

---

[19] As a consequence, the integral between $-R$ and R of the even function $\cos \eta$ multiplied by the odd function $\frac{1}{\eta}$ vanishes.

[20] We define the sign function as follows

$$\mathrm{sgn}(\lambda) = \begin{cases} 1 & \text{if } \lambda > 0, \\ 0 & \text{if } \lambda = 0, \\ -1 & \text{if } \lambda < 0. \end{cases}$$

[21] By (2.5.16), the modulus of the integrand in (2.5.15) is bounded by $2\sup_{r>0}\int_0^r \frac{\sin \eta}{\eta}d\eta < +\infty$

But the inequality in (2.5.17), again by the dominated convergence theorem and taking the limit as $a \to b^-$, implies that $\mu(\{b\}) = 0$ for every $b \in \mathbb{R}$ and therefore

$$\mu(]a, b[) = \frac{1}{2\pi} \int_{\mathbb{R}} g_{a,b}(\eta)\varphi(\eta)d\eta =$$

(using the second equality in (2.5.14) and Fubini's theorem)

$$= \int_a^b \left(\frac{1}{2\pi} \int_{\mathbb{R}} e^{-ix\eta}\varphi(\eta)d\eta \right) dx = \int_a^b \gamma(x)dx,$$

and so γ in (2.5.12) is a density of μ. $\square$

The CHF of $X = (X_1, \ldots, X_n)$ is called the *joint CHF of the random variables* $X_1, \ldots, X_n$; conversely, $\varphi_{X_1}, \ldots, \varphi_{X_n}$ are called *marginal CHFs of X*.

Proposition 2.5.11 *Let $X_1, \ldots, X_n$ be random variables on $(\Omega, \mathcal{F}, P)$ with values respectively in $\mathbb{R}^{d_1}, \ldots, \mathbb{R}^{d_n}$. Let $X = (X_1, \ldots, X_n)$, then:*

(i) $\varphi_{X_i}(\eta_i) = \varphi_X(0, \ldots, 0, \eta_i, 0, \ldots, 0)$;
(ii) $X_1, \ldots, X_n$ are independent if and only if

$$\varphi_X(\eta) = \prod_{i=1}^n \varphi_{X_i}(\eta_i), \qquad \eta = (\eta_1, \ldots, \eta_n).$$

Proof Property (i) is an immediate consequence of the definition of characteristic function. We prove (ii) only in the case $n = 2$. If X_1, X_2 are independent then so are the random variables $e^{i\eta_1 \cdot X_1}$, $e^{i\eta_2 \cdot X_2}$ and therefore we have

$$\varphi_X(\eta_1, \eta_2) = E\left[e^{i\eta_1 \cdot X_1 + i\eta_2 \cdot X_2}\right] = E\left[e^{i\eta_1 \cdot X_1}\right] E\left[e^{i\eta_2 \cdot X_2}\right] = \varphi_{X_1}(\eta_1)\varphi_{X_2}(\eta_2).$$

Conversely, consider two independent random variables $\widetilde{X}_1, \widetilde{X}_2$ such that $\widetilde{X}_1 \overset{d}{=} X_1$ and $\widetilde{X}_2 \overset{d}{=} X_2$. Then we have

$$\varphi_{(\widetilde{X}_1, \widetilde{X}_2)}(\eta_1, \eta_2) = \varphi_{\widetilde{X}_1}(\eta_1)\varphi_{\widetilde{X}_2}(\eta_2) = \varphi_{X_1}(\eta_1)\varphi_{X_2}(\eta_2) = \varphi_{(X_1, X_2)}(\eta_1, \eta_2).$$

Since (X_1, X_2) and $(\widetilde{X}_1, \widetilde{X}_2)$ have the same CHF, by Theorem 2.5.6, they also have the same law: this implies that X_1, X_2 are independent. $\square$

2.5.2 Multivariate Normal Distribution

Given $\mu \in \mathbb{R}^d$ and a $d \times d$ matrix C, *symmetric and positive definite*, we define the Gaussian d-dimensional density function with parameters μ and C as follows:

$$\Gamma(x) = \frac{1}{\sqrt{(2\pi)^d \det C}} e^{-\frac{1}{2}\langle C^{-1}(x-\mu), x-\mu\rangle}, \qquad x \in \mathbb{R}^d. \qquad (2.5.18)$$

A direct calculation shows that

$$\int_{\mathbb{R}^d} \Gamma(x)dx = 1, \qquad\qquad\qquad\qquad (2.5.19)$$

$$\int_{\mathbb{R}^d} x_i \Gamma(x)dx = \mu_i, \qquad\qquad\qquad (2.5.20)$$

$$\int_{\mathbb{R}^d} (x_i - \mu_i)(x_j - \mu_j)\, \Gamma(x)dx = C_{ij}, \qquad (2.5.21)$$

for every $i, j = 1, \ldots, d$. Equation (2.5.19) simply shows that Γ is a density; Eqs. (2.5.20) and (2.5.21) motivate the following

Definition 2.5.12 (Gaussian Vector, I) If X is a d-dimensional r.v. with density Γ in (2.5.18) then we say that X has a multi-normal distribution with mean μ and (positive definite) covariance matrix C. We write $X \sim \mathcal{N}_{\mu,C}$ and we say that X is a *Gaussian vector*.

If $X \sim \mathcal{N}_{\mu,C}$ then $E[X] = \mu$ by (2.5.20) and $\mathrm{cov}(X) = C$ by (2.5.21).

Proposition 2.5.13 [!] *The CHF of $X \sim \mathcal{N}_{\mu,C}$ is given by*

$$\varphi_X(\eta) = e^{i\langle\mu,\eta\rangle - \frac{1}{2}\langle C\eta,\eta\rangle}, \qquad \eta \in \mathbb{R}^d. \qquad (2.5.22)$$

Proof The computation of the Fourier transform of Γ in (2.5.18) is analogous to the one-dimensional case (cf. formula (2.5.7)). $\qquad\square$

We observe that the CHF in (2.5.22) is an exponential function comprising two distinct components: *a linear term* in η dependent on the *expectation* μ and *a quadratic term* in η dependent on the *covariance matrix C*.

It is remarkable that, unlike the density Γ in which the inverse of C appears, in the characteristic function φ_X the quadratic form of the matrix C itself appears. Therefore, for φ_X to be well-defined, it is not necessary that C is strictly positive definite. In fact, in many applications, degenerate covariance matrices occur, and it is useful to extend Definition 2.5.12 in the following way:

Definition 2.5.14 (Gaussian Vector, II) Given $\mu \in \mathbb{R}^d$ and a $d \times d$ symmetric and *positive semi-definite* matrix C, we say that X has a multi-normal distribution with mean μ and covariance matrix C, and we write $X \sim \mathcal{N}_{\mu,C}$, if the CHF of X is the φ_X in (2.5.22). We also say that X is a *Gaussian vector*.

By Theorem 2.5.6, the previous definition is well-posed and coincides with Definition 2.5.12 if $C > 0$, since the CHF uniquely identifies the distribution. Moreover, Definition 2.5.14 is not empty in the sense that a r.v. X, which has φ_X in (2.5.22) as a characteristic function, *exists*: in fact, by Remark 2.2.44, given C, a $d \times d$ symmetric and positive semi-definite matrix, there exists α such that $C = \alpha\alpha^*$; then it suffices to set $X = \alpha Z + \mu$ where Z is a standard multi-normal random variable, i.e., $Z \sim \mathcal{N}_{0,I}$ with I the $d \times d$ identity matrix. Indeed, by (2.5.2) we have

$$\varphi_{\alpha Z + \mu}(\eta) = e^{i\eta \cdot \mu}\varphi_Z(\alpha^*\eta) = e^{i\eta \cdot \mu - \frac{|\alpha^*\eta|^2}{2}} = e^{i\langle\mu,\eta\rangle - \frac{1}{2}\langle C\eta,\eta\rangle}.$$

Using the characteristic function, it is easy to prove some fundamental properties of the normal distribution, such as invariance for linear transformations. Recall that, when we use matrix notation, the d-dimensional random vector X is identified with the $d \times 1$ column matrix.

Proposition 2.5.15 *[!] A random vector X is a Gaussian vector if and only if the real r.v. $\langle\eta, X\rangle$ has normal distribution for any $\eta \in \mathbb{R}^d$. Moreover, let α be a $N \times d$ constant matrix and $\beta \in \mathbb{R}^N$ with $N, d \in \mathbb{N}$. If $X \sim \mathcal{N}_{\mu,C}$ then $\alpha X + \beta$ is a Gaussian vector with distribution*

$$\alpha X + \beta \sim \mathcal{N}_{\alpha\mu+\beta,\alpha C\alpha^*}. \tag{2.5.23}$$

Proof We prove that if $\langle\eta, X\rangle$ has normal distribution for any $\eta \in \mathbb{R}^d$ then X is a Gaussian vector. In fact, it suffices to find the CHF of X:

$$\varphi_X(\eta) = E\left[e^{i\langle\eta,X\rangle}\right] = \varphi_{\langle\eta,X\rangle}(1) = e^{iE[\langle\eta,X\rangle] - \frac{1}{2}\mathrm{var}(\langle\eta,X\rangle)}$$

where the last equality holds by assumption. It remains to calculate the mean and variance of $\langle\eta, X\rangle$:

$$E[\langle\eta, X\rangle] = \langle\eta, E[X]\rangle,$$

$$\mathrm{var}(\langle\eta, X\rangle) = \mathrm{var}\left(\sum_{i=1}^d \eta_i X_i\right) = \sum_{i,j=1}^d \eta_i\eta_j\mathrm{cov}(X_i, X_j) = \langle C\eta, \eta\rangle$$

where $C = \big(\mathrm{cov}(X_i, X_j)\big)$. Therefore, we have $X \sim \mathcal{N}_{E[X],C}$.

Conversely, let us assume $X \sim \mathcal{N}_{\mu,C}$ and compute the CHF of $\alpha X + \beta$: by Proposition 2.5.3-(iv) we have

$$\varphi_{\alpha X + \beta}(\eta) = e^{i\langle\eta,\beta\rangle}\varphi_X(\alpha^*\eta) =$$

(by the expression (2.5.22) of the CHF of X evaluated at $\alpha^* \eta$)

$$= e^{i \langle \eta, \beta \rangle} e^{i \langle \mu, \alpha^* \eta \rangle - \frac{1}{2} \langle C \alpha^* \eta, \alpha^* \eta \rangle}$$

$$= e^{i \langle \alpha \mu + \beta, \eta \rangle - \frac{1}{2} \langle \alpha C \alpha^* \eta, \eta \rangle},$$

which proves (2.5.23).   □

For example, as a consequence of (2.5.23), we have that if (X, Y) has a bidimensional normal distribution then X and $X + Y$ are random variables with a normal distribution.

Example 2.5.16 Let $X, Y \sim \mathcal{N}_{0,1}$ be independent and $(u, v) \in \mathbb{R}^2$ such that $u^2 + v^2 = 1$. The we have

$$Z := uX + vY \sim \mathcal{N}_{0,1}.$$

Indeed, by Theorem 2.3.23-(iv) $(X, Y) \sim \mathcal{N}_{0,I}$ where I indicates the 2×2 identity matrix; then since

$$uX + vY = \alpha \begin{pmatrix} X \\ Y \end{pmatrix}, \qquad \text{with } \alpha = \begin{pmatrix} u & v \end{pmatrix},$$

the thesis follows from (2.5.23), being

$$\operatorname{var}(Z) = \alpha \alpha^* = u^2 + v^2 = 1.$$

Example 2.5.17 Let $(X, Y, Z) \sim \mathcal{N}_{\mu, C}$ with

$$\mu = (\mu_X, \mu_Y, \mu_Z), \qquad C = \begin{pmatrix} 1 & -1 & 1 \\ -1 & 2 & -2 \\ 1 & -2 & 2 \end{pmatrix}.$$

Note that $C \geq 0$ and $\det C = 0$ (the last two rows of C are linearly dependent): thus (X, Y, Z) does not have a density. However, $Y \sim \mathcal{N}_{\mu_Y, 2}$ and $(X, Z) \sim \mathcal{N}_{(\mu_X, \mu_Z), \hat{C}}$ with

$$\hat{C} = \begin{pmatrix} 1 & 1 \\ 1 & 2 \end{pmatrix},$$

and therefore Y and (X, Z) have Gaussian densities. For completeness, we provide the matrix α of the Cholesky factorization $C = \alpha \alpha^*$ (cf. Remark 2.2.44):

$$\alpha = \begin{pmatrix} 1 & -1 & 1 \\ 0 & 1 & -1 \\ 0 & 0 & 0 \end{pmatrix}.$$

Proposition 2.5.18 *[!] Let* $X = (X_1, \ldots, X_d)$ *be a Gaussian vector. The real random variables* $X_1, \ldots, X_d$ *are independent if and only if they are uncorrelated, that is,* cov $(X_h, X_k) = 0$ *for every* $h, k = 1, \ldots, d$ *with* $h \neq k$.

Proof If $X_1, \ldots, X_d$ are independent random variables, then cov $(X_h, X_k) = 0$ for any $h \neq k$ by Theorem 2.3.28. Conversely, let $\mu_h = E[X_h]$ and $C_{hk} = $ cov (X_h, X_k): by Proposition 2.5.15, the r.v. X_h has a normal distribution with CHF given by

$$\varphi_{X_h}(\eta_h) = e^{i\mu_h \eta_h - \frac{1}{2} C_{hh} \eta_h^2}, \qquad \eta_h \in \mathbb{R}.$$

On the other hand, by assumption $C_{hk} = 0$ for $h \neq k$ and therefore

$$\varphi_X(\eta) = e^{i\mu \cdot \eta - \frac{1}{2} \sum_{h=1}^{d} C_{hh} \eta_h^2} = \prod_{h=1}^{d} \varphi_{X_h}(\eta_h), \qquad \eta = (\eta_1, \ldots, \eta_d) \in \mathbb{R}^d,$$

and thus the thesis follows from Proposition 2.5.11. $\qquad\qquad\qquad\qquad \square$

Example 2.5.19 [!] The assumption in Proposition 2.5.18 that $X_1, \ldots, X_d$ have a *joint* multi-variate normal distribution cannot be removed. Indeed, there exist random variables with normal marginal distributions that are uncorrelated but not independent. For example, consider two *independent* random variables, respectively with standard normal distribution, $X \sim \mathcal{N}_{0,1}$, and Bernoulli distribution, $Z \sim \mu_Z := \frac{1}{2} (\delta_{-1} + \delta_1)$. Letting $Y = ZX$, we prove that $Y \sim \mathcal{N}_{0,1}$: indeed, due to the independence assumption, the joint distribution of X and Z is the product distribution

$$\mathcal{N}_{0,1} \otimes \mu_Z$$

and thus for every $f \in m\mathcal{B}$ and bounded we have

$$E[f(ZX)] = \int_{\mathbb{R}^2} f(zx) \left(\mathcal{N}_{0,1} \otimes \mu_Z \right) (dx, dz) =$$

(by Fubini's theorem)

$$= \int_{\mathbb{R}} \left(\int_{\mathbb{R}} f(zx) \mathcal{N}_{0,1}(dx) \right) \mu_Z(dz)$$

$$= \frac{1}{2} \int_{\mathbb{R}} f(-x) \mathcal{N}_{0,1}(dx) + \frac{1}{2} \int_{\mathbb{R}} f(x) \mathcal{N}_{0,1}(dx)$$

$$= \int_{\mathbb{R}} f(x) \mathcal{N}_{0,1}(dx).$$

In particular, if $f = \mathbb{1}_H$ with $H \in \mathcal{B}$, we obtain

$$P(Y \in H) = \mathcal{N}_{0,1}(H),$$

that is, $Y \sim \mathcal{N}_{0,1}$.

We now prove that $\text{cov}(X, Y) = 0$ but X, Y *are not independent*. We have:

$$\text{cov}(X, Y) = E[XY] = E\left[ZX^2\right] =$$

(due to the independence of X and Z)

$$= E[Z]E\left[X^2\right] = 0.$$

We verify that X, Y are not independent:

$$P((X \in [0, 1]) \cap (Y \in [0, 1])) = P((X \in [0, 1]) \cap (ZX \in [0, 1])) =$$

(since on the event $(X \in [0, 1])$ we have $(ZX \in [0, 1]) = (Z = 1) \cap (X \in [0, 1])$)

$$= P((X \in [0, 1]) \cap (Z = 1)) =$$

(due to the independence of X and Z)

$$= \frac{1}{2} P(X \in [0, 1]).$$

On the other hand, since $Y \sim \mathcal{N}_{0,1}$, we have $P(Y \in [0, 1]) < \frac{1}{2}$ and thus $P((X \in [0, 1]) \cap (Y \in [0, 1])) < P(X \in [0, 1])P(Y \in [0, 1])$.

This example does not contradict Proposition 2.5.18 since (X, Y) is not a Gaussian vector, even if its components have normal distribution. In fact, the joint CHF is given by

$$\varphi_{(X,Y)}(\eta_1, \eta_2) = E\left[e^{i(\eta_1 X + \eta_2 Y)}\right]$$

$$= E\left[e^{iX(\eta_1 - \eta_2)}\mathbb{1}_{(Z=-1)}\right] + E\left[e^{iX(\eta_1 + \eta_2)}\mathbb{1}_{(Z=1)}\right] =$$

(due to the independence of X and Z)

$$= \frac{1}{2} E\left[e^{iX(\eta_1 - \eta_2)}\right] + \frac{1}{2} E\left[e^{iX(\eta_1 + \eta_2)}\right] =$$

(since $X \sim \mathcal{N}_{0,1}$)

$$= \frac{1}{2} \left(e^{-\frac{(\eta_1 - \eta_2)^2}{2}} + e^{-\frac{(\eta_1 + \eta_2)^2}{2}} \right) = \frac{e^{\eta_1 \eta_2} + e^{-\eta_1 \eta_2}}{2} e^{-\frac{\eta_1^2 + \eta_2^2}{2}},$$

which is not the CHF of a two-dimensional normal distribution. Incidentally, this also proves that $\varphi_{(X,Y)}(\eta_1, \eta_2) \neq \varphi_X(\eta_1)\varphi_Y(\eta_2)$, that is, it confirms that X, Y are not independent.

2.5.3 Series Expansion of CHF and Moments

We prove an interesting result that shows that the *moments of a random variable* $X \in L^p(\Omega, P)$, that is, the expected values $E[X^k]$ of the powers of X with $k \leq p$, can be expressed in terms of derivatives the CHF of X (see in particular Remark 2.5.21).

Theorem 2.5.20 *[!] Let X be a real r.v. in $L^p(\Omega, P)$ for some $p \in \mathbb{N}$. Then we have the following expansion of the CHF of X around the origin:*

$$\varphi_X(\eta) = \sum_{k=0}^{p} \frac{E[(iX)^k]}{k!} \eta^k + o(\eta^p) \qquad \text{as } \eta \to 0. \tag{2.5.24}$$

Proof Recall the Taylor formula with Lagrange remainder for $f \in C^p(\mathbb{R})$: for every $\eta \in \mathbb{R}$ there exists $\lambda \in [0, 1]$ such that

$$f(\eta) = \sum_{k=0}^{p-1} \frac{f^{(k)}(0)}{k!} \eta^k + \frac{f^{(p)}(\lambda \eta)}{p!} \eta^p.$$

Applying this formula to the smooth function $f(\eta) = e^{i\eta X}$, we obtain

$$e^{i\eta X} = \sum_{k=0}^{p} \frac{(iX)^k}{k!} \eta^k + \frac{(iX)^p \left(e^{i\lambda \eta X} - 1 \right)}{p!} \eta^p,$$

where in this case $\lambda \in [0, 1]$ depends on X and is therefore random. Applying the expected value to the last identity, we get

$$\varphi_X(\eta) = \sum_{k=0}^{p} \frac{E[(iX)^k]}{k!} \eta^k + R(\eta)\eta^p$$

where

$$R(\eta) = \frac{1}{p!} E\left[(iX^p)\left(e^{i\lambda\eta X} - 1\right)\right] \longrightarrow 0 \qquad \text{as } \eta \to 0,$$

by the dominated convergence theorem, since by assumption

$$\left|(iX^p)\left(e^{i\lambda\eta X} - 1\right)\right| \le 2|X|^p \in L^1(\Omega, P).$$

$\square$

Remark 2.5.21 [!] If $X \in L^p(\Omega, P)$ then, by (2.5.24), φ_X is differentiable p times at the origin with

$$\varphi_X^{(k)}(0) = E\left[(iX)^k\right], \qquad k = 0, \ldots, p. \tag{2.5.25}$$

Remark 2.5.22 Suppose that $X \in L^p(\Omega, P)$ for every $p \in \mathbb{N}$ and that φ_X is an analytic function. Then, starting from the moments of X, it is possible to derive φ_X and therefore the law of X.

Example 2.5.23 Let X be a r.v. with Cauchy distribution as in (2.5.5). Then $X \notin L^1(\Omega, P)$ and the CHF φ_X in (2.5.6) is not differentiable at the origin.

Example 2.5.24 Given $X \sim \mathcal{N}_{\mu,\sigma^2}$, we have that $X \in L^p(\Omega, P)$ for every $p \in \mathbb{N}$. Since

$$\varphi_X(\eta) = e^{i\mu\eta - \frac{\sigma^2\eta^2}{2}}$$

then through careful perseverance (or with the aid of symbolic computation software), we can derive the following expressions:

$$\varphi'(\eta) = i\left(\mu + i\eta\sigma^2\right)\varphi(\eta),$$

$$\varphi^{(2)}(\eta) = i^2\left(\sigma^2 + \left(\mu + i\eta\sigma^2\right)^2\right)\varphi(\eta),$$

$$\varphi^{(3)}(\eta) = i^3\left(\mu + i\eta\sigma^2\right)\left(3\sigma^2 + \left(\mu + i\eta\sigma^2\right)^2\right)\varphi(\eta),$$

$$\varphi^{(4)}(\eta) = i^4\left(\mu^4 + 2\mu^2\sigma^2(3 + 2i\mu\eta) + 2\eta^2\sigma^6(-3 - 2i\mu\eta)\right.$$

$$\left. + 3\sigma^4(1 - 2\mu\eta(\mu\eta - 2i)) + \eta^4\sigma^8\right)\varphi(\eta),$$

so that

$$\varphi'(0) = i\mu,$$

$$\varphi^{(2)}(0) = -\left(\mu^2 + \sigma^2\right),$$

$$\varphi^{(3)}(0) = -i\left(\mu^3 + 3\mu\sigma^2\right),$$

$$\varphi^{(4)}(0) = \mu^4 + 6\mu^2\sigma^2 + 3\sigma^4.$$

Then, by (2.5.25) we have

$$E[X] = \mu,$$

$$E\left[X^2\right] = \mu^2 + \sigma^2,$$

$$E\left[X^3\right] = \mu^3 + 3\mu\sigma^2,$$

$$E\left[X^4\right] = \mu^4 + 6\mu^2\sigma^2 + 3\sigma^4.$$

Example 2.5.25 Given $X \sim \text{Exp}_\lambda$, we have that $X \in L^p(\Omega, P)$ for every $p \in \mathbb{N}$. Since

$$\varphi_X(\eta) = \frac{\lambda}{\lambda - i\eta}$$

then we have:

$$\varphi^{(k)}(\eta) = \frac{i^k k! \lambda}{(\lambda - i\eta)^{k+1}}, \qquad k \in \mathbb{N},$$

and in particular

$$\varphi^{(k)}(0) = \frac{i^k k!}{\lambda^k}.$$

Then, by (2.5.25) we get

$$E\left[X^k\right] = \frac{k!}{\lambda^k}.$$

2.6 Complements

2.6.1 Sum of Random Variables

Theorem 2.6.1 *Let $X, Y \in AC$ on $(\Omega, \mathcal{F}, P)$ with values in $\mathbb{R}^d$, with joint density $\gamma_{(X,Y)}$. Then $X + Y \in AC$ and has density*

$$\gamma_{X+Y}(z) = \int_{\mathbb{R}^d} \gamma_{(X,Y)}(x, z - x)dx, \qquad z \in \mathbb{R}^d. \tag{2.6.1}$$

Moreover, if X, Y are independent then

$$\gamma_{X+Y}(z) = (\gamma_X * \gamma_Y)(z) := \int_{\mathbb{R}^d} \gamma_X(x)\gamma_Y(z - x)dx, \qquad z \in \mathbb{R}^d, \tag{2.6.2}$$

that is, the density of $X + Y$ is the convolution *of the densities of X and Y.*

Similarly, if X, Y are discrete random variables on (Ω, P) with values in $\mathbb{R}^d$, with joint distribution function $\bar{\mu}_{(X,Y)}$, then $X + Y$ is a discrete r.v. with distribution function

$$\bar{\mu}_{X+Y}(z) = \sum_{x \in X(\Omega)} \bar{\mu}_{(X,Y)}(x, z - x), \qquad z \in \mathbb{R}^d.$$

In particular, if X, Y are independent then

$$\bar{\mu}_{X+Y}(z) = (\bar{\mu}_X * \bar{\mu}_Y)(z) := \sum_{x \in X(\Omega)} \bar{\mu}_X(x)\bar{\mu}_Y(z - x), \tag{2.6.3}$$

that is, $\bar{\mu}_{X+Y}$ is the discrete convolution of the distribution functions $\bar{\mu}_X$ of X and $\bar{\mu}_Y$ of Y.

Proof For every $H \in \mathscr{B}_d$ we have

$$P(X + Y \in H) = E[\mathbb{1}_H(X + Y)] = \int_{\mathbb{R}^d \times \mathbb{R}^d} \mathbb{1}_H(x + y)\gamma_{(X,Y)}(x, y)dxdy =$$

(by the change of variables $z = x + y$)

$$= \int_{\mathbb{R}^d \times \mathbb{R}^d} \mathbb{1}_H(z)\gamma_{(X,Y)}(x, z - x)dxdz =$$

(by Fubini's theorem)

$$= \int_H \left(\int_{\mathbb{R}^d} \gamma_{(X,Y)}(x, z - x)dx \right) dz,$$

and this proves that the function γ_{X+Y} in (2.6.1) is a density of $X + Y$. Finally, (2.6.2) follows from (2.6.1) and (2.3.5).

As for the discrete case, we have

$$\bar{\mu}_{X+Y}(z) = P\,(X + Y = z) = P\left(\bigcup_{x \in X(\Omega)} ((X, Y) = (x, z - x))\right) =$$

(by the σ-additivity of P)

$$= \sum_{x \in X(\Omega)} \bar{\mu}_{(X,Y)}(x, z - x) =$$

(if X, Y are independent, by (2.3.6))

$$= \sum_{x \in X(\Omega)} \bar{\mu}_X(x) \bar{\mu}_Y(z - x).$$

$\square$

Example 2.6.2 Let X, Y be independent random variables on $(\Omega, \mathcal{F}, P)$ with values in $\mathbb{R}^d$. Proceeding as in the proof of Theorem 2.6.1, we can show that if $X \in AC$ then also $(X + Y) \in AC$ and has density

$$\gamma_{X+Y}(z) = \int_{\mathbb{R}^d} \gamma_X(z - y)\mu_Y(dy), \qquad z \in \mathbb{R}^d. \tag{2.6.4}$$

For example, let $X \sim \mathcal{N}_{\mu,\sigma^2}$ and $Y \sim Be_p$ be independent. Then $X+Y$ is absolutely continuous and, denoting

$$\Gamma_{\mu,\sigma^2}(x) = \frac{1}{\sqrt{2\pi\sigma^2}} e^{-\frac{1}{2}\left(\frac{x-\mu}{\sigma}\right)^2},$$

by (2.6.4), $X + Y$ has density

$$\gamma_{X+Y}(z) = \int_{\mathbb{R}^d} \Gamma_{\mu,\sigma^2}(z - y)Be_p(dy)$$

$$= p\Gamma_{\mu,\sigma^2}(z - 1) + (1 - p)\Gamma_{\mu,\sigma^2}(z)$$

$$= p\Gamma_{\mu+1,\sigma^2}(z) + (1 - p)\Gamma_{\mu,\sigma^2}(z)$$

More generally, if Y is a discrete r.v. with distribution of the form (2.1.4), that is,

$$\sum_{n \geq 1} p_n \delta_{y_n},$$

then $X + Y$ has a density which is a linear combination of Gaussian functions with the same variance and with the poles shifted by y_n:

$$\gamma_{X+Y}(z) = \sum_{n \geq 1} p_n \Gamma_{\mu+y_n,\sigma^2}(z).$$

2.6.2 Examples

Proposition 2.6.3 (Sum of Independent Bernoulli Variables) *Let $(X_i)_{i=1,\ldots,n}$ be a family of independent Bernoulli random variables, $X_i \sim \mathrm{Be}_p$. Then*

$$S := X_1 + \cdots + X_n \sim \mathrm{Bin}_{n,p}. \tag{2.6.5}$$

As a consequence, for $X \sim \mathrm{Bin}_{n,p}$ we have

$$E[X] = np, \qquad \mathrm{var}(X) = np(1-p). \tag{2.6.6}$$

Moreover, if $X \sim \mathrm{Bin}_{n,p}$ and $Y \sim \mathrm{Bin}_{m,p}$ are independent random variables then $X + Y \sim \mathrm{Bin}_{n+m,p}$.

Proof Denoting

$$C_i = (X_i = 1), \qquad i = 1, \ldots, n,$$

we have that $(C_i)_{i=1,\ldots,n}$ is a family of n repeated independent trials with probability p. The r.v. S in (2.6.5) indicates the number of successes among the n trials (as in Example 2.1.7-(iii)) and therefore, as we have already proven, $S \sim \mathrm{Bin}_{n,p}$. Alternatively, one can calculate the distribution function of S as a discrete convolution using (2.6.3), but the calculations are a bit tedious. Formulas (2.6.6) are an immediate consequence of the linearity of the integral and the fact that the variance of independent random variables is equal to the sum of the individual variances (cf. formula (2.3.9)):

$$E[X] = E[S] = nE[X_1] = np, \qquad \mathrm{var}(X) = \mathrm{var}(S) = n\,\mathrm{var}(X_1) = np(1-p).$$

To prove the second part of the statement, let us first consider the case where

$$X = X_1 + \cdots + X_n, \qquad Y = Y_1 + \cdots + Y_m$$

with $X_1, \ldots, X_n, Y_1, \ldots, Y_m \sim \mathrm{Be}_p$ independent. Then, as previously proven, we have

$$X + Y = X_1 + \cdots + X_n + Y_1 + \cdots + Y_m \sim \mathrm{Bin}_{n+m,p}.$$

Now let us consider the general case where $X' \sim \mathrm{Bin}_{n,p}$ and $Y' \sim \mathrm{Bin}_{m,p}$ are independent: then $X' \stackrel{d}{=} X$, $Y' \stackrel{d}{=} Y$ and the thesis follows from (2.6.3) since

$$\bar{\mu}_{X'+Y'} = \bar{\mu}_{X'} * \bar{\mu}_{Y'} = \bar{\mu}_X * \bar{\mu}_Y = \bar{\mu}_{X+Y}.$$

$\square$

Example 2.6.4 (Binomial Model) One of the most classic models used in finance to describe the evolution of the price of a risky asset is the so-called *binomial model*. We introduce a sequence (X_k) of random variables where X_k represents the price of the asset at time k, for $k = 0, 1, \ldots, n$: we assume that $X_0 \in \mathbb{R}_{>0}$ and, given two parameters $0 < d < u$, we recursively define

$$X_k = u^{\alpha_k} d^{1-\alpha_k} X_{k-1}, \qquad k = 1, \ldots, n,$$

where the α_k are independent Bernoulli random variables, $\alpha_k \sim \mathrm{Be}_p$. In the end, we have

$$X_k = \begin{cases} u X_{k-1} & \text{with probability } p, \\ d X_{k-1} & \text{with probability } 1 - p, \end{cases}$$

and

$$X_n = u^{Y_n} d^{n-Y_n} S_0$$

where $Y_n = \sum_{k=1}^{n} \alpha_k \sim \mathrm{Bin}_{n,p}$ by Proposition 2.6.3. Then we have

$$P(X_n = u^k d^{n-k} X_0) = P(Y_n = k) = \binom{n}{k} p^k (1-p)^{n-k}, \qquad k = 0, \ldots, n,$$

which are the probabilities of the possible prices at time n. Figure 2.9 depicts one of the possible paths (or trajectories) of the binomial asset price.

Example 2.6.5 (Sum of Independent Poisson Variables) Let $\lambda_1, \lambda_2 > 0$ and $X_1 \sim \mathrm{Poisson}_{\lambda_1}$, $X_2 \sim \mathrm{Poisson}_{\lambda_2}$ be independent. Then $X_1 + X_2 \sim \mathrm{Poisson}_{\lambda_1+\lambda_2}$. Indeed, if $\bar{\mu}_1, \bar{\mu}_2$ are the distribution functions of X_1, X_2, by Theorem 2.6.1 we have

$$\bar{\mu}_{X_1+X_2}(n) = (\bar{\mu}_1 * \bar{\mu}_2)(n) = \sum_{k=0}^{n} \bar{\mu}_1(k)\bar{\mu}_2(n-k) =$$

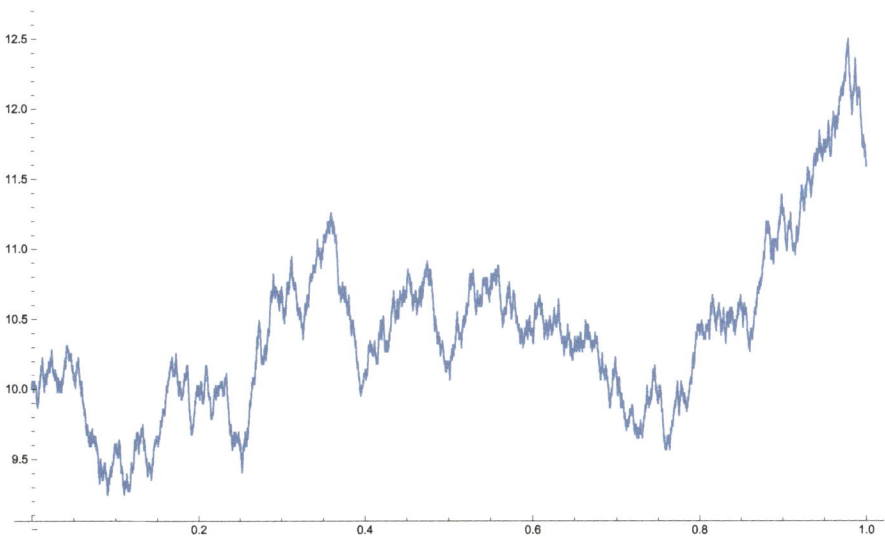

Fig. 2.9 Plot of a trajectory of the binomial process

(the limits in which k varies in the summation are determined by the fact that $\bar\mu_1(k) \neq 0$ only if $k \in \mathbb{N}_0$ and $\bar\mu_2(n-k) \neq 0$ only if $n - k \in \mathbb{N}_0$)

$$= \sum_{k=0}^{n} e^{-\lambda_1} \frac{\lambda_1^k}{k!} e^{-\lambda_2} \frac{\lambda_2^{n-k}}{(n-k)!} = \frac{e^{-\lambda_1-\lambda_2}}{n!} \sum_{k=0}^{n} \binom{n}{k} \lambda_1^k \lambda_2^{n-k} = \frac{e^{-(\lambda_1+\lambda_2)}}{n!} (\lambda_1 + \lambda_2)^n.$$

Example 2.6.6 (Sum of Independent Normal Variables) If $X \sim \mathcal{N}_{\mu,\sigma^2}$ and $Y \sim \mathcal{N}_{\nu,\delta^2}$ are independent real random variables, then

$$X + Y \sim \mathcal{N}_{\mu+\nu,\sigma^2+\delta^2}.$$

Indeed, by (2.6.2) and setting

$$\gamma_{\mu,\sigma^2}(x) := \frac{1}{\sigma\sqrt{2\pi}} e^{-\frac{1}{2}\left(\frac{x-\mu}{\sigma}\right)^2}, \qquad x \in \mathbb{R},$$

a direct calculation shows that

$$\gamma_{\mu,\sigma^2} * \gamma_{\nu,\delta^2} = \gamma_{\mu+\nu,\sigma^2+\delta^2}.$$

Example 2.6.7 (Chi-Square Distribution with n Degrees of Freedom) Another consequence of Theorem 2.6.1 is that, if $X \sim \mathrm{Gamma}_{\alpha,\lambda}$ and $Y \sim \mathrm{Gamma}_{\beta,\lambda}$ are independent real random variables, then

$$X + Y \sim \mathrm{Gamma}_{\alpha+\beta,\lambda}. \qquad (2.6.7)$$

As a particular case, we have that if $X, Y \sim \mathrm{Exp}_\lambda = \mathrm{Gamma}_{1,\lambda}$ are independent random variables, then

$$X + Y \sim \mathrm{Gamma}_{2,\lambda}$$

with density $\gamma_{X+Y}(t) = \lambda^2 t e^{-\lambda t} \mathbb{1}_{\mathbb{R}_{>0}}(t)$.

In Example 2.1.38 we introduced the *chi-square* distribution $\chi^2 := \mathrm{Gamma}_{\frac{1}{2},\frac{1}{2}}$ which is the distribution of the r.v. X^2 where $X \sim \mathcal{N}_{0,1}$ is a standard normal r.v. More generally, given $X_1, \dots, X_n$ independent random variables with distribution $\mathcal{N}_{0,1}$ then by (2.6.7) we have

$$Z := X_1^2 + \cdots + X_n^2 \sim \Gamma_{\frac{n}{2},\frac{1}{2}}. \qquad (2.6.8)$$

Random variables of type (2.6.8) appear in many applications and in particular in statistics (see, for example, Chapter 8 in [8]). The distribution $\Gamma_{\frac{n}{2},\frac{1}{2}}$ is called *chi-square distribution with n degrees of freedom* and is denoted by $\chi^2(n)$: thus $Z \sim \chi^2(n)$ if it has density

$$\gamma_n(x) = \frac{1}{2^{\frac{n}{2}} \Gamma\left(\frac{n}{2}\right)} \frac{e^{-\frac{x}{2}}}{x^{1-\frac{n}{2}}} \mathbb{1}_{\mathbb{R}_{>0}}(x). \qquad (2.6.9)$$

Notice that γ_n in (2.6.9) is a density if n is any positive real number, without the necessity of it being an integer.

Example 2.6.8 Consider the r.v. Z that represents the "sum of the roll of two dice". The random variables that indicate the result of the roll of each of the two dice have a uniform distribution Unif_6 and are independent. Then if $\bar{\mu}$ indicates the distribution function of Unif_6, i.e., $\bar{\mu}(n) = \frac{1}{6}$ for $n \in I_6 = \{1, \dots, 6\}$, by (2.6.3) the distribution function of Z is given by the convolution $\bar{\mu} * \bar{\mu}$:

$$(\bar{\mu} * \bar{\mu})(n) = \sum_k \bar{\mu}(k)\bar{\mu}(n - k), \qquad 2 \le n \le 12,$$

where, in order for $\bar{\mu}(k)$ and $\bar{\mu}(n - k)$ to be non-null, we must have $k \in I_6$ and $n - k \in I_6$ that is

$$(n - 6) \vee 1 \le k \le (n - 1) \wedge 6.$$

Hence

$$P(Z=n)=(\bar{\mu}*\bar{\mu})(n)=\sum_{k=(n-6)\vee 1}^{(n-1)\wedge 6}\bar{\mu}(k)\bar{\mu}(n-k)=\frac{(n-1)\wedge 6-(n-6)\vee 1+1}{36}.$$

Proposition 2.6.9 (Maximum and Minimum of Independent Variables) *Let* $X_1,\ldots,X_n$ *be independent real random variables. Setting*

$$X = \max\{X_1,\ldots,X_n\} \quad \text{and} \quad Y = \min\{X_1,\ldots,X_n\},$$

we have the following relation between the distribution functions[22]

$$F_X(x) = \prod_{k=1}^{n} F_{X_k}(x), \qquad x \in \mathbb{R}, \tag{2.6.10}$$

$$F_Y(y) = 1 - \prod_{k=1}^{n} \left(1 - F_{X_k}(y)\right), \qquad y \in \mathbb{R}.$$

Proof It is sufficient to observe that

$$(X \le x) = \bigcap_{k=1}^{n}(X_k \le x), \qquad x \in \mathbb{R},$$

and therefore, exploiting the independence assumption,

$$F_X(x) = P(X \le x) = P\left(\bigcap_{k=1}^{n}(X_k \le x)\right) = \prod_{k=1}^{n} P(X_k \le x) = \prod_{k=1}^{n} F_{X_k}(x).$$

For the second identity, we proceed in a similar way using the relation

$$(Y > x) = \bigcap_{k=1}^{n}(X_k > x), \qquad x \in \mathbb{R}.$$

$\square$

Example 2.6.10 If $X_k \sim \text{Exp}_{\lambda_k}$, $k = 1,\ldots,n$, are independent random variables, then

$$Y := \min\{X_1,\ldots,X_n\} \sim \text{Exp}_{\lambda_1+\cdots+\lambda_n}.$$

---

[22] Be careful not to confuse (2.6.10) and (2.3.8)!

In fact, we recall that the density and distribution functions of the Exp_λ are respectively

$$\gamma(t) = \lambda e^{-\lambda t} \qquad \text{and} \qquad F(t) = 1 - e^{-\lambda t}, \qquad t \geq 0,$$

and are null for $t < 0$. Then, by Proposition 2.6.9, we have that

$$F_Y(t) = 1 - \prod_{k=1}^{n} \left(1 - F_{X_k}(t)\right) = 1 - \prod_{k=1}^{n} e^{-\lambda_k t}, \qquad t \geq 0,$$

which is precisely the CDF of $\text{Exp}_{\lambda_1 + \cdots + \lambda_n}$.

Exercise 2.6.11 Let X be the maximum between the results of rolling two dice. Determine $P(X \geq 4)$.

Solution Consider the independent random variables $X_i \sim \text{Unif}_6$, $i = 1, 2$, of the results of the two dice rolls. Then $X = \max\{X_1, X_2\}$ and we have

$$P(X \geq 4) = 1 - P(X \leq 3) = 1 - F_X(3) =$$

(by Proposition 2.6.9)

$$= 1 - F_{X_1}(3) F_{X_1}(3) =$$

(recalling (1.4.7))

$$= 1 - \frac{3}{6} \cdot \frac{3}{6} = \frac{3}{4}.$$

Exercise 2.6.12 Prove that if $X_i \sim \text{Geom}_{p_i}$, $i = 1, 2$, are independent, then $\min\{X_1, X_2\} \sim \text{Geom}_p$ with $p = p_1 + p_2 - p_1 p_2$. Generalize the result to the case of n independent geometric random variables.

Exercise 2.6.13 Determine the distribution of $\max\{X, Y\}$ and $\min\{X, Y\}$ where X, Y are independent random variables with distribution $X \sim \text{Unif}_{[0,2]}$ and $Y \sim \text{Unif}_{[1,3]}$.

Chapter 3
Sequences of Random Variables

> *The new always happens against the overwhelming odds of statistical laws and their probability, which for all practical, everyday purposes amounts to certainty; the new therefore always appears in the guise of a miracle.*
>
> *Hannah Arendt*

The primary focus of this chapter lies in the study of sequences of random variables. However, addressing the existence and construction of such sequences presents a non-trivial challenge, which we defer to the second volume [34]. Consequently, assuming their existence for the time being, we delve into exploring different concepts of convergence for sequences of random variables and establish some fundamental results, notably the Law of Large Numbers and the Central Limit Theorem. Moreover, we explore diverse applications, one of which is the prominent stochastic numerical technique recognized as the Monte Carlo method.

3.1 Convergence for Sequences of Random Variables

In this section, we summarize and compare various definitions of convergence of sequences of random variables. We consider a probability space $(\Omega, \mathscr{F}, P)$ on which a sequence of random variables $(X_n)_{n \in \mathbb{N}}$ and a r.v. X with values in $\mathbb{R}^d$ are defined:

A. Pascucci, *Probability Theory I*, La Matematica per il 3+2 165, https://doi.org/10.1007/978-3-031-63190-0_3

(i) $(X_n)_{n\in\mathbb{N}}$ converges *almost surely* to X if[1]

$$P\left(\lim_{n\to\infty} X_n = X\right) = 1,$$

that is, if

$$\lim_{n\to\infty} X_n(\omega) = X(\omega)$$

for almost every $\omega \in \Omega$. In this case, we write

$$X_n \xrightarrow{\text{a.s.}} X.$$

(ii) Let $(X_n)_{n\in\mathbb{N}}$ and X be respectively a sequence and a r.v. in $L^p(\Omega, P)$ with $p \geq 1$. We say that $(X_n)_{n\in\mathbb{N}}$ converges to X in L^p if

$$\lim_{n\to\infty} E\left[|X_n - X|^p\right] = 0.$$

In this case, we write

$$X_n \xrightarrow{L^p} X.$$

(iii) (X_n) converges *in probability* to X if, for every $\varepsilon > 0$, we have

$$\lim_{n\to\infty} P\left(|X_n - X| \geq \varepsilon\right) = 0.$$

In this case, we write

$$X_n \xrightarrow{P} X.$$

(iv) (X_n) converges *in distribution* (or *in law* or *weakly*) to X if

$$\lim_{n\to\infty} E\left[f(X_n)\right] = E\left[f(X)\right]$$

---

[1] By Remark 2.1.9, the set

$$\left(\lim_{n\to\infty} X_n = X\right) := \{\omega \in \Omega \mid \lim_{n\to\infty} X_n(\omega) = X(\omega)\}$$

is an event.

for every $f \in bC$ where $bC = bC(\mathbb{R}^d)$ denotes the family of continuous and bounded functions from $\mathbb{R}^d$ to $\mathbb{R}$. In this case, we write

$$X_n \xrightarrow{d} X.$$

Remark 3.1.1 (Weak Convergence of Distributions) Weak convergence does not require that the variables X_n are defined on the same probability space, but depends only on the distributions of the variables themselves. More precisely, we say that a sequence $(\mu_n)_{n \in \mathbb{N}}$ of distributions on $\mathbb{R}^d$ converges *weakly* to the distribution μ and write

$$\mu_n \xrightarrow{d} \mu,$$

if

$$\lim_{n \to \infty} \int_{\mathbb{R}^d} f d\mu_n = \int_{\mathbb{R}^d} f d\mu \qquad \text{for every } f \in bC. \tag{3.1.1}$$

Since

$$E[f(X_n)] = \int_{\mathbb{R}^d} f d\mu_{X_n},$$

the convergence in distribution of $(X_n)_{n \in \mathbb{N}}$ is equivalent to the weak convergence of the sequence $(\mu_{X_n})_{n \in \mathbb{N}}$ of the corresponding distributions: in other words, $X_n \xrightarrow{d}$ X if and only if $\mu_{X_n} \xrightarrow{d} \mu_X$.

Example 3.1.2 [!] Let $(x_n)_{n \in \mathbb{N}}$ be a sequence of real numbers converging to $x \in \mathbb{R}$. Then $\delta_{x_n} \xrightarrow{d} \delta_x$ since, for every $f \in bC$, we have

$$\int_{\mathbb{R}} f d\delta_{x_n} = f(x_n) \xrightarrow[n \to \infty]{} f(x) = \int_{\mathbb{R}} f d\delta_x.$$

However, it is not true that

$$\lim_{n \to \infty} \delta_{x_n}(H) = \delta_x(H)$$

for every $H \in \mathscr{B}$: for example, if $x_n = \frac{1}{n}$ and $H = \mathbb{R}_{>0}$. This clarifies why, in the definition (3.1.1) of convergence of distributions, it is natural to assume $f \in bC$ rather than what might seem like a naive definition, namely $f = \mathbb{1}_H$ for every $H \in \mathscr{B}$.

Example 3.1.3 Consider two sequences of real numbers $(a_n)_{n\in\mathbb{N}}$ and $(\sigma_n)_{n\in\mathbb{N}}$ such that $a_n \longrightarrow a \in \mathbb{R}$ and $0 < \sigma_n \longrightarrow 0$ as $n \to \infty$. If $X_n \sim \mathcal{N}_{a_n,\sigma_n^2}$ then $X_n \xrightarrow{d} X \sim \delta_a$. In fact, for every $f \in bC(\mathbb{R})$, we have

$$E[f(X_n)] = \int_{\mathbb{R}} f \, d\mathcal{N}_{a_n,\sigma_n^2} = \int_{\mathbb{R}} f(x) \frac{1}{\sqrt{2\pi\sigma_n^2}} e^{-\frac{1}{2}\left(\frac{x-a_n}{\sigma_n}\right)^2} dx =$$

(by the change of variables $z = \frac{x-a_n}{\sigma_n\sqrt{2}}$)

$$= \int_{\mathbb{R}} f\left(a_n + z\sigma_n\sqrt{2}\right) \frac{e^{-z^2}}{\sqrt{\pi}} dz,$$

which converges $f(a) = E[f(X)]$ by the dominated convergence theorem.

Further, if the variables X and X_n, for each $n \in \mathbb{N}$, are defined on the same probability space $(\Omega, \mathscr{F}, P)$, we also have convergence in L^2: in fact $X_n, X \in L^2(\Omega, P)$ and we have

$$E\left[|X_n - X|^2\right] \leq 2E\left[|X_n - a_n|^2\right] + 2E\left[|a_n - X|^2\right]$$

$$= 2E\left[|X_n - a_n|^2\right] + 2|a_n - a|^2$$

$$= 2\sigma_n^2 + 2|a_n - a|^2 \xrightarrow[n\to\infty]{} 0.$$

3.1.1 Markov's Inequality

Theorem 3.1.4 (Markov's Inequality) *[!] For every random variable X with values in $\mathbb{R}^d$, and for every $\lambda > 0$ and $p \in [0, +\infty[$, the following inequality, known as* Markov's inequality, *holds true:*

$$P(|X| \geq \lambda) \leq \frac{E[|X|^p]}{\lambda^p}. \tag{3.1.2}$$

In particular, if $Y \in L^2(\Omega, P)$ is a real-valued random variable, Chebyshev's inequality *holds true:*

$$P(|Y - E[Y]| \geq \lambda) \leq \frac{\text{var}(Y)}{\lambda^2}. \tag{3.1.3}$$

Proof As for (3.1.2), if $E[|X|^p] = +\infty$ there is nothing to prove, otherwise by the monotonicity property of expectation we have

$$E\left[|X|^p\right] \geq E\left[|X|^p \mathbb{1}_{(|X|\geq\lambda)}\right] \geq$$

(since $p \geq 0$)

$$\geq \lambda^p E\left[\mathbb{1}_{(|X|\geq\lambda)}\right] = \lambda^p P\left(|X| \geq \lambda\right).$$

(3.1.3) follows from (3.1.2) by setting $p = 2$ and $X = Y - E[Y]$, indeed

$$P\left(|Y - E[Y]| \geq \lambda\right) \leq \frac{E\left[|Y - E[Y]|^2\right]}{\lambda^2} = \frac{\text{var}(Y)}{\lambda^2}.$$

□

Remark 3.1.5 Similarly, the following extension of Markov's inequality can proven. Let X be a r.v. with values in $\mathbb{R}^d$, f be an increasing function on $[0, +\infty[$ and $\lambda > 0$. Then we have

$$P(|X| \geq \lambda)f(\lambda) \leq E[f(|X|)].$$

An interesting example is $f(\lambda) = e^{\alpha\lambda^2}$ with $\alpha > 0$.

The Markov's inequality provides an estimate for the extreme values of X in terms of its L^p norm. Conversely, we have the following

Proposition 3.1.6 *[!] Let X be a r.v. and $f \in C^1(\mathbb{R}_{\geq 0})$ such that $f' \geq 0$ or $f' \in L^1(\mathbb{R}_{\geq 0}, \mu_{|X|})$. Then*

$$E[f(|X|)] = f(0) + \int_0^{+\infty} f'(t)P(|X| \geq t)dt. \tag{3.1.4}$$

Proof We have

$$E[f(|X|)] = \int_0^{+\infty} f(y)\mu_{|X|}(dy) =$$

$$= \int_0^{+\infty} \left(f(0) + \int_0^y f'(t)dt\right)\mu_{|X|}(dy) =$$

(by Fubini's theorem)

$$= f(0) + \int_0^{+\infty} f'(t)\int_t^{+\infty} \mu_{|X|}(dy)dt =$$

$$= f(0) + \int_0^{+\infty} f'(t)P(|X| \geq t)dt.$$

□

Example 3.1.7 For $f(t) = t^p$, $p \geq 1$, from (3.1.4) we have

$$E\left[|X|^p\right] = p \int_0^{+\infty} \lambda^{p-1} P\left(|X| \geq t\right) dt.$$

Consequently, to prove that $X \in L^p(\Omega, P)$ it is sufficient to have a good estimate of $P\left(|X| \geq t\right)$, at least for $t \gg 1$. Similarly, for $f(t) = e^{\alpha t^2}$, $\alpha > 0$, we have

$$E\left[e^{\alpha|X|^2}\right] = 1 + \int_0^{+\infty} 2\alpha t e^{\alpha t^2} P\left(|X| \geq t\right) dt.$$

3.1.2 Relations Between Different Definitions of Convergence

Lemma 3.1.8 *Let $(a_n)_{n\in\mathbb{N}}$ be a sequence in a topological space $(E, \mathscr{T})$. If every subsequence $(a_{n_k})_{k\in\mathbb{N}}$ admits a subsequence $(a_{n_{k_i}})_{i\in\mathbb{N}}$ converging to the same $a \in E$, then also $(a_n)_{n\in\mathbb{N}}$ converges to a.*

Proof By contradiction, if $(a_n)_{n\in\mathbb{N}}$ does not converge to a, then there exists $U \in \mathscr{T}$ such that $a \in U$ and a subsequence $(a_{n_k})_{k\in\mathbb{N}}$ such that $a_{n_k} \notin U$ for every $k \in \mathbb{N}$. In this case, no subsequence of $(a_{n_k})_{k\in\mathbb{N}}$ would converge to a, contradicting the hypothesis. □

The following result summarizes the relations between the various types of convergence of sequences of random variables: these are schematically represented in Fig. 3.1.

Theorem 3.1.9 *Let $(X_n)_{n\in\mathbb{N}}$ be a sequence of random variables and X be a r.v. defined on the same probability space $(\Omega, \mathscr{F}, P)$, with values in $\mathbb{R}^d$. The following implications hold true:*

(i) *if $X_n \xrightarrow{a.s.} X$ then $X_n \xrightarrow{P} X$;*

(ii) *if $X_n \xrightarrow{L^p} X$ for some $p \geq 1$ then $X_n \xrightarrow{P} X$;*

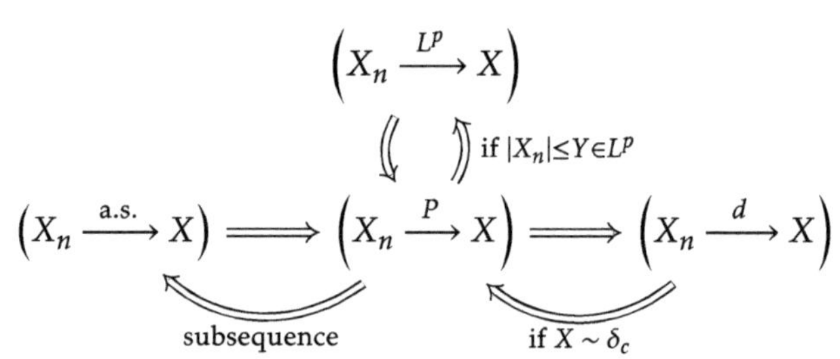

$$\left(X_n \xrightarrow{L^p} X\right)$$

$$\Updownarrow \text{ if } |X_n| \leq Y \in L^p$$

$$\left(X_n \xrightarrow{a.s.} X\right) \implies \left(X_n \xrightarrow{P} X\right) \implies \left(X_n \xrightarrow{d} X\right)$$

$$\underbrace{\qquad}_{\text{subsequence}} \qquad \underbrace{\qquad}_{\text{if } X \sim \delta_c}$$

Fig. 3.1 Relations between the various types of convergence of random variables

(iii) *if* $X_n \xrightarrow{P} X$ *then there exists a subsequence* $(X_{n_k})_{k \in \mathbb{N}}$ *such that* $X_{n_k} \xrightarrow{a.s.} X$;

(iv) *if* $X_n \xrightarrow{P} X$ *then* $X_n \xrightarrow{d} X$;

(v) *if* $X_n \xrightarrow{P} X$ *and there exists* $Y \in L^p(\Omega, P)$ *such that* $|X_n| \leq Y$ *a.s., for every* $n \in \mathbb{N}$, *then* $X_n, X \in L^p(\Omega, P)$ *and* $X_n \xrightarrow{L^p} X$;

(vi) *if* $X_n \xrightarrow{d} X$, *with* $X \sim \delta_c$, $c \in \mathbb{R}^d$, *then* $X_n \xrightarrow{P} X$.

Proof

(i) Fixed $\varepsilon > 0$, if $X_n \xrightarrow{a.s.} X$ then

$$\mathbb{1}_{(|X_n - X| \geq \varepsilon)} \xrightarrow{a.s.} 0$$

and therefore, by the dominated convergence theorem, we have

$$P(|X_n - X| \geq \varepsilon) = E\left[\mathbb{1}_{(|X_n - X| \geq \varepsilon)}\right] \longrightarrow 0.$$

(ii) Fixed $\varepsilon > 0$, by Markov's inequality (3.1.2), we have

$$P(|X_n - X| \geq \varepsilon) \leq \frac{E\left[|X_n - X|^p\right]}{\varepsilon^p}$$

which proves the thesis.

(iii) By assumption, there exists a sequence of indices $(n_k)_{k \in \mathbb{N}}$, with $n_k \to +\infty$, such that $P(A_k) \leq \frac{1}{k^2}$ where

$$A_k := \left(|X - X_{n_k}| \geq 1/k\right).$$

Since

$$\sum_{k \geq 1} P(A_k) < \infty,$$

by Borel-Cantelli's Lemma 1.3.28-(i), we have $P(A_k \text{ i.o.}) = 0$. Hence, the event $(A_k \text{ i.o.})^c$ has probability one: by definition,[2] for each $\omega \in (A_k \text{ i.o.})^c$ there exists $\bar{k} = \bar{k}(\omega) \in \mathbb{N}$ such that

$$|X(\omega) - X_{n_k}(\omega)| < \frac{1}{k}, \qquad k \geq \bar{k}$$

---

[2] The elements of $(A_k \text{ i.o.})^c$ are those that belong to only a finite number of A_k.

and consequently, we have

$$\lim_{k \to \infty} X_{n_k}(\omega) = X(\omega)$$

which proves the thesis.

(iv) Let $f \in bC$. By point iii), every subsequence $(X_{n_k})_{k \in \mathbb{N}}$ admits a subsequence $(X_{n_{k_i}})_{i \in \mathbb{N}}$ such that $X_{n_{k_i}} \xrightarrow{\text{a.s.}} X$. Since f is continuous, we also have $f(X_{n_{k_i}}) \xrightarrow{\text{a.s.}} f(X)$ and since f is bounded, we apply the dominated convergence theorem to obtain

$$\lim_{i \to \infty} E\big[f(X_{n_{k_i}})\big] = E[f(X)].$$

Now, by Lemma 3.1.8 (applied to the sequence $a_n := E[f(X_n)]$ in $\mathbb{R}$ endowed with the Euclidean topology), we also have

$$\lim_{n \to \infty} E[f(X_n)] = E[f(X)]$$

which proves the thesis.

(v) Given that $|X_n| \le Y$ a.s. and $Y \in L^p(\Omega, P)$, it is clear that $X_n \in L^p(\Omega, P)$. As for X, from point iii) we know that there exists a subsequence $(X_{n_k})_{k \in \mathbb{N}}$ such that $X_{n_k} \xrightarrow{\text{a.s.}} X$. Since $|X_{n_k}| \le Y$ a.s., for $k \to \infty$ we obtain $|X| \le Y$ a.s., thus $X \in L^p(\Omega, P)$. Finally, we show that $X_n \xrightarrow{L^p} X$. Again, by point (iii), every subsequence $(X_{n_k})_{k \in \mathbb{N}}$ admits a subsequence $(X_{n_{k_i}})_{i \in \mathbb{N}}$ such that $X_{n_{k_i}} \xrightarrow{\text{a.s.}} X$. By the dominated convergence theorem, we have $X_{n_{k_i}} \xrightarrow{L^p} X$. From Lemma 3.1.8 it follows that $X_n \xrightarrow{L^p} X$.

(vi) Given $c \in \mathbb{R}^d$ and $\varepsilon > 0$, let $f_\varepsilon \in bC$, non-negative and such that $f_\varepsilon(x) \ge 1$ if $|x - c| > \varepsilon$ and $f_\varepsilon(c) = 0$. We have

$$P(|X_n - X| \ge \varepsilon) = P(|X_n - c| \ge \varepsilon)$$
$$= E\big[\mathbb{1}_{(|X_n - c| \ge \varepsilon)}\big] \le E[f_\varepsilon(X_n)] \xrightarrow[n \to \infty]{} f_\varepsilon(c) = 0.$$

$\square$

We provide some counterexamples concerning the implications of Theorem 3.1.9. In the first two examples, we consider the sample space $\Omega = [0, 1]$ equipped with the Lebesgue measure.

Example 3.1.10 The sequence $X_n(\omega) = n^2 \mathbb{1}_{\left[0, \frac{1}{n}\right]}(\omega)$, with $\omega \in [0, 1]$, converges to zero almost surely (and consequently also in probability), but $E[|X_n|^p] = n^{2p-1}$ diverges for every $p \ge 1$.

Example 3.1.11 We give an example of a sequence (X_n) that converges in L^p (and therefore also in probability) with $1 \le p < \infty$, but not almost surely. Represent each positive integer n as $n = 2^k + \ell$, with $k = 0, 1, 2, \ldots$ and $\ell = 0, \ldots, 2^k - 1$. Note that the representation is unique. We set

$$J_n = \left[\frac{\ell}{2^k}, \frac{\ell+1}{2^k}\right] \subseteq [0, 1] \qquad \text{and} \qquad X_n(\omega) = \mathbb{1}_{J_n}(\omega), \qquad \omega \in [0, 1].$$

For every $p \ge 1$, we have

$$E\left[|X_n|^p\right] = E[X_n] = \text{Leb}(J_n) = \frac{1}{2^k},$$

and therefore $X_n \xrightarrow{L^p} 0$ since $k \to \infty$ when $n \to \infty$. On the other hand, each $\omega \in [0, 1]$ belongs to an infinite number of intervals J_n and therefore the real sequence $X_n(\omega)$ does not converge for every $\omega \in [0, 1]$.

Example 3.1.12 Given a r.v. $X \sim \text{Be}_{\frac{1}{2}}$, we set

$$X_n = \begin{cases} X, & \text{if } n \text{ is even}, \\ 1 - X, & \text{if } n \text{ is odd}. \end{cases}$$

Since $(1 - X) \sim \text{Be}_{\frac{1}{2}}$, it is clear that $X_n \xrightarrow{d} X$. However, $|X_{n+1} - X_n| = |2X - 1| = 1$ for every $n \in \mathbb{N}$: therefore, $P(|X_{n+1} - X_n| \ge 1/2) = 1$ for every n and thus X_n does not converge to X in probability (and, consequently, neither in L^p or almost surely).

Remark 3.1.13 There is no metric (nor even a topology) that induces the almost sure convergence of random variables: otherwise, one could combine Lemma 3.1.8 with point (iii) of Theorem 3.1.9 to conclude that if $X_n \xrightarrow{P} X$ then $X_n \xrightarrow{\text{a.s.}} X$, contradicting Example 3.1.11.

On the contrary, the convergences in L^p and in probability are "metrizable". In fact, the convergence in L^p is simply the convergence relative to the norm $\|X\|_p = E\left[|X|^p\right]^{\frac{1}{p}}$ in the space $L^p(\Omega, P)$: it is therefore a type of convergence defined only for integrable variables of order p. Instead, the convergence in probability is defined for any variables and we have that $X_n \xrightarrow{P} X$ if and only if

$$\lim_{n \to \infty} E\left[\frac{|X - X_n|}{1 + |X - X_n|}\right] = 0. \tag{3.1.5}$$

We prove this fact under the (not restrictive) assumption that $X \equiv 0$. We note that for every $\varepsilon > 0$ we have

$$\frac{|x|}{1+|x|} \leq \frac{|x|}{1+|x|} \mathbb{1}_{|x|\geq\varepsilon} + \varepsilon \mathbb{1}_{|x|<\varepsilon} \leq \mathbb{1}_{|x|\geq\varepsilon} + \varepsilon \mathbb{1}_{|x|<\varepsilon}.$$

Applying the expected value, we get

$$E\left[\frac{|X_n|}{1+|X_n|}\right] \leq P(|X_n| \geq \varepsilon) + \varepsilon P(|X_n| < \varepsilon) \leq P(|X_n| \geq \varepsilon) + \varepsilon.$$

Then, when $X_n \xrightarrow{P} 0$, we have

$$\lim_{n\to\infty} E\left[\frac{|X_n|}{1+|X_n|}\right] \leq \varepsilon$$

and (3.1.5) follows from the arbitrariness of ε.

Conversely, we note that

$$\frac{\varepsilon}{1+\varepsilon} \mathbb{1}_{x>\varepsilon} \leq \frac{x}{1+x} \mathbb{1}_{x>\varepsilon} \leq \frac{x}{1+x}$$

and therefore

$$\frac{\varepsilon}{1+\varepsilon} \mathbb{1}_{|X_n|>\varepsilon} \leq \frac{|X_n|}{1+|X_n|}.$$

Applying the expected value, we obtain

$$\frac{\varepsilon}{1+\varepsilon} P(|X_n| > \varepsilon) \leq E\left[\frac{|X_n|}{1+|X_n|}\right]$$

and, by (3.1.5), this implies $X_n \xrightarrow{P} 0$.

Even the weak convergence is metrizable in the space of distributions: for further details see, for example, the monographs [7] and [22].

3.2 Law of Large Numbers

In this section, we prove two versions of the Law of large numbers. This law concerns sequences of real random variables $(X_n)_{n\in\mathbb{N}}$, defined on a probability space $(\Omega, \mathscr{F}, P)$, with the additional assumption that they are *independent and identically distributed* (abbreviated as *i.i.d.*). We denote by

$$S_n = X_1 + \cdots + X_n, \qquad M_n = \frac{S_n}{n}, \qquad (3.2.1)$$

the sum and the arithmetic mean of $X_1, \ldots, X_n$ respectively.

Theorem 3.2.1 (Weak Law of Large Numbers) *[!!] Let $(X_n)_{n \in \mathbb{N}}$ be a sequence of i.i.d. real random variables in $L^2(\Omega, P)$, with expected value $\mu := E[X_1]$ and variance $\sigma^2 := \text{var}(X_1)$. Then we have*

$$E\left[(M_n - \mu)^2\right] = \frac{\sigma^2}{n} \qquad (3.2.2)$$

and consequently, the arithmetic mean M_n *converges in* $L^2(\Omega, P)$ *norm to the constant r.v. equal to* μ:

$$M_n \xrightarrow{L^2} \mu.$$

Proof By linearity, we have

$$E[M_n] = \frac{1}{n} \sum_{k=1}^{n} E[X_k] = \mu,$$

and therefore

$$E\left[(M_n - \mu)^2\right] = \text{var}(M_n) = \frac{\text{var}(X_1 + \cdots + X_n)}{n^2} =$$

(by independence, recalling (2.2.21))

$$= \frac{\text{var}(X_1) + \cdots + \text{var}(X_n)}{n^2} = \frac{\sigma^2}{n}. \qquad (3.2.3)$$

$\square$

Remark 3.2.2 Combining (3.2.2) with Markov's inequality, we get

$$P(|M_n - \mu| \geq \varepsilon) \leq \frac{\sigma^2}{n\varepsilon^2}, \qquad \varepsilon > 0, \ n \in \mathbb{N},$$

and thus M_n also converges *in probability* to μ. Moreover, from Theorem 3.1.9-(iv) it follows that M_n also converges *weakly*:

$$M_n \xrightarrow{d} \mu.$$

The convergence of M_n in $L^2(\Omega, P)$ implies the almost sure convergence of a subsequence of M_n, by Theorem 3.1.9-(iii). Actually, with some additional effort, it can be proven that the entire sequence M_n converges almost surely.

Theorem 3.2.3 (Strong Law of Large Numbers, [21]) *Under the assumptions of Theorem 3.2.1, we also have*

$$M_n \xrightarrow{\text{a.s.}} \mu.$$

Proof Without loss of generality, we may assume $\mu = 0$. We begin by proving that the subsequence M_{n^2} converges almost surely: indeed, by (3.2.3), we have

$$E\left[\sum_{n=1}^{N} M_{n^2}^2\right] = \sum_{n=1}^{N} E\left[M_{n^2}^2\right] = \sum_{n=1}^{N} \frac{\sigma^2}{n^2}, \qquad N \in \mathbb{N},$$

and by Beppo Levi's theorem

$$E\left[\sum_{n=1}^{\infty} M_{n^2}^2\right] = \sum_{n=1}^{\infty} \frac{\sigma^2}{n^2} < \infty$$

from which

$$M_{n^2} \xrightarrow{\text{a.s.}} 0. \tag{3.2.4}$$

Next, we control *all* the terms of the sequence M_n with terms of the type M_{n^2}. For each $n \in \mathbb{N}$ we denote by $k_n = [\sqrt{n}]$ the integer part of the square root of n, so that

$$k_n^2 \leq n < (k_n + 1)^2.$$

By definition of M_n we have

$$M_n - \frac{k_n^2}{n} M_{k_n^2} = \frac{1}{n} \sum_{k=k_n^2+1}^{n} X_k$$

from which, as by (3.2.3), we have

$$E\left[\left(M_n - \frac{k_n^2}{n} M_{k_n^2}\right)^2\right] = \frac{n - k_n^2}{n^2} \sigma^2 \leq$$

(since $0 \geq n - (k_n + 1)^2 = n - k_n^2 - 2k_n - 1$)

$$\leq \frac{2k_n + 1}{n^2} \sigma^2 \leq \frac{2\sqrt{n} + 1}{n^2} \sigma^2 \leq \frac{3\sigma^2}{n^{\frac{3}{2}}}.$$

Again by Beppo Levi's theorem we have

$$E\left[\sum_{n=1}^{\infty} \left(M_n - \frac{k_n^2}{n} M_{k_n^2} \right)^2 \right] \leq \sum_{n=1}^{\infty} \frac{3\sigma^2}{n^{\frac{3}{2}}} < \infty$$

from which

$$M_n - \frac{k_n^2}{n} M_{k_n^2} \xrightarrow{\text{a.s.}} 0.$$

Now $M_{k_n^2} \xrightarrow{\text{a.s.}} 0$ by (3.2.4) and on the other hand $\frac{k_n^2}{n} \to 1$ as $n \to \infty$: consequently also $M_n \xrightarrow{\text{a.s.}} 0$ and this concludes the proof. $\qquad\square$

Example 3.2.4 (Doubling Strategy)   In the game of roulette, a ball is spun and can land in one of the 37 possible positions. These positions consist of 18 red numbers, 18 black numbers, and one green zero. Consider the game strategy that consists of betting on red (the win is double the bet) and doubling the bet every time you lose. So at the first bet you place 1 (i.e., 2^0) Euro and, in case of loss, at the second bet you place 2 (i.e., 2^1) Euro and so on up to the n-th bet in which, if you have always lost, you place 2^{n-1} Euro. At this point (i.e., at the n-th bet having always lost), the amount played is equal to[3]

$$1 + 2 + \cdots + 2^{n-1} = 2^n - 1,$$

and there are two cases:

(i) you lose and in this case the total loss is equal to $2^n - 1$;
(ii) you win and collect $2 \cdot 2^{n-1}$ Euro. The total balance is then positive and is equal to the difference between the win and the amount played:

$$2^n - (2^n - 1) = 1.$$

The probability of losing n times in a row is equal to p^n, where $p = \frac{19}{37}$ is the probability that the ball will stop on black or green. Consequently, the probability of winning at least once in n bets is equal to $1 - p^n$.

---

[3] Remember that $\sum_{k=0}^{n} a^k = \frac{a^{n+1} - 1}{a - 1}$ for $a \neq 1$.

Now consider the case where we decide to implement the doubling strategy up to a maximum of 10 bets. Specifically, we denote by X the gain/loss that we obtain by playing the doubling and collecting 1 Euro if we win within the tenth bet or losing $2^{10} - 1 = 1023$ Euro in the case of 10 consecutive losses. Then X is a Bernoulli r.v. that takes the values -1023 with probability $p^{10} \approx 0.13\%$ and 1 with probability $1 - p^{10} \approx 99.87\%$. Therefore, by implementing the doubling strategy, we win 1 Euro with high probability in exchange for a significant loss (1023 Euro) in very rare cases.

We might then think of implementing the doubling strategy repeatedly for N times: to understand if it is convenient, we can calculate the expectation

$$E[X] \approx -1023 \cdot \frac{0.13}{100} + 1 \cdot \frac{99.87}{100} \approx -0.3$$

and interpret this result in light of the Law of large numbers. The fact that $E[X]$ is equal to -0.3 means that if $X_1, \ldots, X_N$ indicate the individual gains/losses, then overall

$$X_1 + \cdots + X_N$$

will most likely be close to $-0.3N$. This is due to the fact that the *game is not fair* because of the presence of the zero (green) for which the probability of winning by betting on red is slightly less than $\frac{1}{2}$. Actually, it can be shown that even if it were $p = \frac{1}{2}$, then the doubling strategy, with the constraint of doubling at most n times, would produce an average gain of zero. The study of this type of problems related to gambling is at the origin of a large sector of Probability, the so-called *martingale theory*, which, together with numerous applications, has fundamental and profound theoretical results.

3.2.1   An Overview of the Monte Carlo Method

The Law of large numbers is the basis of a very important probabilistic numerical method known as the Monte Carlo method. In many applications, we are interested in calculating (or at least numerically approximating) the expected value $E[f(X)]$, where X is a random variable in $\mathbb{R}^d$ and $f \in L^2(\mathbb{R}^d, \mu_X)$ (thus $f(X) \in L^2(\Omega, P)$). For example, in the case of $d = 1$, if $X \sim \text{Unif}_{[0,1]}$ and $f \in L^2([0, 1])$, then

$$E[f(X)] = \int_0^1 f(x)dx.$$

Hence, an integral (even in multiple dimensions) admits a probabilistic representation, reducing its calculation to that of an expected value.

Now suppose that $(X_n)_{n\in\mathbb{N}}$ is a sequence of real i.i.d. random variables with the same distribution[4] of X. By the strong Law of large numbers, we have

$$E[f(X)] = \lim_{m\to\infty} \frac{f(X_1) + \cdots + f(X_m)}{m} \qquad \text{a.s.}$$

This result can be translated into "practical" terms as follows. Suppose we can randomly draw a value x_n from the r.v. X_n, for each $n = 1, \ldots, m$ with $m \in \mathbb{N}$ fixed, sufficiently large: we say that $x_n \in \mathbb{R}^d$ is a *realization* or *simulation* of the r.v. X_n. Then an approximation of $E[f(X)]$ is given by the arithmetic mean

$$\frac{1}{m} \sum_{n=1}^{m} f(x_n). \tag{3.2.5}$$

In (3.2.5) $x_1, \ldots, x_m$ represent m *independent realizations (simulations) of* X: in other words, x_n is a number (not a random variable) that is *a particular value of the r.v.* X_n *generated independently of* X_h *for* $h \neq n$. Most scientific computing software packages have *random number generators* for the main distributions (uniform, exponential, normal etc...). In conclusion, *the Monte Carlo method allows to numerically approximate the expected value of a function of a r.v. for which we are able to generate (simulate) random values independently.*

The main advantages over *deterministic* numerical integration methods are the following:

(i) for the convergence of the method *no regularity assumptions are required* on the function f other than integrability;

(ii) the convergence rate of the method is *independent of the dimension d*, with no added complexity in implementation for dimensions greater than one.

The issues of convergence and the estimation of the numerical error of the Monte Carlo method will be briefly discussed in Remark 3.4.7. The Monte Carlo method is useful for solving a range of problems, including partial differential equations and optimization problems. It is especially valuable in scenarios where deterministic algorithms struggle due to the so-called *curse of dimensionality*. At the moment, Monte Carlo is the only known numerical method for solving large-scale problems that typically arise in real applications. There are many monographs dedicated to Monte Carlo, including [17]; a concise presentation of the method can also be found in [33].

Figure 3.2 represents the histogram of a vector of 10,000 random numbers generated by the standard normal distribution $\mathcal{N}_{0,1}$: the figure illustrates how the histogram "approximates" the graph (depicted as the continuous blue line) of the Gaussian density function of $\mathcal{N}_{0,1}$.

---

[4] Usually one says that $(X_n)_{n\in\mathbb{N}}$ is a sequence of independent copies of X.

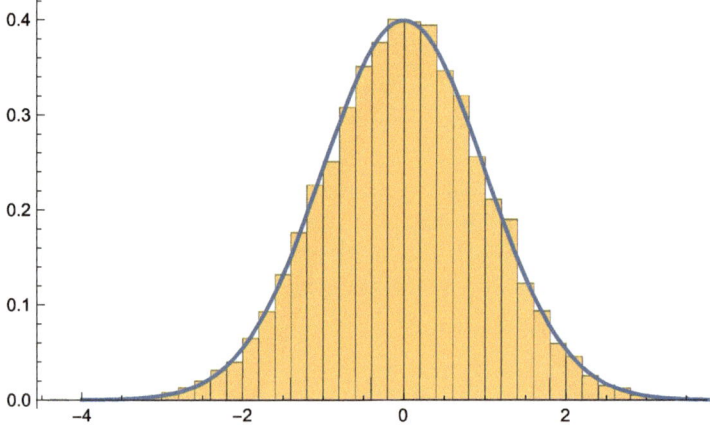

Fig. 3.2 Histogram of 10,000 random numbers drawn from the distribution $\mathcal{N}_{0,1}$ and graph (blue line) of the Gaussian density of $\mathcal{N}_{0,1}$

3.2.2 Bernstein Polynomials

We present a probabilistic proof of the well-known density result of polynomials in the space $C([0, 1])$ of continuous functions on the interval $[0, 1]$, with respect to the uniform norm

$$\|f\|_\infty = \max_{p \in [0,1]} |f(p)|.$$

Proposition 3.2.5 *The* Bernstein polynomial *of degree n of $f \in C([0, 1])$ is defined as*

$$f_n(p) = \sum_{k=0}^{n} \binom{n}{k} p^k (1-p)^{n-k} f(k/n), \qquad p \in [0, 1]. \qquad (3.2.6)$$

We have

$$\lim_{n \to \infty} \|f - f_n\|_\infty = 0.$$

Proof Let $(X_n)_{n \in \mathbb{N}}$ be a sequence of i.i.d. real random variables with distribution Be_p, and $M_n := \frac{X_1 + \cdots + X_n}{n}$. Recall that, by Proposition 2.6.3, $X_1 + \cdots + X_n \sim \text{Bin}_{n,p}$. Then the probabilistic interpretation of formula (3.2.6) is

$$f_n(p) = E[f(M_n)], \qquad p \in [0, 1].$$

Now observe that

$$\text{var}\,(M_n) = \frac{p(1-p)}{n} \le \frac{1}{4n},$$

and since $E\,[M_n] = p$, by Markov's inequality (3.1.3) we have

$$P\,(|M_n - p| \ge \delta) \le \frac{1}{4n\delta^2}, \qquad \delta > 0. \qquad (3.2.7)$$

Since f is *uniformly continuous on* $[0, 1]$, for every $\varepsilon > 0$ there exists δ_ε such that $|f(x) - f(y)| \le \varepsilon$ if $|x - y| \le \delta_\varepsilon$. Then we have

$$|f(p) - f_n(p)| = |f(p) - E\,[f\,(M_n)]| \le$$

(by Jensen's inequality)

$$\le E\,[|f(p) - f\,(M_n)|]$$
$$\le \varepsilon + E\,\Big[|f(p) - f\,(M_n)|\,\mathbb{1}_{(|M_n - p| \ge \delta_\varepsilon)}\Big]$$
$$\le \varepsilon + 2\|f\|_\infty P\,(|M_n - p| \ge \delta_\varepsilon)\,.$$

Using (3.2.7), we obtain

$$\limsup_{n \to \infty} \|f - f_n\|_\infty \le \varepsilon$$

and the thesis follows from the arbitrariness of ε. $\qquad\qquad\qquad\qquad\qquad\Box$

3.3 Necessary and Sufficient Conditions for Weak Convergence

We present two necessary and sufficient conditions for the weak convergence of a sequence $(X_n)_{n \in \mathbb{N}}$ of real random variables: the first condition is expressed in terms of the CDFs $(F_{X_n})_{n \in \mathbb{N}}$, and the second condition in terms of the CHFs $(\varphi_{X_n})_{n \in \mathbb{N}}$.

3.3.1 Convergence of Distribution Functions

Given that each distribution is uniquely characterized by its CDF, it is natural to inquire about the potential connection between weak convergence and pointwise convergence of their respective CDFs. Let us explore this relationship through a couple of straightforward examples.

Fig. 3.3 CDF of the distributions $\text{Unif}_{[0,1]}$ (solid line), $\text{Unif}_{[0,\frac{1}{2}]}$ (dashed line) and $\text{Unif}_{[0,\frac{1}{3}]}$ (dotted line)

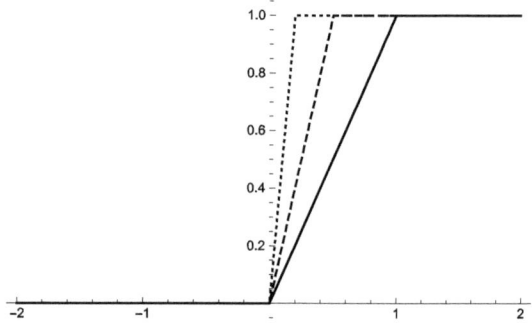

Example 3.3.1 The sequence of uniform distributions $\text{Unif}_{[0,\frac{1}{n}]}$, with $n \in \mathbb{N}$, converges weakly to the Dirac delta δ_0 since, for every $f \in bC$, we have

$$\int_{\mathbb{R}} f \, d\text{Unif}_{[0,\frac{1}{n}]} = n \int_0^{\frac{1}{n}} f(x) \, dx \xrightarrow[n \to \infty]{} f(0) = \int_{\mathbb{R}} f \, d\delta_0.$$

On the other hand, the sequence of CDFs $F_{\text{Unif}_{[0,\frac{1}{n}]}}$, represented in Fig. 3.3, converges pointwise to F_{δ_0} only on $\mathbb{R} \setminus \{0\}$: we note that 0 is the only point of discontinuity of F_{δ_0}.

Example 3.3.2 It is not difficult to check that:

- if $x_n \nearrow x_0$ then $F_{\delta_{x_n}}(x) \longrightarrow F_{\delta_{x_0}}(x)$ for every $x \in \mathbb{R}$;
- if $x_n \searrow x_0$ then $F_{\delta_{x_n}}(x) \longrightarrow F_{\delta_{x_0}}(x)$ for every $x \in \mathbb{R} \setminus \{x_0\}$.

Theorem 3.3.3 *Let* $(\mu_n)_{n \in \mathbb{N}}$ *be a sequence of real distributions and* μ *be a real distribution. The following statements are equivalent:*

(i) $\mu_n \xrightarrow{d} \mu$;
(ii) $F_{\mu_n}(x) \xrightarrow[n \to \infty]{} F_\mu(x)$ *for every* x *point of continuity of* F_μ.

Proof Let us first note that the statement has the following equivalent formulation in terms of random variables: let $(X_n)_{n \in \mathbb{N}}$ be a sequence of real random variables and X be a real random variable. The following statements are equivalent:

(i) $X_n \xrightarrow{d} X$;
(ii) $F_{X_n}(x) \xrightarrow[n \to \infty]{} F_X(x)$ for every x point of continuity of F_X.

[(i) $\Longrightarrow$ (ii)] Fix x, a point of continuity of F_X: then for every $\varepsilon > 0$ there exists $\delta > 0$ such that $|F_X(x) - F_X(y)| \leq \varepsilon$ if $|x - y| \leq \delta$. Let $f \in bC$ be such that $|f| \leq 1$ and

$$f(y) = \begin{cases} 1 & \text{for } y \leq x, \\ 0 & \text{for } y \geq x + \delta. \end{cases}$$

Note that

$$E\left[f(X_n)\right] \geq E\left[f(X_n)\mathbb{1}_{(X_n \leq x)}\right] = P(X_n \leq x) = F_{X_n}(x).$$

Then we have

$$\limsup_{n \to \infty} F_{X_n}(x) \leq \limsup_{n \to \infty} E\left[f(X_n)\right] =$$

(by assumption, since $X_n \xrightarrow{d} X$)

$$= E\left[f(X)\right] \leq F_X(x + \delta) \leq F_X(x) + \varepsilon.$$

Similarly, if $f \in bC$ is such that $|f| \leq 1$ and

$$f(y) = \begin{cases} 1 & \text{for } y \leq x - \delta, \\ 0 & \text{for } y \geq x, \end{cases}$$

then

$$E\left[f(X_n)\right] \leq E\left[\mathbb{1}_{\{X_n \leq x\}}\right] = F_{X_n}(x).$$

So we have

$$\liminf_{n \to \infty} F_{X_n}(x) \geq \liminf_{n \to \infty} E\left[f(X_n)\right] =$$

(by assumption)

$$= E\left[f(X)\right] \geq F_X(x - \delta) \geq F_X(x) - \varepsilon.$$

The thesis follows from the arbitrariness of ε.

[(ii) $\Longrightarrow$ (i)] Given a, b points of continuity of F_X, by assumption we have

$$E\left[\mathbb{1}_{]a,b]}(X_n)\right] = F_{X_n}(b) - F_{X_n}(a) \xrightarrow[n \to \infty]{} F_X(b) - F_X(a) = E\left[\mathbb{1}_{]a,b]}(X)\right].$$

Fix $R > 0$ and $f \in bC$ with support contained in the compact interval $[-R, R]$. Since the points of discontinuity of F_X are at most countably infinite, f can be approximated *uniformly* (in L^∞ norm) by linear combinations of functions of the type $\mathbb{1}_{]a,b]}$ with a, b points of continuity of F_X. It follows that for such f we have

$$\lim_{n \to \infty} E\left[f(X_n)\right] = E\left[f(X)\right].$$

Finally, fix $\varepsilon > 0$ and consider R large enough so that $F_X(-R) \leq \varepsilon$ and $F_X(R) \geq 1 - \varepsilon$: also assume that R and $-R$ are points of continuity of F_X. Then for every $f \in bC$ we have

$$E\left[f(X_n) - f(X)\right] = J_{1,n} + J_{2,n} + J_3$$

where

$$J_{1,n} = E\left[f(X_n)\mathbb{1}_{]-R,R]}(X_n)\right] - E\left[f(X)\mathbb{1}_{]-R,R]}(X)\right],$$
$$J_{2,n} = E\left[f(X_n)\mathbb{1}_{]-R,R]^c}(X_n)\right],$$
$$J_3 = -E\left[f(X)\mathbb{1}_{]-R,R]^c}(X)\right].$$

Now, as proven above, we have

$$\lim_{n\to\infty} J_{1,n} = 0$$

while, by assumption,

$$\left|J_{2,n}\right| \leq \|f\|_\infty \left(F_{X_n}(-R) + (1 - F_{X_n}(R))\right) \xrightarrow[n\to\infty]{} \|f\|_\infty \left(F_X(-R)\right.$$
$$+ (1 - F_X(R)))$$
$$\leq 2\varepsilon\|f\|_\infty,$$

and

$$|J_3| \leq \|f\|_\infty \left(F_X(-R) + (1 - F_X(R))\right) \leq 2\varepsilon\|f\|_\infty.$$

This concludes the proof. $\square$

It is not sufficient that the CDFs F_{μ_n} converge to a continuous function to conclude that μ_n converges weakly, as shown by the following

Example 3.3.4 The sequence of Dirac deltas δ_n does not converge weakly, however

$$F_{\delta_n}(x) = \mathbb{1}_{[n,+\infty[}(x) \xrightarrow[n\to\infty]{} 0, \qquad x \in \mathbb{R},$$

that is, F_{δ_n} converges pointwise to the identically null function which, obviously, is continuous on $\mathbb{R}$ (but is not a CDF).

Example 3.3.4 does not contradict Theorem 3.3.3 since the limit function of the F_{δ_n} is not a distribution function. This example also shows that it is possible for a sequence of CDFs to converge to a function that is not a CDF.

3.3.2 Compactness in the Space of Distributions

In this section, we introduce the property of *tightness* which provides a characterization of *relative compactness* in the space of distributions: it guarantees that from a sequence of distributions one can extract a subsequence converging weakly. In particular, tightness avoids situations like that of Example 3.3.4.

Definition 3.3.5 (Tightness) A family of real distributions $(\mu_i)_{i \in I}$ is *tight* if for every $\epsilon > 0$ there exists $M > 0$ such that

$$\mu_i \big(] - \infty, -M] \cup [M, +\infty[\big) \leq \epsilon \quad \text{for every } i \in I.$$

Exercise 3.3.6 Prove that the family consisting of a single real distribution is tight.[5]

The tightness property can also be attributed to families of random variables $(X_i)_{i \in I}$ or of CDFs $(F_i)_{i \in I}$: they are tight if their corresponding families of distributions are tight, that is

$$P(|X_i| \geq M) \leq \varepsilon \quad \text{for every } i \in I,$$

and

$$F_i(-M) \leq \varepsilon, \qquad F_i(M) \geq 1 - \varepsilon \quad \text{for every } i \in I.$$

Theorem 3.3.7 (Helly's Theorem) *[!!]* *Every tight sequence of real distributions* $(\mu_n)_{n \in \mathbb{N}}$ *admits a subsequence weakly convergent to a distribution* μ.

Proof Let $(\mu_n)_{n \in \mathbb{N}}$ be a tight sequence of distributions and let $(F_n)_{n \in \mathbb{N}}$ be the sequence of their corresponding CDFs. Based on Theorem 3.3.3, it is sufficient to prove that there exists a CDF F and a subsequence F_{n_k} that converges to F at the points of continuity of F.

The construction of F is based on Cantor's diagonal argument. We consider an enumeration $(q_h)_{h \in \mathbb{N}}$ of the rational numbers. Since $(F_n(q_1))_{n \in \mathbb{N}}$ is a sequence in $[0, 1]$, it admits a subsequence $\big(F_{1,n}(q_1)\big)_{n \in \mathbb{N}}$ converging to a value that we denote by $F(q_1) \in [0, 1]$. Now $\big(F_{1,n}(q_2)\big)_{n \in \mathbb{N}}$ is a sequence in $[0, 1]$ that admits a subsequence $\big(F_{2,n}(q_2)\big)_{n \in \mathbb{N}}$ converging to a value that we denote by $F(q_2) \in [0, 1]$: note that we also have

$$F_{2,n}(q_1) \xrightarrow[n \to \infty]{} F(q_1)$$

---

[5] More generally, every distribution μ on a *separable and complete* metric space $(\mathbb{M}, \rho)$, is tight in the following sense: for every $\epsilon > 0$ there exists a compact set K such that $\mu(\mathbb{M} \setminus K) < \epsilon$. For the proof, see for instance Theorem 1.4 in [7].

since $F_{2,n}$ is a subsequence of $F_{1,n}$. We repeat the argument until we construct, for each $k \in \mathbb{N}$, a sequence $(F_{k,n})_{n\in\mathbb{N}}$ such that

$$F_{k,n}(q_h) \xrightarrow[n\to\infty]{} F(q_h), \qquad \forall h \leq k.$$

Using the diagonal argument, we consider the subsequence $F_{n_k} := F_{k,k}$ that is such that

$$F_{n_k}(q) \xrightarrow[n\to\infty]{} F(q), \qquad q \in \mathbb{Q}.$$

We complete the definition of F by setting

$$F(x) := \inf_{x<q\in\mathbb{Q}} F(q), \qquad x \in \mathbb{R} \setminus \mathbb{Q}.$$

By construction, F takes values in $[0, 1]$, is increasing and right continuous. To prove that F is a distribution function, it remains to verify that

$$\lim_{x\to-\infty} F(x) = 0, \qquad \lim_{x\to+\infty} F(x) = 1. \qquad (3.3.1)$$

Only at this point[6] and only to prove (3.3.1), we use the hypothesis that $(F_n)_{n\in\mathbb{N}}$ is a *tight* sequence: given $\varepsilon > 0$, there exists M (it is not restrictive to assume $M \in \mathbb{Q}$) such that $F_{n_k}(-M) \leq \varepsilon$ for every $k \in \mathbb{N}$. Therefore, for every $x \leq -M$, we have

$$F(x) \leq F(-M) = \lim_{k\to\infty} F_{n_k}(-M) \leq \varepsilon.$$

Similarly, for every $x \geq M$, we have

$$1 \geq F(x) \geq F(M) = \lim_{k\to\infty} F_{n_k}(M) \geq 1 - \varepsilon.$$

(3.3.1) follows from the arbitrariness of ε.

Finally, we conclude by proving that F_{n_k} converges to F at its points of continuity. In fact, if F is continuous at x then for every $\varepsilon > 0$ there exist $a, b \in \mathbb{Q}$ such that $a < x < b$ and

$$F(x) - \varepsilon \leq F(y) \leq F(x) + \varepsilon, \qquad y \in [a, b].$$

---

[6] Think back to the sequence of Example 3.3.4, defined by $X_n \equiv n$ for $n \in \mathbb{N}$: it does not admit weakly convergent subsequences and yet we have $\lim_{n\to\infty} F_{X_n}(x) = F(x) \equiv 0$ for every $x \in \mathbb{R}$. Indeed, $(X_n)_{n\in\mathbb{N}}$ is not a tight sequence of random variables.

Then we have

$$\liminf_{k\to\infty} F_{n_k}(x) \geq \liminf_{k\to\infty} F_{n_k}(a) = F(a) \geq F(x) - \varepsilon,$$

$$\limsup_{k\to\infty} F_{n_k}(x) \leq \limsup_{k\to\infty} F_{n_k}(b) = F(b) \leq F(x) + \varepsilon,$$

which proves the thesis due to the arbitrariness of ε.   □

3.3.3 Convergence of Characteristic Functions and Lévy's Continuity Theorem

In this section, we explore the connection between weak convergence of distributions and the pointwise convergence of their respective characteristic functions. We focus on the case where $d = 1$, although the principles discussed can be readily extended to higher dimensions.

Theorem 3.3.8 (Lévy's Continuity Theorem) *[!!] Let $(\mu_n)_{n\in\mathbb{N}}$ be a sequence of real distributions and let $(\varphi_n)_{n\in\mathbb{N}}$ be the sequence of the corresponding characteristic functions. Then we have:*

(i) *if $\mu_n \xrightarrow{d} \mu$ then φ_n converges pointwise to the CHF φ of μ, that is $\varphi_n(\eta) \xrightarrow[n\to\infty]{} \varphi(\eta)$ for every $\eta \in \mathbb{R}$;*

(ii) *conversely, if φ_n converges pointwise to a function φ continuous at 0, then φ is the CHF of a distribution μ and $\mu_n \xrightarrow{d} \mu$.*

Proof

(i) For every fixed η, the function $f(x) := e^{ix\eta}$ is continuous and bounded: therefore, if $\mu_n \xrightarrow{d} \mu$ then

$$\varphi_n(\eta) = \int_{\mathbb{R}} f \, d\mu_n \xrightarrow[n\to\infty]{} \int_{\mathbb{R}} f \, d\mu = \varphi(\eta).$$

(ii) We prove that if φ_n converges pointwise to φ, with φ being a continuous function at 0, then $(\mu_n)_{n\in\mathbb{N}}$ is tight. We observe that $\varphi(0) = 1$ and, due to the continuity of φ at 0, we have

$$\frac{1}{t} \int_{-t}^{t} (1 - \varphi(\eta)) \, d\eta \xrightarrow[t\to0^+]{} 0. \tag{3.3.2}$$

Now let $t > 0$: we have

$$J_1(x, t) := \int_{-t}^{t} \left(1 - e^{i\eta x}\right) d\eta = 2t - \int_{-t}^{t} (\cos(x\eta) + i \sin(x\eta)) \, d\eta$$

$$= 2t - \frac{2 \sin(xt)}{xt} =: J_2(x, t).$$

We note that $J_2(x, t) \geq 0$ since

$$|\sin x| = \left| \int_{0}^{x} \cos t \, dt \right| \leq |x|.$$

Therefore, integrating with respect to μ_n, on one hand we have

$$\int_{\mathbb{R}} J_2(x, t) \mu_n(dx) \geq \int_{t|x| \geq 2} J_2(x, t) \mu_n(dx) \geq$$

(since $\left| \frac{\sin(tx)}{tx} \right| \leq \frac{1}{t|x|} \leq \frac{1}{2}$ if $t|x| \geq 2$)

$$\geq \int_{t|x| \geq 2} \mu_n(dx) = \mu_n \left(\left] -\infty, -\frac{2}{t} \right] \cup \left[\frac{2}{t}, +\infty \right[\right). \qquad (3.3.3)$$

On the other hand, by Fubini's and the dominated convergence theorems we have

$$\int_{\mathbb{R}} J_1(x, t) \mu_n(dx) = \frac{1}{t} \int_{-t}^{t} (1 - \varphi_n(\eta)) \xrightarrow[n \to \infty]{} \frac{1}{t} \int_{-t}^{t} (1 - \varphi(\eta)) \, d\eta.$$

From (3.3.2) it follows that, for every $\varepsilon > 0$, there exist $t > 0$ and $\bar{n} = \bar{n}(\varepsilon, t) \in \mathbb{N}$ such that

$$\left| \int_{\mathbb{R}} J_1(x, t) \mu_n(dx) \right| \leq \varepsilon, \qquad n \geq \bar{n}.$$

Combining this estimate with (3.3.3), we conclude that

$$\mu_n \left(\left] -\infty, -\frac{2}{t} \right] \cup \left[\frac{2}{t}, +\infty \right[\right) \leq \varepsilon, \qquad n \geq \bar{n},$$

and therefore $(\mu_n)_{n \in \mathbb{N}}$ is tight.

We just proved that any subsequence μ_{n_k} is tight and therefore, by Helly's theorem, admits a further subsequence $\mu_{n_{k_j}}$ that converges weakly to a distribution μ. By point i), $\varphi_{n_{k_j}}$ converges pointwise to the CHF of μ: on the other hand, by assumption, $\varphi_{n_{k_j}}$ converges pointwise to φ and therefore φ is the CHF of μ. In summary, every subsequence μ_{n_k} admits a subsequence that converges weakly to the distribution μ that has CHF equal to φ.

Let now $f \in bC$: as just proved, every subsequence of $\int_{\mathbb{R}} f d\mu_n$ admits a subsequence that converges to $\int_{\mathbb{R}} f d\mu$. By Lemma 3.1.8, $\int_{\mathbb{R}} f d\mu_n$ converges to $\int_{\mathbb{R}} f d\mu$ and the thesis follows from the arbitrariness of f.   □

Example 3.3.9 The continuity hypothesis at 0 of Lévy's theorem is necessary. In fact, consider $X_n \sim \mathcal{N}_{0,n}$ with $n \in \mathbb{N}$. Then

$$\varphi_{X_n}(\eta) = e^{-\frac{n\eta^2}{2}}$$

converges to zero as $n \to \infty$ for every $\eta \neq 0$ and $\varphi_{X_n}(0) = 1$. On the other hand, for every $x \in \mathbb{R}$ we have

$$F_{X_n}(x) = \int_{-\infty}^{x} \frac{1}{\sqrt{2\pi n}} e^{-\frac{y^2}{2n}} dy =$$

(setting $z = \frac{y}{\sqrt{2n}}$)

$$= \int_{-\infty}^{\frac{x}{\sqrt{2n}}} \frac{1}{\sqrt{\pi}} e^{-z^2} dz \xrightarrow[n\to\infty]{} \frac{1}{2},$$

and therefore, by Theorem 3.3.3, X_n does not converge weakly.

3.3.4 Examples of Weak Convergence

In this section we exhibit some remarkable examples of weak convergence. We will see sequences of discrete random variables that converge to absolutely continuous random variables and, conversely, sequences of absolutely continuous that converge to discrete random variables. Weak convergence is established through Lévy's continuity theorem, that is, by studying the pointwise convergence of the corresponding sequence of characteristic functions.

Example 3.3.10 (From Geometric to Exponential) Consider a sequence of random variables with geometric distribution

$$X_n \sim \text{Geom}_{p_n}, \qquad n \in \mathbb{N},$$

where $0 < p_n < 1$, so that

$$P(X_n = k) = p_n (1 - p_n)^{k-1}, \qquad k \in \mathbb{N}.$$

The CHF of X_n is easily calculated:

$$\varphi_{X_n}(\eta) = \sum_{k=1}^{\infty} e^{i\eta k} p_n (1-p_n)^{k-1} = e^{i\eta} p_n \sum_{k=1}^{\infty} \left(e^{i\eta}(1-p_n)\right)^{k-1}$$

$$= \frac{e^{i\eta} p_n}{1 - e^{i\eta}(1-p_n)} = \frac{p_n}{e^{-i\eta} - 1 + p_n}.$$

Now let us verify that if $np_n \xrightarrow[n\to\infty]{} \lambda$ for a certain $\lambda \in \mathbb{R}_{>0}$ then $\frac{X_n}{n} \xrightarrow{d} X \sim$ Exp_λ. In fact, we have

$$\varphi_{\frac{X_n}{n}}(\eta) = E\left[e^{i\eta \frac{X_n}{n}}\right] = \varphi_{X_n}\left(\frac{\eta}{n}\right) = \frac{p_n}{e^{-i\frac{\eta}{n}} - 1 + p_n} =$$

(expanding the exponential in Taylor series, as $n \to \infty$)

$$= \frac{p_n}{-i\frac{\eta}{n} + o\left(\frac{1}{n}\right) + p_n} = \frac{np_n}{-i\eta + o(1) + np_n} \xrightarrow[n\to\infty]{} \frac{\lambda}{\lambda - i\eta} = \varphi_{\mathrm{Exp}_\lambda}(\eta).$$

Example 3.3.11 (From Normal to Dirac) Let us take up Example 3.1.3 and consider a sequence $(X_n)_{n\in\mathbb{N}}$ of random variables with normal distribution $X_n \sim \mathcal{N}_{a_n, \sigma_n^2}$ where $a_n \longrightarrow a \in \mathbb{R}$ and $\sigma_n \longrightarrow 0$. Thanks to Lévy's continuity theorem, we can easily confirm that $X_n \xrightarrow{d} X \sim \delta_a$, that is, X_n converges weakly to a r.v. with Dirac delta distribution centered at a. In fact, we have

$$\varphi_{X_n}(\eta) = e^{ia_n\eta - \frac{\eta^2 \sigma_n^2}{2}} \xrightarrow[n\to\infty]{} e^{ia\eta}, \qquad \eta \in \mathbb{R}.$$

Example 3.3.12 (From Binomial to Poisson) Consider a sequence of random variables with binomial distribution

$$X_n \sim \mathrm{Bin}_{n, p_n}, \qquad n \in \mathbb{N}.$$

If $np_n \xrightarrow[n\to\infty]{} \lambda$ for some $\lambda \in \mathbb{R}_{>0}$ then $X_n \xrightarrow{d} X \sim \mathrm{Poisson}_\lambda$: in fact, by (2.5.3) and Lemma 3.4.1, we have

$$\varphi_{X_n}(\eta) = \left(1 + p_n\left(e^{i\eta} - 1\right)\right)^n = \left(1 + \frac{np_n}{n}\left(e^{i\eta} - 1\right)\right)^n \xrightarrow[n\to\infty]{} e^{\lambda(e^{i\eta} - 1)}$$

$$= \varphi_{\mathrm{Poisson}_\lambda}(\eta).$$

Example 3.3.13 (From Binomial to Normal) Let $X_n \sim \mathrm{Bin}_{n,p}$. Recall (cf. Proposition 2.6.3) that the distribution of X_n coincides with the distribution of the sum of n independent Bernoulli random variables. We anticipate that, by the Central

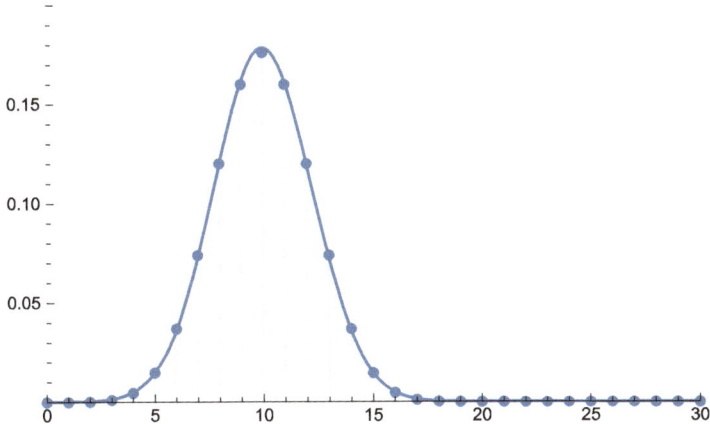

Fig. 3.4 Plots of the density of the normal distribution $\mathcal{N}_{np,np(1-p)}$ and of the binomial distribution function $\mathrm{Bin}_{n,p}$ for $p = 0.5$ and $n = 20$

limit theorem (Theorem 3.4.4, which we will prove shortly as a direct consequence of Lévy's continuity theorem), we have

$$Z_n \xrightarrow{\ d\ } X \sim \mathcal{N}_{0,1},$$

where

$$Z_n = \frac{X_n - \mu_n}{\sigma_n}, \qquad \mu_n = E[X_n] = np, \quad \sigma_n^2 = \mathrm{var}(X_n) = np(1-p).$$

This result can be informally expressed by saying that for every $p \in\,]0, 1[$, the distribution $\mathcal{N}_{np,np(1-p)}$ is a good approximation of $\mathrm{Bin}_{n,p}$ for n large enough: see for example Fig. 3.4 for a comparison between the graphs of the normal density of $\mathcal{N}_{np,np(1-p)}$ and the binomial distribution function $\mathrm{Bin}_{n,p}$, for $p = 0.5$ and $n = 20$. This result will be resumed and explained with greater precision in Remark 3.4.8.

3.4   Law of Large Numbers and Central Limit Theorem

We present a unified approach to the proofs of the weak Law of large numbers and the Central limit theorem. This approach is based on Lévy's continuity theorem and Theorem 2.5.20 on the Taylor series expansion of the characteristic function. The name *Central limit theorem* was first used by the Hungarian mathematician George Pólya in his paper [37], to emphasize how this theorem has a *central* role in probability. The book [16] examines the historical development of the central limit theorem and other related probabilistic limit theorems spanning the period from approximately 1810 to 1950.

We recall the notation

$$S_n = X_1 + \cdots + X_n, \qquad M_n = \frac{S_n}{n} \qquad (3.4.1)$$

for the sum and the arithmetic average of the random variables $X_1, \ldots, X_n$, respectively. The following result, well-known in the case of real sequences, holds true.

Lemma 3.4.1 *Let $(z_n)_{n \in \mathbb{N}}$ be a sequence of complex numbers converging to $z \in \mathbb{C}$. Then we have*

$$\lim_{n \to \infty} \left(1 + \frac{z_n}{n}\right)^n = e^z.$$

Proof We follow the proof of [12], Theorem 3.4.2. First, we prove that for every $w_1, \ldots, w_n, \zeta_1, \ldots, \zeta_n \in \mathbb{C}$, with modulus less than or equal to $c > 0$, we have

$$\left| \prod_{k=1}^{n} w_k - \prod_{k=1}^{n} \zeta_k \right| \le c^{n-1} \sum_{k=1}^{n} |w_k - \zeta_k|. \qquad (3.4.2)$$

(3.4.2) is true for $n = 1$ and in general can be proven by induction observing that

$$\left| \prod_{k=1}^{n} w_k - \prod_{k=1}^{n} \zeta_k \right| \le \left| w_n \prod_{k=1}^{n-1} w_k - z_n \prod_{k=1}^{n-1} \zeta_k \right| + \left| w_n \prod_{k=1}^{n-1} \zeta_k - \zeta_n \prod_{k=1}^{n-1} \zeta_k \right|$$

$$\le c \left| \prod_{k=1}^{n-1} w_k - \prod_{k=1}^{n-1} \zeta_k \right| + c^{n-1} |w_n - \zeta_n|.$$

Then we notice that for every $w \in \mathbb{C}$ with $|w| \le 1$ we have $|e^w - (1 + w)| \le |w|^2$ since

$$\left| e^w - (1 + w) \right| = \left| \sum_{k \ge 0} \frac{w^k}{k!} - (1 + w) \right| \le \sum_{k \ge 2} \frac{|w|^k}{k!} = |w|^2 \sum_{k \ge 2} \frac{1}{k!} \le |w|^2. \qquad (3.4.3)$$

To prove the thesis, we fix $R > |z|$: for every $n \in \mathbb{N}$ large enough, we also have $R > |z_n|$. We apply (3.4.2) with

$$w_k = 1 + \frac{z_n}{n}, \qquad \zeta_k = e^{\frac{z_n}{n}}, \qquad k = 1, \ldots, n;$$

observing that $|w_k| \leq 1 + \frac{|z_n|}{n} \leq e^{\frac{R}{n}}$, we have

$$\left|\left(1 + \frac{z_n}{n}\right)^n - e^{z_n}\right| \leq \left(e^{\frac{R}{n}}\right)^{n-1} \sum_{k=1}^{n} \left|1 + \frac{z_n}{n} - e^{\frac{z_n}{n}}\right| \leq$$

(by (3.4.3))

$$\leq e^{\frac{R(n-1)}{n}} n \left|\frac{z_n}{n}\right|^2 \leq e^R \frac{R^2}{n}$$

which proves the thesis.   □

Theorem 3.4.2 (Weak Law of Large Numbers) *Let $(X_n)_{n\in\mathbb{N}}$ be a sequence of i.i.d. real random variables in $L^1(\Omega, P)$, with expected value $\mu := E[X_1]$. Then the arithmetic mean M_n converges weakly to the constant r.v. equal to μ:*

$$M_n \xrightarrow{d} \mu.$$

Proof By Lévy's continuity Theorem 3.3.8, it is sufficient to prove that the sequence of characteristic functions φ_{M_n} converges pointwise to the CHF of the distribution δ_μ:

$$\lim_{n\to\infty} \varphi_{M_n}(\eta) = e^{i\mu\eta}, \qquad \eta \in \mathbb{R}. \tag{3.4.4}$$

We have

$$\varphi_{M_n}(\eta) = E\left[e^{i\frac{\eta}{n}S_n}\right] =$$

(since the X_n are i.i.d.)

$$= \left(E\left[e^{i\frac{\eta}{n}X_1}\right]\right)^n =$$

(by Theorem 2.5.20 and the summability assumption)

$$= \left(1 + \frac{i\mu\eta}{n} + o\left(\frac{1}{n}\right)\right)^n \xrightarrow[n\to\infty]{} e^{i\mu\eta}$$

thanks to Lemma 3.4.1. This proves (3.4.4) and concludes the proof.   □

Remark 3.4.3 (Kolmogorov's Strong Law of Large Numbers) The integrability assumptions of Theorem 3.4.2 are weaker than the Law of large numbers in the version of Theorem 3.2.1 where we assumed that $X_n \in L^2(\Omega, P)$. With more sophisticated methods, it is also possible to extend Theorem 3.2.3 and obtain the so-called *Kolmogorov's strong Law of large numbers*: if $(X_n)_{n\in\mathbb{N}}$ is a sequence of real i.i.d. random variables in $L^1(\Omega, P)$ with expected value $\mu := E[X_1]$, then M_n converges *almost surely* to μ. For more details, see, for example, [21].

Suppose now that $(X_n)_{n \in \mathbb{N}}$ is a sequence of real i.i.d. random variables in $L^2(\Omega, P)$ and let

$$\mu := E[X_1] \quad \text{and} \quad \sigma^2 := \text{var}(X_1).$$

Recall that the expected value and variance of the arithmetic mean M_n in (3.2.1) are given by

$$E[M_n] = \mu \quad \text{and} \quad \text{var}(M_n) = \frac{\sigma^2}{n}.$$

Consider then the *normalized arithmetic mean*, defined as

$$\tilde{M}_n := \frac{M_n - E[M_n]}{\sqrt{\text{var}(M_n)}} = \frac{M_n - \mu}{\frac{\sigma}{\sqrt{n}}}$$

and note that

$$\tilde{M}_n = \frac{S_n - \mu n}{\sigma \sqrt{n}} = \frac{1}{\sqrt{n}} \sum_{k=1}^{n} \frac{X_k - \mu}{\sigma}. \tag{3.4.5}$$

The Central limit theorem states that, regardless of the distribution of the X_n, the sequence of normalized arithmetic means $\tilde{M}_n$ converges weakly to the standard normal distribution.

Theorem 3.4.4 (Central Limit Theorem) *[!!!] Let $(X_n)_{n \in \mathbb{N}}$ be a sequence of real i.i.d. random variables in $L^2(\Omega, P)$, with expectation μ and positive standard deviation σ. Then, for $\tilde{M}_n$ as in (3.4.5), we have*

$$\tilde{M}_n \xrightarrow{d} Z \sim \mathcal{N}_{0,1}. \tag{3.4.6}$$

Proof By Lévy's continuity Theorem 3.3.8, it is sufficient to prove that the sequence of characteristic functions $\varphi_{\tilde{M}_n}$ converges pointwise to the CHF of the distribution $\mathcal{N}_{0,1}$:

$$\lim_{n \to \infty} \varphi_{\tilde{M}_n}(\eta) = e^{-\frac{\eta^2}{2}}, \qquad \eta \in \mathbb{R}. \tag{3.4.7}$$

By (3.4.5) we have

$$\varphi_{\tilde{M}_n}(\eta) = E\left[e^{i \frac{\eta}{\sqrt{n}} \sum_{k=1}^{n} \frac{X_k - \mu}{\sigma}} \right] =$$

(since the X_n are i.i.d.)

$$= \left(E \left[e^{i \frac{\eta}{\sqrt{n}} \frac{X_1 - \mu}{\sigma}} \right] \right)^n =$$

(by Theorem 2.5.20, being by assumption $\frac{X_1 - \mu}{\sigma} \in L^2(\Omega, P)$ with zero mean and unit variance)

$$= \left(1 + \frac{(i\eta)^2}{2n} + o\left(\frac{1}{n} \right) \right)^n \xrightarrow[n \to \infty]{} e^{-\frac{\eta^2}{2}}$$

thanks to Lemma 3.4.1. This proves (3.4.7) and concludes the proof. $\qquad\square$

Remark 3.4.5 In the particular case when $\mu = 0$ and $\sigma = 1$, (3.4.6) becomes

$$\frac{S_n}{\sqrt{n}} \xrightarrow{d} Z \sim \mathcal{N}_{0,1}.$$

Remark 3.4.6 (Central Limit Theorem and Law of Large Numbers) Given the expression of $\tilde{M}_n$ in (3.4.5), the Central limit theorem can be reformulated in the following way:

$$M_n \simeq \mu + \frac{\sigma}{\sqrt{n}} Z \sim \mathcal{N}_{\mu, \frac{\sigma^2}{n}}, \qquad \text{for } n \gg 1, \qquad (3.4.8)$$

where the symbol $\simeq$ indicates that the distributions of M_n and $\mu + \frac{\sigma}{\sqrt{n}} Z$ are "asymptotically equivalent". Formula (3.4.8) offers an approximation of the distribution of the random variable M_n, elucidating the convergence result of the Law of large numbers.

Remark 3.4.7 (Central Limit Theorem and Monte Carlo Method) Averages M_n of i.i.d. variables, defined as in (3.4.1), appear naturally in the Monte Carlo method that we introduced in Sect. 3.2.1. Under the assumptions of the Central limit theorem, given

$$p_\lambda := P\left(|M_n - \mu| \leq \lambda \frac{\sigma}{\sqrt{n}} \right) = P\left(|\tilde{M}_n| \leq \lambda \right), \qquad \lambda > 0,$$

we have the estimate

$$p_\lambda \simeq P\left(|Z| \leq \lambda \right), \qquad Z \sim \mathcal{N}_{0,1}.$$

(continued)

Remark 3.4.7 (continued)
Now, recall (cf. (2.1.12)) that

$$P(|Z| \leq \lambda) = 2F(\lambda) - 1, \qquad \lambda > 0,$$

with F in (3.4.10). For the estimation of the numerical error of the Monte Carlo method, we start from the values of p most commonly used, namely $p = 95\%$ and $p = 99\%$: setting $\lambda = F^{-1}\left(\frac{p+1}{2}\right)$, we obtain

$$P\left(|M_n - \mu| \leq 1.96\frac{\sigma}{\sqrt{n}}\right) \simeq 95\% \quad \text{and} \quad P\left(|M_n - \mu| \leq 2.57\frac{\sigma}{\sqrt{n}}\right) \simeq 99\%.$$

For this reason

$$r_{95} := 1.96\frac{\sigma}{\sqrt{n}} \qquad \text{and} \qquad r_{99} := 2.57\frac{\sigma}{\sqrt{n}}$$

are commonly called *radii of the confidence intervals* at 95% and 99%, respectively. In other words, if M_n represents the (random) Monte Carlo approximation of the value μ, then

$$[M_n - r_{95}, M_n + r_{95}] \qquad \text{and} \qquad [M_n - r_{99}, M_n + r_{99}]$$

are the intervals (with random endpoints) to which μ (which is the unknown value that is intended to be approximated) belongs with probability equal to 95% and 99%, respectively. In this perspective, it is natural to interpret the result of a Monte Carlo approximation as *a confidence interval*, rather than a single value.

Remark 3.4.8 (Central Limit Theorem and Sums of i.i.d. Random Variables)
As already anticipated in Example 3.3.13, the Central limit theorem is a valid tool for approximating the law of random variables defined as sums of i.i.d. variables. For example, we know (cf. Proposition 2.6.3) that $X \sim \text{Bin}_{n,p}$ is equal in law to $X_1 + \cdots + X_n$ with $X_j \sim \text{Be}_p$ i.i.d. Then we have the following asymptotic approximation of the CDF of X for $n \to +\infty$:

$$P(X \leq k) \approx P\left(Z \leq \frac{k - pn}{\sqrt{np(1-p)}}\right), \qquad Z \sim \mathscr{N}_{0,1}. \tag{3.4.9}$$

Formula (3.4.9) simply follows from the fact that, given $\mu = E[X_1] = p$ and $\sigma^2 = \text{var}(X_1) = p(1 - p)$, by the Central limit theorem we have

$$P(X \leq k) = P\left(\frac{X - \mu n}{\sigma \sqrt{n}} \leq \frac{k - \mu n}{\sigma \sqrt{n}}\right) \approx P\left(Z \leq \frac{k - \mu n}{\sigma \sqrt{n}}\right).$$

Formula (3.4.9) is equivalent to

$$F_X(k) \approx F\left(\frac{k - pn}{\sqrt{np(1 - p)}}\right)$$

where F_X denotes the CDF of $X \sim \text{Bin}_{n,p}$ and

$$F(x) = \int_{-\infty}^{x} \frac{e^{-\frac{z^2}{2}}}{\sqrt{2\pi}} dz \qquad (3.4.10)$$

is the standard normal CDF.

Under stronger assumptions, the Berry-Esseen theorem provides an explicit estimate of the rate of convergence in the Central limit theorem.

Theorem 3.4.9 (Berry-Esseen Theorem) *There exists a constant[7] $C < 1$ such that, if (X_n) is a sequence of i.i.d. random variables in $L^3(\Omega, P)$ with*

$$E[X_1] = 0, \qquad \text{var}(X_1) := \sigma^2, \qquad E\left[|X_1|^3\right] =: \varrho,$$

then we have

$$|F_n(x) - F(x)| \leq \frac{C\varrho}{\sigma^3 \sqrt{n}}, \qquad x \in \mathbb{R}, \ n \in \mathbb{N},$$

where F_n denotes the CDF of the normalized mean $\widetilde{M}_n$ in (3.4.5) and F is the standard normal CDF in (3.4.10).

For the proof of the Berry-Esseen theorem, we refer to, for instance, [12].

---

[7] The optimal value of C is not known: as reported in [47] it is known that $0.4097 < C < 0.469$.

Chapter 4
Conditional Probability

We have not succeeded in answering all our problems - indeed we sometimes feel we have not completely answered any of them. The answers we have found have only served to raise a whole set of new questions. In some ways we feel that we are as confused as ever, but we think we are confused on a higher level, and about more important things.

Earl C. Kelley

In a probability space $(\Omega, \mathscr{F}, P)$, let X be a random variable and $\mathscr{G}$ a sub-σ-algebra of $\mathscr{F}$. In this chapter, we introduce the concepts of conditional distribution and expectation of X given $\mathscr{G}$. Recalling that a σ-algebra can be interpreted as a set of "information", *the conditional expectation of X given $\mathscr{G}$ represents the best estimate of the random value X based on the information contained in $\mathscr{G}$*. The larger $\mathscr{G}$ is, the better and more detailed is the estimate of X given by the conditional expectation: from a mathematical standpoint, a conditional expectation is a *random variable* that enjoys certain properties. The concepts of conditional expectation and distribution are fundamental in the theory of stochastic processes and underpin all applications of probability theory aimed at modeling random phenomena evolving over time: in this case, it is necessary to describe not only the evolution of the random value X but also the evolution of the information that, over time, becomes available and enables better estimation of X. Throughout this chapter, unless stated otherwise, X denotes a $\mathbb{R}^d$-valued random variable.

4.1 The Discrete Case

We introduce the concept of conditioning on the σ-algebra generated by a *discrete* random variable. We treat this very particular case with a merely introductory purpose to the general definition which is technically more complex and will be introduced in the following sections.

A. Pascucci, *Probability Theory I*, La Matematica per il 3+2 165, https://doi.org/10.1007/978-3-031-63190-0_4

Consider a r.v. Y defined on the space $(\Omega, \mathscr{F}, P)$ and assume that Y is *discrete*[1] in the following sense:

(i) the distinct values assumed by Y form a set of at most countable cardinality: in other words, the image of Ω through Y is of the form $Y(\Omega) = (y_n)_{n \in \mathbb{N}}$ with distinct y_n;

(ii) for each $n \in \mathbb{N}$, the event $B_n := (Y = y_n)$ is not negligible, i.e., $P(B_n) > 0$.

Under these assumptions, *the family $(B_n)_{n \in \mathbb{N}}$ forms a finite or countable partition of Ω, whose elements are non-negligible events*. Note that $\sigma(Y)$, the σ-algebra generated by Y, consists of the empty set, the elements of the partition $(B_n)_{n \in \mathbb{N}}$ and their unions.

Definition 4.1.1 (Conditional Probability Given a Discrete r.v.)   In the space $(\Omega, \mathscr{F}, P)$, the conditional probability given the discrete r.v. Y is the family $P(\cdot \mid Y) = \big(P_\omega(\cdot \mid Y)\big)_{\omega \in \Omega}$ of probability measures on $(\Omega, \mathscr{F})$ defined by

$$P_\omega(A \mid Y) := P(A \mid Y = Y(\omega)), \qquad A \in \mathscr{F}, \tag{4.1.1}$$

where $P(\cdot \mid Y = Y(\omega))$ indicates the conditional probability given the event $(Y = Y(\omega))$ (cf. Definition 1.3.2).

Remark 4.1.2   For each $A \in \mathscr{F}$, $P(A \mid Y)$ is a *random variable* that is constant on the elements of the partition $(B_n)_{n \in \mathbb{N}}$:

$$P(A \mid Y) = \sum_{n \geq 1} P(A \mid B_n) \mathbb{1}_{B_n}.$$

Since $P_\omega(\cdot \mid Y)$ is a probability measure for each $\omega \in \Omega$, the concepts of distribution and conditional expectation on Y are naturally defined.

Definition 4.1.3 (Conditional Distribution and Expectation)   Let X be a r.v. on $(\Omega, \mathscr{F}, P)$ with values in $\mathbb{R}^d$:

(i) the conditional distribution (or law) of X given Y, denoted by $\mu_{X \mid Y}$, is the distribution of X with respect to the conditional probability $P(\cdot \mid Y)$:

$$\mu_{X \mid Y}(H) := P(X \in H \mid Y), \qquad H \in \mathscr{B}_d; \tag{4.1.2}$$

(ii) if $X \in L^1(\Omega, P)$, the conditional expectation of X given Y, denoted by $E[X \mid Y]$, is the expected value of X under the conditional probability $P(\cdot \mid Y)$:

$$E[X \mid Y] := \int_\Omega X \, dP(\cdot \mid Y). \tag{4.1.3}$$

---

[1] Assumption (ii) is not really restrictive: if Z verifies (i) then there exists a discrete r.v. Y such that $P(Y = y) > 0$ for each $y \in Y(\Omega)$ and $Z = Y$ a.s.

Remark 4.1.4 Note that conditional distribution and expectation depend on ω and are therefore *random quantities*, in fact:

(i) the meaning of definition (4.1.2) is

$$\mu_{X|Y}(H; \omega) := P_\omega(X \in H \mid Y), \qquad H \in \mathscr{B}_d, \ \omega \in \Omega.$$

Consequently:

(i-a) for each $\omega \in \Omega$, $\mu_{X|Y}(\cdot; \omega)$ is a *distribution* on $(\mathbb{R}^d, \mathscr{B}_d)$: hence we say that $\mu_{X|Y}$ is a *random distribution*;

(i-b) for each $H \in \mathscr{B}_d$, $\mu_{X|Y}(H)$ is a *random variable* that is constant on the elements of the partition $(B_n)_{n \in \mathbb{N}}$:

$$\mu_{X|Y}(H) = \sum_{n \geq 1} P(X \in H \mid B_n) \mathbb{1}_{B_n}; \qquad (4.1.4)$$

(ii) the meaning of definition (4.1.3) is

$$E[X|Y](\omega) := \int_\Omega X \, dP_\omega(\cdot|Y), \qquad \omega \in \Omega.$$

Consequently, $E[X|Y]$ is a *random variable* that is constant on the elements of the partition $(B_n)_{n \in \mathbb{N}}$:

$$E[X|Y] = \sum_{n \geq 1} E[X|B_n] \mathbb{1}_{B_n}, \qquad (4.1.5)$$

where, by Proposition 2.4.2,

$$E[X|B_n] = \frac{1}{P(B_n)} \int_{B_n} X \, dP.$$

Equivalently,

$$E[X|Y](\omega) = E[X|Y = Y(\omega)] = \frac{1}{P(Y = Y(\omega))} \int_{(Y = Y(\omega))} X \, dP, \qquad \omega \in \Omega.$$

Example 4.1.5 Let us take up Example 2.4.5: from an urn containing $n \geq 2$ numbered balls, two balls are drawn in sequence and without replacement. Let X_1 and X_2 be the random variables that indicate respectively the number of the first and second ball drawn. Then, for each $k \in I_n$, we have

$$\mu_{X_2|X_1=k}(\{h\}) = \begin{cases} \frac{1}{n-1}, & \text{if } h \in I_n \setminus \{k\}, \\ 0 & \text{otherwise,} \end{cases}$$

or equivalently

$$\mu_{X_2|X_1} = \mathrm{Unif}_{I_n \setminus \{X_1\}}.$$

We now generalize two well-known fundamental tools for calculating expectation.

Theorem 4.1.6 [!] *Let X and Y be random variables on $(\Omega, \mathscr{F}, P)$ with Y discrete. If $f \in m\mathscr{B}_d$ and $f(X) \in L^1(\Omega, P)$ then*

$$E[f(X)|Y] = \int_{\mathbb{R}^d} f \, d\mu_{X|Y}.$$

Proof For each $\omega \in \Omega$ we have

$$E[f(X)|Y](\omega) = \int_\Omega f(X) dP_\omega(\cdot|Y) =$$

(by Theorem 2.2.26)

$$= \int_{\mathbb{R}^d} f(x)\mu_{X|Y}(dx; \omega).$$

$\square$

Theorem 4.1.7 (Law of Total Probability) [!] *Let X and Y be random variables on $(\Omega, \mathscr{F}, P)$ with Y discrete. We have*

$$\mu_X = E[\mu_{X|Y}]. \tag{4.1.6}$$

Proof For each $H \in \mathscr{B}_d$, by (4.1.4) we have

$$E[\mu_{X|Y}(H)] = \sum_{n \geq 1} P(X \in H|B_n)P(B_n) = \sum_{n \geq 1} P((X \in H) \cap B_n)$$

$$= P(X \in H) = \mu_X(H).$$

$\square$

Example 4.1.8 The number of spam emails received daily by a mailbox is a r.v. with a Poisson$_{10}$ distribution. By installing an anti-spam software, it is possible to halve the average number of spam emails received. Knowing that this software protects only 80% of a company's mailboxes, let us determine the distribution and average number of spam emails received daily by each mailbox in the company.

Let $Y \sim \mathrm{Be}_p$, with $p = 80\%$, be the r.v. that is 1 if a mailbox is protected and 0 otherwise. If X indicates the number of spam emails received, we have by hypothesis

$$\mu_{X|Y} = Y\mathrm{Poisson}_5 + (1 - Y)\mathrm{Poisson}_{10}.$$

Then, by the Law of total probability (4.1.6), we have

$$\mu_X = E\left[\mu_{X|Y}\right] = p\mu_{X|Y=1} + (1 - p)\mu_{X|Y=0} = p\mathrm{Poisson}_5 + (1 - p)\mathrm{Poisson}_{10}$$

from which

$$E[X] = pE[X|Y = 1] + (1 - p)E[X|Y = 0] = 80\% \cdot 5 + 20\% \cdot 10 = 6.$$

Finally, we have

$$\begin{aligned} E[X|Y] &= \int_{\mathbb{R}} x\mu_{X|Y}(dx) \\ &= Y\int_{\mathbb{R}} x\mathrm{Poisson}_5(dx) + (1 - Y)\int_{\mathbb{R}} x\mathrm{Poisson}_{10}(dx) \\ &= 5Y + 10(1 - Y). \end{aligned}$$

Example 4.1.9 [!] Consider the *random exponential distribution* $\mu_{X|Y} = \mathrm{Exp}_Y$ where $Y \sim \mathrm{Geom}_p$: then we have

$$P(X \geq x|Y) = \mathrm{Exp}_Y([x, +\infty[) = \int_x^{+\infty} Ye^{-tY}\,dt = \left[-e^{-tY}\right]_{t=x}^{t=+\infty} = e^{-xY},$$

for each $x \geq 0$. Therefore, we have

$$E[P(X \geq x|Y)] = E\left[e^{-xY}\right] = \sum_{n \in \mathbb{N}} e^{-nx}p(1 - p)^{n-1} = \frac{p}{p - 1 + e^x}$$

and on the other hand, by the Law of total probability, we have

$$E[P(X \geq x|Y)] = P(X \geq x)$$

which provides the expression of the CDF (and therefore the distribution) of X. In fact, noting that clearly $P(X \geq x|Y) = 1$ if $x < 0$, we have

$$P(X \geq x) = \begin{cases} 1 & \text{if } x < 0, \\ \frac{p}{p-1+e^x} & \text{if } x \geq 0, \end{cases}$$

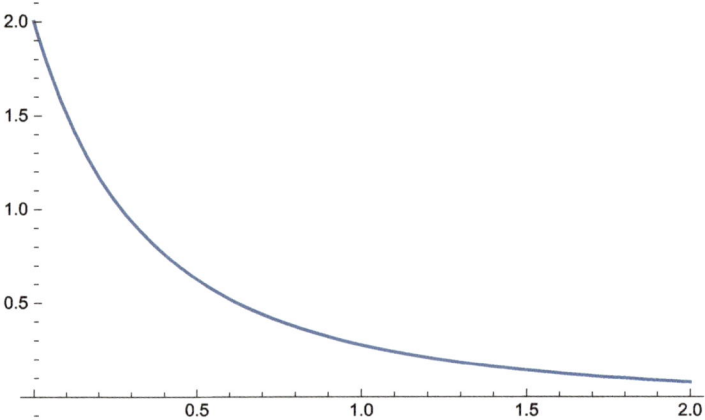

Fig. 4.1 Graph of the density in (4.1.7) for p=0.5

from which it follows that X is an absolutely continuous r.v. with density (see Fig. 4.1)

$$\gamma_X(x) = \frac{d}{dx}(1 - P(X \geq x)) = \begin{cases} 0 & \text{if } x < 0, \\ \frac{pe^x}{(p-1+e^x)^2} & \text{if } x \geq 0. \end{cases} \tag{4.1.7}$$

We can think of X as an exponential-type r.v. with stochastic intensity.[2] This example shows that through the concept of conditional distribution, it is possible to consider probabilistic models in which the value of the parameters is uncertain or stochastic. Hence arises the fundamental importance of conditional distribution in numerous applications, particularly in statistics.

Conditional expectation has two properties that uniquely characterize it.

Proposition 4.1.10 *[!] Given two random variables X and Y on $(\Omega, \mathscr{F}, P)$, with $X \in L^1(\Omega, P)$ and Y discrete, let $Z = E[X \mid Y]$. Then we have:*

(i) $Z \in m\sigma(Y)$;
(ii) for every $W \in b\sigma(Y)$ we have

$$E[ZW] = E[XW].$$

Moreover, if Z' is a r.v. that satisfies properties (i) and (ii), then $Z'(\omega) = Z(\omega)$ for every $\omega \in \Omega$.

---

[2] In the exponential distribution Exp_λ, the parameter $\lambda > 0$ is usually called *intensity*.

Proof Property (i) is an immediate consequence of (4.1.5). As for (ii), by Doob's Theorem 2.3.3 there exists a measurable and bounded function f such that $W = f(Y)$ or, more explicitly

$$W = \sum_{n \geq 1} f(y_n) \mathbb{1}_{B_n}. \tag{4.1.8}$$

Then by (4.1.5) we have

$$E[WZ] = E\left[f(Y) \sum_{n \geq 1} E[X \mid B_n] \mathbb{1}_{B_n} \right]$$

$$= \sum_{n \geq 1} f(y_n) E[X \mid B_n] E\left[\mathbb{1}_{B_n} \right] =$$

(by (2.4.1))

$$= \sum_{n \geq 1} f(y_n) E\left[X \mathbb{1}_{B_n} \right] = E[XW].$$

Finally, if Z' satisfies properties (i) and (ii), then Z' is of the form (4.1.8) and, by ii) with $W = \mathbb{1}_{B_n}$, we have

$$f(y_n) P(B_n) = E\left[Z' \mathbb{1}_{B_n} \right] = E\left[X \mathbb{1}_{B_n} \right]$$

and therefore $f(y_n) = E[X \mid B_n]$.   □

Remark 4.1.11 (Conditional Probability Function) Let Y be a discrete r.v. with values in a measurable space $(E, \mathscr{E})$. According to definition (4.1.1), the conditional probability is a family of probability measures $(P_\omega(\cdot \mid Y))_{\omega \in \Omega}$ and in this sense can be interpreted as a *random probability*. It is possible to give an alternative definition of conditional probability in which $P(\cdot \mid Y)$ depends on $y \in Y(\Omega)$ instead of $\omega \in \Omega$: precisely, we say that $y \mapsto P(\cdot \mid Y = y)$ is the *conditional probability function*[3] given Y. Note that $P(\cdot \mid Y) = (P(\cdot \mid Y = y))_{y \in Y(\Omega)}$ is a finite or at most countable family of probability measures on $(\Omega, \mathscr{F})$, because Y is discrete. Similarly, the

(continued)

---

[3] We use the term conditional probability *function* given Y to emphasize the fact that, according to this definition, $P(\cdot \mid Y)$ is a function that associates each $y \in Y(\Omega)$ with the probability measure $P(\cdot \mid Y = y)$.

Remark 4.1.11 (continued)
conditional distribution function of X given Y is the function

$$y \longmapsto \mu_{X|Y}(H; y) = P(X \in H \mid Y = y), \qquad H \in \mathscr{B}, \ y \in Y(\Omega),$$

and the *conditional expectation function* as

$$y \longmapsto E[X \mid Y = y] = \int_\Omega X \, dP(\cdot \mid Y = y) =$$

(by Proposition 2.4.2)

$$= \frac{1}{P(Y = y)} \int_{(Y=y)} X \, dP, \qquad y \in Y(\Omega).$$

Let

$$\bar{\mu}_X(x) = P(X = x), \qquad x \in X(\Omega),$$

be the distribution function of the *discrete* r.v. X (cf. Remark 1.4.16). By analogy, we denote by $\bar{\mu}_{X|Y}(x, y) = P(X = x \mid Y = y)$ the *conditional distribution function* of X given Y. Then, we have the interesting formula

$$\bar{\mu}_{X|Y}(x, y) = \frac{P((X = x) \cap (Y = y))}{P(Y = y)} = \frac{\bar{\mu}_{(X,Y)}(x, y)}{\bar{\mu}_Y(y)}, \qquad x \in X(\Omega), \ y \in Y(\Omega),$$

$$(4.1.9)$$

that is, $\bar{\mu}_{(X,Y)} = \bar{\mu}_Y \bar{\mu}_{X|Y}$, resembling the multiplication rule (1.3.5) for events.

Example 4.1.12 The number of emails received each day is a r.v. $Y \sim \text{Poisson}_\lambda$ with $\lambda = 20$. Each email has a probability $p = 15\%$ of being spam, independently of the others. We determine the distribution of the r.v. X that indicates the number of spam emails received each day.

Intuitively, we expect that $X \sim \text{Poisson}_{\lambda p}$. In fact, by hypothesis, we have

$$P(X = k \mid Y = n) = \begin{cases} \text{Bin}_{n,p}(\{k\}) & \text{if } k \leq n, \\ 0 & \text{if } k > n, \end{cases}$$

is the probability that, out of n emails received, exactly k of them are spam. By the Law of total probability, we have

$$P(X = k) = \sum_{n \geq 0} P(X = k \mid Y = n) P(Y = n)$$

$$= \sum_{n \geq k} \binom{n}{k} p^k (1-p)^{n-k} e^{-\lambda} \frac{\lambda^n}{n!}$$

$$= \frac{e^{-\lambda}(\lambda p)^k}{k!} \sum_{n \geq k} \frac{(1-p)^{n-k}\lambda^{n-k}}{(n-k)!} =$$

(setting $h = n - k$)

$$= \frac{e^{-\lambda}(\lambda p)^k}{k!} \sum_{h \geq 0} \frac{(1-p)^h \lambda^h}{h!} = e^{-\lambda p} \frac{(\lambda p)^k}{k!} = \text{Poisson}_{\lambda p}(\{k\}).$$

Remark 4.1.13 Consider $Y = \mathbb{1}_B$ with $B \in \mathscr{F}$ such that $0 < P(B) < 1$: the σ-algebra generated by Y is

$$\sigma(Y) = \{\emptyset, \Omega, B, B^c\}$$

and can be interpreted as "the information regarding whether event B has occurred or not". Notice the conceptual difference between:

(i) conditioning on B, in the sense of conditioning on the fact *that B has occurred*;
(ii) conditioning on Y, in the sense of conditioning on the information *whether B has occurred or not*.

For this reason the conditional expectation $E[X \mid Y]$ is defined as in (4.1.5), that is:

$$E[X \mid Y](\omega) := \begin{cases} E[X \mid B] & \text{if } \omega \in B, \\ E[X \mid B^c] & \text{if } \omega \in B^c. \end{cases}$$

Intuitively, $E[X \mid B]$ represents the expectation of X estimated based on the observation that B has occurred: therefore, $E[X \mid B]$ is a number, a *deterministic value*. On the contrary, one can think of $E[X \mid Y]$ as a future estimate of X that will depend on observing whether B has occurred or not (or the estimate of X that is given by an individual who knows whether B has occurred or not): for this reason, $E[X \mid Y]$ is defined as a *random variable*.

4.1.1 Examples

Example 4.1.14 We determine $E[X_1 \mid Y]$ where $X_1, \ldots, X_n \sim \text{Be}_p$, with $0 < p < 1$, are independent and $Y = X_1 + \cdots + X_n$. Since $Y \sim \text{Bin}_{n,p}$, we have

$$E[X_1 \mid Y = k] = 0 \cdot P(X_1 = 0 \mid Y = k) + 1 \cdot P(X_1 = 1 \mid Y = k) =$$

(setting $Z = X_2 + \cdots + X_n \sim \text{Bin}_{n-1,p}$)

$$= \frac{P((X_1 = 1) \cap (Z = k - 1))}{P(Y = k)} =$$

(due to the independence of X_1 and Z)

$$= \frac{P(X_1 = 1)P(Z = k - 1)}{P(Y = k)}$$

$$= \frac{p\binom{n-1}{k-1}p^{k-1}(1-p)^{n-1-(k-1)}}{\binom{n}{k}p^k(1-p)^{n-k}} = \frac{k}{n}, \qquad k = 0, \ldots, n,$$

is the conditional expectation *function* of X_1 given Y. Similarly, we have

$$E[X_1 \mid Y] = \frac{Y}{n}.$$

Example 4.1.15 Urn A contains $n \in \mathbb{N}$ balls, of which only $k_1 \leq n$ are white. Urn B contains $n \in \mathbb{N}$ balls, of which only $k_2 \leq n$ are white. A urn is chosen at random and a sequence of draws with replacement is performed. We determine the distribution of the number X of draws needed to find the first white ball.

Let $Y \sim \text{Be}_p$, with $p = \frac{1}{2}$, be the r.v. that equals 1 if urn A is chosen and 0 otherwise. Then, recalling Example 2.1.25 on the geometric distribution, we have

$$\mu_{X|Y} = Y\text{Geom}_{\frac{k_1}{n}} + (1 - Y)\text{Geom}_{\frac{k_2}{n}},$$

and by the Law of total probability (4.1.6) we have

$$\mu_X = \frac{1}{2}\left(\text{Geom}_{\frac{k_1}{n}} + \text{Geom}_{\frac{k_2}{n}}\right).$$

Thus, we also have

$$E[X] = \frac{n(k_1 + k_2)}{2k_1 k_2}.$$

Example 4.1.16 Let $X_i \sim \text{Poisson}_{\lambda_i}$, $i = 1, 2$, be independent and $Y := X_1 + X_2$. We know (cf. Example 2.6.5) that $Y \sim \text{Poisson}_{\lambda_1 + \lambda_2}$. Let us prove that

$$\mu_{X_1|Y} = \text{Bin}_{Y, \frac{\lambda_1}{\lambda_1 + \lambda_2}}.$$

We consider the conditional distribution *function* $\mu_{X_1|Y=\cdot}$ of X_1 given Y. For $k \in \{0, 1, \ldots, n\}$, we have

$$\mu_{X_1|Y=n}(\{k\}) = \frac{P((X_1 = k) \cap (Y = n))}{P(Y = n)} =$$

(by the independence of X_1 and X_2)

$$= \frac{P(X_1 = k)P(X_2 = n - k)}{P(Y = n)} = \frac{\frac{e^{-\lambda_1}\lambda_1^k}{k!}\frac{e^{-\lambda_2}\lambda_2^{n-k}}{(n-k)!}}{\frac{e^{-\lambda_1-\lambda_2}(\lambda_1+\lambda_2)^n}{n!}}$$

and on the other hand, $\mu_{X_1|Y=n}(\{k\}) = 0$ for the other values of k. From this, we easily conclude.

Exercise 4.1.17 Let $X_i \sim \text{Geom}_p$, $i = 1, 2$, be independent and $Y := X_1 + X_2$. Prove that

(i) $\mu_Y(\{n\}) = (n - 1)p^2(1 - p)^{n-2}$, for $n \geq 2$;
(ii) $\mu_{X_1|Y} = \text{Unif}_{\{1,2,\ldots,Y-1\}}$.

4.2   Conditional Expectation

In a space $(\Omega, \mathscr{F}, P)$ let X be a integrable r.v. and $\mathscr{G}$ a sub-σ-algebra of $\mathscr{F}$. In this section, we give the definition of conditional expectation of X given $\mathscr{G}$. For a general $\mathscr{G}$, it is not possible to define $E[X \mid \mathscr{G}]$ as in the discrete case because it is not clear how to partition the sample space Ω based on $\mathscr{G}$. The problem is that a σ-algebra can have a very complicated structure: consider, for instance, the Borel σ-algebra on Euclidean space. Moreover, in the case $\mathscr{G} = \sigma(Y)$ with Y absolutely continuous, the definition (4.1.1) loses meaning because each event of the type $(Y = Y(\omega))$ is negligible.

To overcome these problems, the general definition of conditional expectation is given in terms of the two characterizing properties of Proposition 4.1.10. The following result shows that a r.v. that satisfies such properties *always exists and is a.s. unique.*

Theorem 4.2.1 *Let* $X \in L^1(\Omega, \mathscr{F}, P)$ *with values in* $\mathbb{R}^d$ *and* $\mathscr{G}$ *be a sub-σ-algebra of* $\mathscr{F}$. *There exists a* $\mathbb{R}^d$-valued r.v. $Z \in L^1(\Omega, P)$ *that satisfies the following properties:*

(i) $Z \in m\mathscr{G}$;
(ii) *for each bounded r.v.* $W \in m\mathscr{G}$, *we have*

$$E[ZW] = E[XW]. \tag{4.2.1}$$

Moreover, if Z' verifies (i) and (ii) then $Z = Z'$ almost surely. These results also generalize to integrable random variables (see Corollary 4.2.9).

Proof (Uniqueness) Consider the case $d = 1$. We prove a slightly more general result from which uniqueness easily follows: let X, X' be integrable r.v., such that $X \leq X'$ almost surely, and let Z, Z' be r.v. that verify the properties (i) and (ii) respectively for X and X'. Then $Z \leq Z'$ almost surely.

Indeed, let

$$A_n = (Z - Z' \geq 1/n), \qquad n \in \mathbb{N}.$$

Then $A_n \in \mathscr{G}$ by i), and we have

$$0 \geq E\left[(X - X')\mathbb{1}_{A_n}\right] = E\left[X\mathbb{1}_{A_n}\right] - E\left[X'\mathbb{1}_{A_n}\right] =$$

(by (ii))

$$= E\left[Z\mathbb{1}_{A_n}\right] - E\left[Z'\mathbb{1}_{A_n}\right] = E\left[(Z - Z')\mathbb{1}_{A_n}\right] \geq \frac{1}{n}P(A_n)$$

from which $P(A_n) = 0$ and, due to the continuity from below of P, we also have $P(Z > Z') = 0$. The case $d > 1$ follows by reasoning component by component.
(Existence) We give a proof of existence based on functional analysis results, in particular related to orthogonal projection in Hilbert spaces. We first consider the more restrictive hypothesis that X belongs to $L^2(\Omega, \mathscr{F}, P)$ which is a Hilbert space with the scalar product

$$\langle X, Z \rangle = E[XZ].$$

Also $L^2(\Omega, \mathscr{G}, P)$ is a Hilbert space and is a closed subspace of $L^2(\Omega, \mathscr{F}, P)$ since $\mathscr{G} \subseteq \mathscr{F}$. Then there exists the projection Z of X onto $L^2(\Omega, \mathscr{G}, P)$ and by definition we have:

(i) $Z \in L^2(\Omega, \mathscr{G}, P)$ and therefore in particular Z is $\mathscr{G}$-measurable;
(ii) for every $W \in L^2(\Omega, \mathscr{G}, P)$

$$E[(Z - X)W] = 0. \tag{4.2.2}$$

So Z is precisely the r.v. we are looking for: from a geometric standpoint, Z is the $\mathscr{G}$-measurable r.v. that best approximates X in the sense that, among the $\mathscr{G}$-measurable random variables, it is the least distant from X with respect to the L^2 distance.

Now consider $X \in L^1(\Omega, \mathscr{F}, P)$ such that $X \geq 0$ almost surely. The case of X with values in $\mathbb{R}^d$ is proved by reasoning on the positive and negative part of each single component. The sequence defined by

$$X_n = X \wedge n, \qquad n \in \mathbb{N},$$

is increasing, belongs to L^2 and converges pointwise to X: to each X_n, we associate Z_n defined as above, i.e., as the projection of X_n onto $L^2(\Omega, \mathscr{G}, P)$. As seen in the first part of the proof, for every $n \in \mathbb{N}$ we have $0 \leq Z_n \leq Z_{n+1}$ almost surely: consequently, we also have that, except for a negligible event A, we have

$$0 \leq Z_n \leq Z_{n+1}, \qquad \forall n \in \mathbb{N}.$$

We define

$$Z(\omega) = \sup_{n \in \mathbb{N}} Z_n(\omega), \qquad \omega \in \Omega \setminus A,$$

and $Z = 0$ on A. Then $Z \in m\mathscr{G}$ being the pointwise limit of random variables in $m\mathscr{G}$. Moreover, let W be bounded and $\mathscr{G}$-measurable: without loss of generality, we can consider $W \geq 0$. By Beppo Levi's theorem, we have

$$E[XW] = \lim_{n \to \infty} E[X_n W] = \lim_{n \to \infty} E[Z_n W] = E[ZW].$$

$\square$

Remark 4.2.2 [!]   By the second Dynkin's Theorem A.0.8, property (ii) of Theorem 4.2.1 is equivalent to the following property, generally easier to verify:

(ii-b) we have

$$E[Z\mathbb{1}_G] = E[X\mathbb{1}_G]$$

for every $G \in \mathscr{A}$, where $\mathscr{A}$ is a $\cap$-closed family such that $\sigma(\mathscr{A}) = \mathscr{G}$.

Definition 4.2.3 (Conditional Expectation)   Let X be an integrable r.v. and $\mathscr{G}$ be a sub-σ-algebra of $\mathscr{F}$. If Z satisfies the properties (i) and (ii) of Theorem 4.2.1 then we write

$$Z = E[X \mid \mathscr{G}] \tag{4.2.3}$$

and say that Z is *a version of the conditional expectation* of X given $\mathscr{G}$. In particular, if $\mathscr{G} = \sigma(Y)$ for some r.v. Y on $(\Omega, \mathscr{F}, P)$, we write

$$Z = E[X \mid Y]$$

instead of $Z = E[X \mid \sigma(Y)]$.

Remark 4.2.4 Formula (4.2.3) *is not to be understood as an equation*, i.e., as an identity between the members on the right and left of the equality: on the contrary, it is a *notation,* a symbol that indicates that Z satisfies properties (i) and (ii) of Theorem 4.2.1 (and therefore is a version of the conditional expectation of X given $\mathscr{G}$). The conditional expectation is defined implicitly, through properties (i) and (ii), *up to negligible events of $\mathscr{G}$:* in other words, if $Z = E[X \mid \mathscr{G}]$ and Z' differs from Z on a negligible event of $\mathscr{G}$, then also $Z' = E[X \mid \mathscr{G}]$. For this reason, we speak of *version* of the conditional expectation, even though later on, for simplicity, we will improperly say that Z is the conditional expectation of X given $\mathscr{G}$. However, be careful: if $Z = E[X \mid \mathscr{G}]$ and $Z' = Z$ a.s., it is not necessarily true that $Z' = E[X \mid \mathscr{G}]$. This is a subtlety that must be paid attention to: modifying Z on an event C negligible but such that $C \notin \mathscr{G}$, the $\mathscr{G}$-measurability property can be lost.

Convention 4.2.5 [!] Later, it will be useful to consider equalities of conditional expectations. To avoid ambiguity we will use the following convention: if $\mathscr{H} \subseteq \mathscr{G}$, the writing

$$E[X \mid \mathscr{H}] = E[X \mid \mathscr{G}]$$

means that if $Z = E[X \mid \mathscr{H}]$ then $Z = E[X \mid \mathscr{G}]$ (however, there may exist a version Z' of $E[X \mid \mathscr{G}]$ that is not the conditional expectation of X given $\mathscr{H}$, in particular if $Z' \in m\mathscr{G} \setminus m\mathscr{H}$). Note that the notations $E[X \mid \mathscr{H}] = E[X \mid \mathscr{G}]$ and $E[X \mid \mathscr{G}] = E[X \mid \mathscr{H}]$ are not equivalent unless $\mathscr{H} = \mathscr{G}$.

Remark 4.2.6 One may wonder why the conditional expectation is not defined as an *equivalence class*, identifying functions (random variables) that are almost surely equal as is usually done in functional analysis theory. Certainly, the presentation would be more elegant and would avoid having to continuously mention the *version* (i.e., the representative of the equivalence class) of the conditional expectation. This issue is also discussed in the introduction of Williams' book [46]. First of all, it is necessary to consider the fact that the identification by equivalence classes depends on the fixed probability measure: while in functional analysis the measurable space structure is generally fixed once and for all, in probability theory it is normal to work simultaneously with different measures and σ-algebras. Moreover, the typical situation is that such measures, even if defined on the same σ-algebra, *are not equivalent* (i.e., they do not have the same negligible and certain events): think of the case of a probability P and the conditional probability $P(\cdot \mid B)$ with

(continued)

Remark 4.2.6 (continued)
$0 < P(B) < 1$ for which $P(B^c \mid B) = 0$. The situation becomes even more complicated in the theory of stochastic processes, where *uncountable* families of σ-algebras and probability measures are considered: in this context, the use of equivalence classes is simply not feasible.

Remark 4.2.7 [!] Let $X, Y \in L^2(\Omega, P)$ and $Z = E[X \mid Y]$. Then

$$E[X - Z] = 0, \qquad \text{cov}(X - Z, Y) = 0, \qquad (4.2.4)$$

that is, $X - Z$ *has zero mean and is uncorrelated with* Y. The first equation follows from (4.2.2) with $W = 1$. For the second one, we have

$$\text{cov}(X - Z, Y) = E[(X - Z)Y] - E[X - Z]E[Y] = 0$$

since $E[(X - Z)Y] = 0$ by formula[4] (4.2.1) with $W = Y$.

Example 4.2.8 [!] Consider a two-dimensional normal random vector $(X, Y) \sim \mathcal{N}_{\mu,C}$ with

$$\mu = (e_X, e_Y), \qquad C = \begin{pmatrix} \sigma_X^2 & \sigma_{XY} \\ \sigma_{XY} & \sigma_Y^2 \end{pmatrix} \geq 0.$$

We prove that there exist $a, b \in \mathbb{R}$ such that $aY + b = E[X \mid Y]$. If $aY + b = E[X \mid Y]$ then a, b are uniquely determined by the equations in (4.2.4) which here become

$$E[aY + b] = E[X], \qquad \text{cov}(X - (aY + b), Y) = 0.$$

Thus

$$ae_Y + b = e_X, \qquad a\sigma_Y^2 = \sigma_{XY}$$

from which, assuming $\sigma_Y \neq 0$,

$$a = \frac{\sigma_{XY}}{\sigma_Y^2}, \qquad b = e_X - \frac{\sigma_{XY}}{\sigma_Y^2}e_Y,$$

---

[4] More precisely, see (4.2.2).

which provides a further interpretation of the regression line seen in Sect. 2.2.9. On the other hand, if a, b are determined in this way then $Z := aY + b = E[X \mid Y]$ since:

(i) clearly $Z \in m\sigma(Y)$;
(ii) $X - Z$ and Y have a joint normal distribution (since $(X - Z, Y)$ is a linear function of (X, Y)) and therefore are not only uncorrelated but also independent (see Proposition 2.5.18). Consequently, for every $W \in m\sigma(Y)$ (which is therefore independent of $X - Z$), we have

$$E[(X - Z)W] = (E[X] - E[Z]) E[W] = 0.$$

These facts can be summarized as follows: *the multi-normal distribution has the remarkable property of having marginal distributions (μ_X and μ_Y) and conditional marginal distributions (i.e., $\mu_{X|Y}$) that are still normal.*

In the proof of Theorem 4.2.1 we also proved the following result:

Corollary 4.2.9 *Let $X \in m\mathscr{F}^+$ and $\mathscr{G}$ be a sub-σ-algebra of $\mathscr{F}$. There exists a r.v. Z that satisfies the following properties:*

(i) $Z \in m\mathscr{G}^+$;
(ii) for every r.v. $W \in m\mathscr{G}^+$, we have

$$E[ZW] = E[XW].$$

Furthermore, if Z' satisfies (i) and (ii) then $Z = Z'$ almost surely.

Corollary 4.2.9 allows including integrable (not necessarily absolutely integrable) random variables in Definition 4.2.3 of conditional expectation.

4.2.1 Main Properties

In this section, we establish a comprehensive list of significant properties of conditional expectation. We consider two real random variables $X, Y \in L^1(\Omega, \mathscr{F}, P)$ and $\mathscr{G}, \mathscr{H}$ sub-σ-algebras of $\mathscr{F}$.

Theorem 4.2.10 *The following properties hold:*

(1) **law of total probability:**

$$E[X] = E[E[X \mid \mathscr{G}]]; \qquad (4.2.5)$$

(2) if $X \in m\mathscr{G}$ then

$$X = E\left[X \mid \mathscr{G}\right];$$

(3) if X and $\mathscr{G}$ are independent then

$$E\left[X\right] = E\left[X \mid \mathscr{G}\right];$$

*(4) **linearity:** for every $a, b \in \mathbb{R}$ we have*

$$aE\left[X \mid \mathscr{G}\right] + bE\left[Y \mid \mathscr{G}\right] = E\left[aX + bY \mid \mathscr{G}\right];$$

*(5) **monotonicity:** if $P(X \le Y) = 1$ then*

$$E\left[X \mid \mathscr{G}\right] \le E\left[Y \mid \mathscr{G}\right],$$

in the sense that if $Z = E\left[X \mid \mathscr{G}\right]$ and $W = E\left[Y \mid \mathscr{G}\right]$ then $P(Z \le W) = 1$;
*(6) **pullout property:** if X is $\mathscr{G}$-measurable and bounded then*

$$XE\left[Y \mid \mathscr{G}\right] = E\left[XY \mid \mathscr{G}\right]; \qquad\qquad (4.2.6)$$

*(7) **tower property:** if $\mathscr{H} \subseteq \mathscr{G}$, we have[5]*

$$E\left[E\left[X \mid \mathscr{G}\right] \mid \mathscr{H}\right] = E\left[X \mid \mathscr{H}\right];$$

*(8) **Beppo Levi's theorem:** if $0 \le X_n \nearrow X$ a.s. then*

$$\lim_{n \to \infty} E\left[X_n \mid \mathscr{G}\right] = E\left[X \mid \mathscr{G}\right];$$

*(9) **Fatou's lemma:** if $(X_n)_{n \in \mathbb{N}}$ is a sequence of random variables in $m\mathscr{F}^+$, then*

$$E\left[\liminf_{n \to \infty} X_n \mid \mathscr{G}\right] \le \liminf_{n \to \infty} E\left[X_n \mid \mathscr{G}\right];$$

*(10) **dominated convergence theorem:** if $(X_n)_{n \in \mathbb{N}}$ is a sequence that converges a.s. to X and $|X_n| \le Y \in L^1(\Omega, P)$ a.s. for every $n \in \mathbb{N}$, then we have*

$$\lim_{n \to \infty} E\left[X_n \mid \mathscr{G}\right] = E\left[X \mid \mathscr{G}\right];$$

---

[5] We also have

$$E\left[X \mid \mathscr{H}\right] = E\left[E\left[X \mid \mathscr{H}\right] \mid \mathscr{G}\right]$$

which follows directly from property (2) and the fact that $E\left[X \mid \mathscr{H}\right] \in m\mathscr{G}$ since $\mathscr{H} \subseteq \mathscr{G}$.

(11) **Jensen's inequality:** *if φ is a convex function such that $\varphi(X) \in L^1(\Omega, P)$, then*

$$\varphi\left(E\left[X \mid \mathcal{G}\right]\right) \leq E\left[\varphi(X) \mid \mathcal{G}\right];$$

(12) for every $p \geq 1$ we have

$$\|E\left[X \mid \mathcal{G}\right]\|_p \leq \|X\|_p;$$

(13) **freezing lemma:** *let $\mathcal{G}, \mathcal{H}$ be independent, $X \in m\mathcal{G}$ and $f = f(x, \omega) \in m(\mathcal{B} \otimes \mathcal{H})$ such that $f(X, \cdot) \in L^1(\Omega, P)$ or $f \geq 0$. Then we have*

$$F(X) = E\left[f(X, \cdot) \mid \mathcal{G}\right] \quad \text{where} \quad F(x) := E\left[f(x, \cdot)\right] = \int_\Omega f(x, \omega) P(d\omega),$$
$$(4.2.7)$$

or, in a more compact notation,

$$E\left[f(x, \cdot)\right]\big|_{x=X} = E\left[f(X, \cdot) \mid \mathcal{G}\right];$$

(14) **conditional CHF and independence:** *X and $\mathcal{G}$ are independent if and only if*

$$E\left[e^{i\eta X}\right] = E\left[e^{i\eta X} \mid \mathcal{G}\right], \qquad \eta \in \mathbb{R},$$

that is, if the CHF $\varphi_X(\eta)$ is a version of the conditional CHF $\varphi_{X|\mathcal{G}}(\eta)$ for any $\eta \in \mathbb{R}$;
(15) if $Z = E\left[X \mid \mathcal{G}\right]$ and $Z \in m\mathcal{H}$ with $\mathcal{H} \subseteq \mathcal{G}$ then $Z = E\left[X \mid \mathcal{H}\right]$.

Proof

   (1) It suffices to set $W = 1$ in (4.2.1).
   (2) It follows directly from the definition.
   (3) The constant r.v. $Z := E[X]$ is clearly $\mathcal{G}$-measurable (because $\sigma(Z) = \{\emptyset, \Omega\}$) and moreover, for every bounded r.v. $W \in m\mathcal{G}$, due to the independence assumption, we have

$$E[XW] = E[X]E[W] = E[E[X]W] = E[ZW].$$

   This proves that $Z = E[X \mid \mathcal{G}]$.
   (4) We need to show that if $Z = E[X \mid \mathcal{G}]$ and $W = E[Y \mid \mathcal{G}]$, in the sense that they satisfy properties (i) and (ii) of Theorem 4.2.1, then $aZ + bW = E[aX + bY \mid \mathcal{G}]$. This verification is left as a straightforward exercise.
   (5) This property is proven in the first part of the proof of Theorem 4.2.1.
   (6) Let $Z = E[Y \mid \mathcal{G}]$. We have to prove that $XZ = E[XY \mid \mathcal{G}]$:

      (i) $X \in m\mathcal{G}$ by assumption and therefore $XZ \in m\mathcal{G}$;

(ii) given $W \in m\mathscr{G}$ bounded, we have $XW \in b\mathscr{G}$ and thus

$$E\,[(XZ)W] = E\,[Z(XW)] =$$

(since $Z = E\,[Y \mid \mathscr{G}]$)

$$= E\,[Y(XW)] = E\,[(XY)W)]$$

and this proves the claim.

(7) Let $Z = E\,[X \mid \mathscr{H}]$. We have to prove that $Z = E\,[E\,[X \mid \mathscr{G}] \mid \mathscr{H}]$. By definition

(i) $Z \in m\mathscr{H}$;
(ii) given $W \in m\mathscr{H}$ bounded, we have

$$E\,[ZW] = E\,[XW]\,.$$

On the other hand, if $W \in m\mathscr{H}$ then $W \in m\mathscr{G}$ since $\mathscr{H} \subseteq \mathscr{G}$, and thus

$$E\,[E\,[X \mid \mathscr{G}]\,W] = E\,[XW]\,.$$

Then $E\,[ZW] = E\,[E\,[X \mid \mathscr{G}]\,W]$ and this proves the tower property.

(8) Let $Y_n := E\,[X_n \mid \mathscr{G}], n \geq 1$. Due to the monotonicity of conditional expectation, $0 \leq Y_n \leq Y_{n+1}$ a.s. and thus there exists

$$Y := \lim_{n \to \infty} E\,[X_n \mid \mathscr{G}]\,,$$

with $Y \in m\mathscr{G}^+$ because it is the pointwise limit of $\mathscr{G}$-measurable random variables. Moreover, for every $W \in m\mathscr{G}^+$, we have $0 \leq Y_n W \nearrow YW$ and $0 \leq X_n W \nearrow XW$ a.s.; thus, by Beppo Levi's theorem, we have

$$E\,[YW] = \lim_{n \to \infty} E\,[Y_n W] = \lim_{n \to \infty} E\,[X_n X] = E\,[XW]\,,$$

which proves the thesis.

(9)–(10)–(11) The proof is substantially analogous to the deterministic case.

(12) It follows easily from Jensen's inequality with $\varphi(x) = |x|^p$.

(13) Let $\mathscr{M}$ be the family of functions $f \in b(\mathscr{B} \otimes \mathscr{H})$ that satisfy (4.2.7): $\mathscr{M}$ is a monotone family of functions (cf. Definition A.0.7), as can be easily shown using Beppo Levi's theorem for conditional expectation. Moreover, (4.2.7) holds true for functions of the form

$f(x, \omega) = g(x)Y(\omega)$ with $g \in b\mathcal{B}$ and $Y \in b\mathcal{H}$: indeed, in this case, we have $F(x) = g(x)E[Y]$ and, by property (4.2.6),

$$E[g(X)Y \mid \mathcal{G}] = g(X)E[Y \mid \mathcal{G}] = g(X)E[Y] = F(X).$$

Then the thesis follows from Dynkin's Theorem A.0.8.
(14) For every $Y \in m\mathcal{G}$ and $\eta_1, \eta_2 \in \mathbb{R}$, we have

$$\varphi_{(X,Y)}(\eta_1, \eta_2) = E\left[e^{i\eta_1 X} e^{i\eta_2 Y}\right] =$$

(by definition of conditional expectation)

$$= E\left[E\left[e^{i\eta_1 X} \mid \mathcal{G}\right] e^{i\eta_2 Y}\right] =$$

(by hypothesis)

$$= E\left[e^{i\eta_1 X}\right] E\left[e^{i\eta_2 Y}\right] = \varphi_X(\eta_1)\varphi_Y(\eta_2)$$

and the thesis follows from Proposition 2.5.11-(ii).
(15) It is a simple exercise.

$\square$

An immediate consequence of point (13) of Theorem 4.2.10 is the following particular version of the freezing lemma for which we provide an alternative simpler proof.

Lemma 4.2.11 (Freezing Lemma) *[!] Let $\mathcal{G}$ be a sub-σ-algebra of $\mathcal{F}$. If $X \in m\mathcal{G}$, Y is a r.v. independent of $\mathcal{G}$ and $f \in m\mathcal{B}_2$ is such that $f(X, Y) \in L^1(\Omega, P)$, then we have*

$$F(X) = E[f(X, Y) \mid \mathcal{G}] \quad \text{where} \quad F(x) := E[f(x, Y)] = \int_\Omega f(x, Y(\omega))P(d\omega),$$
$$(4.2.8)$$

or, in a more compact notation,

$$E[f(x, Y)]\mid_{x=X} = E[f(X, Y) \mid \mathcal{G}].$$

Proof By Fubini's theorem, the function F in (4.2.8) is Borel measurable and thus $F(X) \in m\mathcal{G}$. Moreover, Y is independent of (W, X) for every $W \in b\mathcal{G}$: then we have

$$E[Wf(X, Y)] = \int_{\mathbb{R}^3} wf(x, y)\mu_{(W,X,Y)}(dw, dx, dy) =$$

(by independence)

$$= \int_{\mathbb{R}^3} wf(x,y)\mu_{(W,X)} \otimes \mu_Y(dw,dx,dy) =$$

(by Fubini's theorem)

$$= \int_{\mathbb{R}^2} w \left(\int_{\mathbb{R}} f(x,y)\mu_Y(dy) \right) \mu_{(W,X)}(dw,dx)$$

$$= \int_{\mathbb{R}^2} wF(x)\mu_{(W,X)}(dw,dx) = E[WF(X)]$$

and this proves the thesis. □

Example 4.2.12 [!] Let us resume Example 2.5.4 and consider N and $Z_1, Z_2, \dots$ independent random variables with $N \sim \text{Poisson}_\lambda$ and Z_n identically distributed for $n \in \mathbb{N}$. We determine the CHF of

$$X := \begin{cases} 0 & \text{if } N = 0, \\ \sum_{k=1}^{N} Z_k & \text{if } N \geq 1. \end{cases}$$

We have

$$\varphi_X(\eta) = E\left[e^{i\eta X}\right] = E\left[\prod_{k=1}^{N} e^{i\eta Z_k}\right] =$$

(by the Law of total probability (4.2.5))

$$= E\left[E\left[\prod_{k=1}^{N} e^{i\eta Z_k} \mid N\right]\right] = E\left[\left(\varphi_{Z_1}(\eta)\right)^N\right]$$

where in the last step we used the freezing lemma and the fact that, due to the independence of the random variables Z_k, we have

$$E\left[\prod_{k=1}^{n} e^{i\eta Z_k}\right] = \varphi_{Z_1}(\eta)^n, \qquad n \in \mathbb{N}.$$

Then we have

$$\varphi_X(\eta) = e^{-\lambda} \sum_{n \geq 0} \frac{\lambda^n}{n!} \varphi_{Z_1}(\eta)^n = e^{\lambda(\varphi_{Z_1}(\eta)-1)}$$

where φ_{Z_1} denotes the CHF of Z_1.

Example 4.2.13 Let X, Y, U, V be independent random variables with $X, Y \sim \mathcal{N}_{0,1}$ and $U^2 + V^2 \neq 0$ a.s. Prove that

$$Z := \frac{XU + YV}{\sqrt{U^2 + V^2}} \sim \mathcal{N}_{0,1}.$$

Indeed, we have

$$\varphi_Z(\eta) = E\left[e^{i\eta \frac{XU+YV}{\sqrt{U^2+V^2}}} \right] =$$

(by the Law of total probability (4.2.5))

$$= E\left[E\left[e^{i\eta \frac{XU+YV}{\sqrt{U^2+V^2}}} \mid (U, V) \right] \right] =$$

(by the freezing lemma and Example 2.5.16)

$$= E\left[e^{-\frac{\eta^2}{2}} \right] = e^{-\frac{\eta^2}{2}}$$

and this proves the thesis.

4.2.2 Changes of Probability Measure

We adopt the notations of Appendix B.1 and write $Q \ll_{\mathscr{F}} P$ to indicate that Q is an absolutely continuous measure with respect to P on the σ-algebra $\mathscr{F}$. Moreover, $E^P[\cdot]$ denotes the expectation under the probability P.

Theorem 4.2.14 (Bayes' Formula) *Let P, Q be probability measures on $(\Omega, \mathscr{F})$ with $Q \ll_{\mathscr{F}} P$. If $X \in L^1(\Omega, Q)$ and $\mathscr{G}$ is a sub-σ-algebra of $\mathscr{F}$, then*

$$E^Q[X \mid \mathscr{G}] = \frac{E^P[XL \mid \mathscr{G}]}{E^P[L \mid \mathscr{G}]} \tag{4.2.9}$$

where $L = \frac{dQ}{dP} \mid_{\mathscr{F}}$ is the Radon-Nikodym derivative of Q with respect to P on $\mathscr{F}$.

Proof Let $Z = E^Q[X \mid \mathscr{G}]$ and $L^{\mathscr{G}} = E^P[L \mid \mathscr{G}]$. Observe that $Q(L^{\mathscr{G}} > 0) = 1$ since

$$Q(L^{\mathscr{G}} = 0) = E^Q\left[\mathbb{1}_{(L^{\mathscr{G}}=0)} \right] = E^P\left[\mathbb{1}_{(L^{\mathscr{G}}=0)} L \right] =$$

(by property ii) of the definition of conditional expectation and since $(L^{\mathcal{G}} = 0) \in \mathcal{G}$

$$= E^P \left[\mathbb{1}_{(L^{\mathcal{G}}=0)} L^{\mathcal{G}} \right] = 0.$$

Hence, Eq. (4.2.9) is equivalent to $ZL^{\mathcal{G}} = E^P[XL \mid \mathcal{G}]$ To show this last equation, we observe that $ZL^{\mathcal{G}}$ is obviously $\mathcal{G}$-measurable and to conclude we use Remark 4.2.2: for every $G \in \mathcal{G}$ we have

$$\int_G ZL^{\mathcal{G}} dP = \int_G E^P[ZL \mid \mathcal{G}] dP = \int_G ZL \, dP$$

$$= \int_G E^Q[X \mid \mathcal{G}] dQ = \int_G X \, dQ = \int_G XL \, dP.$$

$\square$

Remark 4.2.15 Let us denote with $L^{\mathcal{F}}$ and $L^{\mathcal{G}}$ the Radon-Nikodym derivatives of Q with respect to P on $\mathcal{F}$ and on $\mathcal{G}$ respectively: note that $L^{\mathcal{F}}$, unlike $L^{\mathcal{G}}$, is not necessarily $\mathcal{G}$-measurable. On the other hand, we have

$$L^{\mathcal{G}} = E^P \left[L^{\mathcal{F}} \mid \mathcal{G} \right],$$

since $L^{\mathcal{G}}$ is integrable and $\mathcal{G}$-measurable and

$$\int_G L^{\mathcal{G}} dP = Q(G) = \int_G L^{\mathcal{F}} dP, \qquad G \in \mathcal{G},$$

being $\mathcal{G} \subseteq \mathcal{F}$.

4.2.3   Conditional Expectation Function

In this section, we consider the case $\mathcal{G} = \sigma(Y)$ where Y is a r.v. on $(\Omega, \mathcal{F}, P)$ taking values in a measurable space $(E, \mathcal{E})$. In analogy with Remark 4.1.11, we give an alternative definition of conditional expectation as a *function*.

Let $X \in L^1(\Omega, \mathcal{F}, P)$ take values in $\mathbb{R}^d$. If $Z = E[X \mid Y]$ then $Z \in m\sigma(Y)$ and therefore, by Doob's Theorem 2.3.3, there exists (and in general is not unique) a function $\Phi \in m\mathcal{E}$ such that $Z = \Phi(Y)$: to fix ideas, consider the following diagram

$$(\Omega, \mathcal{F}) \xrightarrow{\;E[X \mid Y]\;} \left(\mathbb{R}^d, \mathcal{B}_d \right)$$
$$Y \searrow \qquad \nearrow \Phi$$
$$(E, \mathcal{E})$$

Definition 4.2.16 (Conditional Expectation Function) Let

$$\Phi : (E, \mathcal{E}) \longrightarrow \left(\mathbb{R}^d, \mathscr{B}_d \right)$$

be any function such that

(i) $\Phi \in m\mathcal{E}$;
(ii) $\Phi(Y) = E[X \mid Y]$.

Then we say that Φ is a *version of the conditional expectation function* of X given Y and we write

$$\Phi(y) = E[X \mid Y = y]. \tag{4.2.10}$$

Notice that, if $f \in b\mathscr{B}_d$ and Y is a r.v. in $\mathbb{R}^d$, then

$$f(y) = E[f(Y) \mid Y = y], \qquad y \in \mathbb{R}^d.$$

Remark 4.2.17 Notation $E[X \mid Y = y]$ in (4.2.10) *does not indicate the conditional expectation of X given $(Y = y)$ in the sense of Definition 1.3.2.* Indeed, such a definition requires that $(Y = y)$ is not negligible while in (4.2.10) Y is a generic random variable: for instance, if Y is absolutely continuous, then the probability of the event $(Y = y)$ is zero for every y. Therefore, (4.2.10) should not be understood as an equation and does not uniquely identify Φ: rather, it is a notation to indicate that Φ is any function that satisfies the two properties i) and ii) of Definition 4.2.16. In other words, a measurable *function* Φ is a version of the conditional expectation function of X given Y if and only if the *random variable* $\Phi(Y)$ is a version of the conditional expectation of X given Y.

Also, note that the "relevant" values of $\Phi(y)$ are those it takes on for $y \in Y(\Omega)$. In general, $Y(\Omega) \subseteq E$: therefore, provided $\mathcal{E}$-measurability is preserved, modifying Φ on $E \setminus Y(\Omega)$ is sufficient to obtain an alternative version of the conditional expectation.

In conclusion, the conditional expectation given Y can be interpreted either as a *random variable* or as a *function*. These two perspectives are essentially equivalent, and the decision of which to employ typically relies on the context.

Example 4.2.18 In Example 4.2.8 we saw that if (X, Y) has a bivariate normal distribution then there exist $a, b \in \mathbb{R}$ such that $ay + b = E[X \mid Y = y]$, that is, the linear function $\Phi(y) = ay + b$ is a version of the *conditional expectation function* of X given Y.

4.2.4 Least Square Monte Carlo

As seen in the proof of Theorem 4.2.1, in the space of square integrable random variables, the conditional expectation can be defined as an orthogonal projection and therefore is expressed as the solution of a *least squares problem*. Precisely, we have the following

Proposition 4.2.19 (Characterization of Conditional Expectation in L^2) *Let $Z = E[X \mid \mathcal{G}]$ with $X \in L^2(\Omega, \mathcal{F}, P)$ and $\mathcal{G}$ sub-σ-algebra of $\mathcal{F}$. Then*

$$E\left[|X - Z|^2\right] \le E\left[|X - W|^2\right], \qquad W \in L^2(\Omega, \mathcal{G}, P). \qquad (4.2.11)$$

Proof We have

$$E\left[|X - W|^2 \mid \mathcal{G}\right] = E\left[|X - Z + Z - W|^2 \mid \mathcal{G}\right]$$

$$= E\left[|X - Z|^2 \mid \mathcal{G}\right] + E\left[|Z - W|^2 \mid \mathcal{G}\right]$$

$$+ 2E\left[\langle X - Z, Z - W\rangle \mid \mathcal{G}\right] =$$

(since $Z - W \in m\mathcal{G}$ and by (4.2.6))

$$= E\left[|X - Z|^2 \mid \mathcal{G}\right] + |Z - W|^2 + 2\langle E[X - Z \mid \mathcal{G}], Z - W\rangle =$$

(since $E[X - Z \mid \mathcal{G}] = 0$)

$$= E\left[|X - Z|^2 \mid \mathcal{G}\right] + |Z - W|^2 \ge E\left[|X - Z|^2 \mid \mathcal{G}\right].$$

Applying the expected value, we obtain (4.2.11).   □

Given a Borel-measurable function F such that $F(X, Y) \in L^2(\Omega, \mathcal{F}, P)$, a common task is calculating the conditional expectation

$$E[F(X, Y) \mid Y]$$

based on the knowledge of the joint law of X and Y. This problem is equivalent to the calculation of a version Φ of the conditional expectation *function*, i.e., $\Phi(y) = E[F(X, Y) \mid Y = y]$: by (4.2.11) we have[6]

$$E\left[|F(X, Y) - \Phi(Y)|^2\right] = \min_{f \in L^2(\mathbb{R}^n, \mathcal{B}_n, \mu_Y)} E\left[|F(X, Y) - f(Y)|^2\right].$$

In other words, determining Φ is equivalent to solving the least squares problem

$$\Phi = \underset{f \in L^2(\mathbb{R}^n, \mathcal{B}_n, \mu_Y)}{\arg\min} E\left[|F(X, Y) - f(Y)|^2\right]. \tag{4.2.12}$$

Sometimes this problem can be solved exactly: this is the case of Example 4.2.8, in which $F(x, y) = x$ and $(X, Y) \sim \mathcal{N}_{\mu, C}$. Often, however, it is necessary to resort to numerical methods. When X, Y are independent, then by the freezing lemma, we simply have $\Phi(y) = E[F(X, y)]$, $y \in \mathbb{R}$: thus, to determine Φ, it is sufficient to determine an expected value, and this can be done numerically with the Monte Carlo method. More generally, there exists an extension of this method, called Least Square Monte Carlo (LSMC), which is based on a multi-linear regression of the type seen in Sect. 2.2.9.

Let us see how to proceed in the one-dimensional case: consider a basis of $L^2(\mathbb{R}, \mathcal{B}, \mu_Y)$, for example, the polynomial[7] functions $\beta_k(y) := y^k$ with $k = 0, 1, 2, \ldots$. For fixed $n \in \mathbb{N}$, we set

$$\beta = (\beta_0, \beta_1, \ldots, \beta_n).$$

We approximate problem (4.2.12) in finite dimension by looking for a solution $\bar{\lambda} \in \mathbb{R}^{n+1}$ of

$$\min_{\lambda \in \mathbb{R}^{n+1}} E\left[|\langle \beta(Y), \lambda \rangle - F(X, Y)|^2\right]. \tag{4.2.13}$$

Once $\bar{\lambda}$ is determined, the approximation of the conditional expectation function in (4.2.12) is given by

$$\Phi(y) \simeq \langle \beta(y), \bar{\lambda} \rangle.$$

Problem (4.2.13) can be tackled by approximating the expected value using the Monte Carlo method. We construct two vectors $x, y \in \mathbb{R}^M$ whose components are obtained by simulating M values of the variables X and Y, with M sufficiently large. To fix ideas, M can be of the order of 10^5 or more, while on the contrary, it is

---

[6] Recall that, by Doob's theorem, every $W \in L^2(\Omega, \sigma(Y), P)$ is of form $W = f(Y)$ for a certain $f \in L^2(\mathbb{R}^n, \mathcal{B}_n, \mu_Y)$.

[7] Clearly, the fact that polynomials belong to $L^2(\mathbb{R}, \mathcal{B}, \mu_Y)$ depends on the distribution μ_Y.

sufficient that the number of elements of the basis n is small, of the order of a few units (for more details see, for example, [18] or the monograph [17]). Set

$$Q(\lambda) := \sum_{k=1}^{M} \left(\langle \beta(y_k), \lambda \rangle - F(x_k, y_k) \right)^2, \qquad \lambda \in \mathbb{R}^{n+1},$$

the expected value in (4.2.13) is approximated by

$$\frac{Q(\lambda)}{M} \approx E\left[|\langle \lambda, \beta(Y) \rangle - F(X, Y)|^2 \right], \qquad M \gg 1.$$

As in Sect. 2.2.9, *since Q is a quadratic function of λ, the minimum is determined by imposing $\nabla Q(\lambda) = 0$.* In vector notation, we have

$$Q(\lambda) = |\mathbf{B}\lambda - \mathbf{F}|^2$$

where $\mathbf{B} = (b_{ki})$ with $b_{ki} = \beta_i(y_k)$ and $\mathbf{F} = (F(x_k, y_k))$ for $k = 1, \dots, M$ and $i = 0, \dots, n$. Therefore,

$$\nabla Q(\lambda) = 2\mathbf{B}^* (\mathbf{B}\lambda - \mathbf{F})$$

and imposing the condition $\nabla Q(\lambda) = 0$, in the case the matrix $\mathbf{B}^*\mathbf{B}$ is invertible, we obtain

$$\bar{\lambda} = \left(\mathbf{B}^*\mathbf{B} \right)^{-1} \mathbf{B}^*\mathbf{F}.$$

The calculation of $\bar{\lambda}$ requires the inversion of the matrix $\mathbf{B}^*\mathbf{B}$ which has dimensions $(n+1) \times (n+1)$, hence the importance of keeping n small. Note that instead $\mathbf{B}$ is a very large matrix, of dimension $M \times (n+1)$.

As an example, in Fig. 4.2 we depict the plots of the first four LSMC approximations (with a polynomial basis) of the conditional expectation function given Y,

$$\Phi(y) = E[F(X, Y) \mid Y = y], \qquad F(x, y) = \max\{1 - e^{x^2 y}, 0\},$$

with (X, Y) having bivariate normal distribution with zero mean, standard deviations $\sigma_X = 0.8$, $\sigma_Y = 0.5$ and correlation $\varrho = -0.7$.

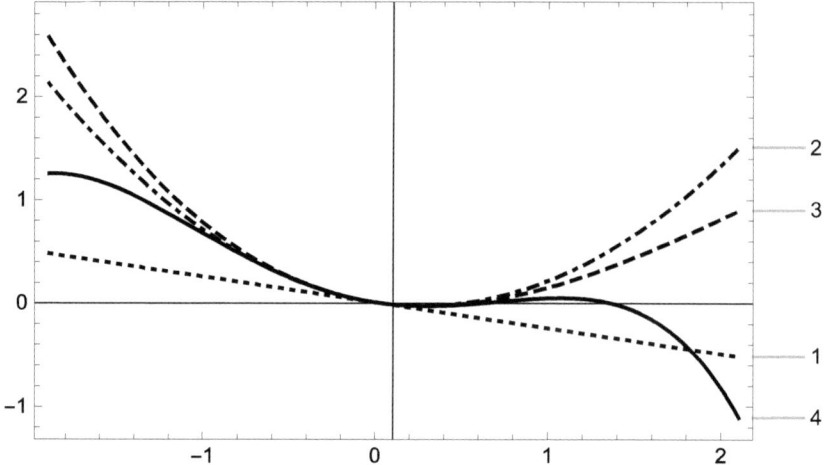

Fig. 4.2 LSMC approximations

4.3 Conditional Probability

In a probability space $(\Omega, \mathscr{F}, P)$, we consider a sub-σ-algebra $\mathscr{G}$ of $\mathscr{F}$. For each $A \in \mathscr{F}$ we fix a version $Z_A = E[\mathbb{1}_A \mid \mathscr{G}]$ of the conditional expectation of $\mathbb{1}_A$ given $\mathscr{G}$. It would seem natural to define the *conditional probability* given $\mathscr{G}$ by setting

$$P_\omega(A \mid \mathscr{G}) = Z_A(\omega), \qquad \omega \in \Omega. \tag{4.3.1}$$

However, since Z_A is determined up to a P-negligible event *that depends on A*, it is not assured (and generally not the case) that $P_\omega(\cdot \mid \mathscr{G})$, as defined, is a probability measure for every $\omega \in \Omega$.

Definition 4.3.1 (Regular Version of Conditional Probability) In the space $(\Omega, \mathscr{F}, P)$, a regular version of the conditional probability given $\mathscr{G}$ is a family $P(\cdot \mid \mathscr{G}) = \big(P_\omega(\cdot \mid \mathscr{G})\big)_{\omega \in \Omega}$ of *probability measures* on $(\Omega, \mathscr{F})$ such that[8]

$$P(A \mid \mathscr{G}) = E[\mathbb{1}_A \mid \mathscr{G}], \qquad A \in \mathscr{F}. \tag{4.3.2}$$

---

[8] We recall that (4.3.2) means that for every $A \in \mathscr{F}$ we have :

i) $\omega \mapsto P_\omega(A \mid \mathscr{G})$ is a $\mathscr{G}$-measurable r.v.;
ii) for every $W \in b\mathscr{G}$ we have

$$E[WP(A \mid \mathscr{G})] = E\big[W\mathbb{1}_A\big].$$

The existence of a regular version of conditional probability is a far from trivial problem: in [10], [11, p. 624], [20, p. 210], examples demonstrating the non-existence are given. Various authors have offered *sufficient*[9] conditions on $(\Omega, \mathscr{F}, P)$ that ensure the existence of a regular version of conditional probability: the most classic result in this regard is the following Theorem 4.3.2. We recall that a *Polish space* is a separable[10] and complete metric space.

Theorem 4.3.2 *Let P be a probability measure defined on a Polish space Ω equipped with the Borel σ-algebra $\mathscr{B}$. For every sub-σ-algebra $\mathscr{G}$ of $\mathscr{B}$, there exists a regular version of the conditional probability $P(\cdot \mid \mathscr{G})$.*

We prove Theorem 4.3.2 in the particular case where $\Omega = \mathbb{R}^d$ (cf. Theorem 4.3.4): for the general proof, see, for example, [44, p. 13] or [11, p. 380]. The idea is to exploit the existence of a countable subset A dense in Ω, to first define a family of probability measures $(P_\omega(\cdot \mid \mathscr{G}))_{\omega \in A}$ that verifies (4.3.1) and then prove the thesis by density of A in Ω.

Example 4.3.3 Assume there exists a regular version of the conditional probability $P(\cdot \mid \mathscr{G})$. If $G \in \mathscr{G}$ then $P(G \mid \mathscr{G})$ takes only the values 0 and 1. In fact, we have

$$P(G \mid \mathscr{G}) = E[\mathbb{1}_G \mid \mathscr{G}] = \mathbb{1}_G.$$

Let now X be a r.v. on $(\Omega, \mathscr{F}, P)$ with values in $\mathbb{R}^d$. If there exists a regular version $P(\cdot \mid \mathscr{G})$ of the conditional probability given $\mathscr{G}$, we set

$$\mu_{X \mid \mathscr{G}}(H) := P(X \in H \mid \mathscr{G}), \qquad H \in \mathscr{B}_d.$$

Note that, by definition, $\mu_{X \mid \mathscr{G}} = (\mu_{X \mid \mathscr{G}}(\cdot; \omega))_{\omega \in \Omega}$ is a family of *distributions* in $\mathbb{R}^d$ and for this reason is called *regular version* of the conditional distribution of X given $\mathscr{G}$.

Even without assuming the existence of $P(\cdot \mid \mathscr{G})$, we can still define a regular version of the conditional distribution of X given $\mathscr{G}$ based on the concept of conditional expectation. This is the content of the following

Theorem 4.3.4 (Regular Version of the Conditional Law) *[!] In a probability space $(\Omega, \mathscr{F}, P)$, let X be a r.v. with values in $\mathbb{R}^d$ and $\mathscr{G}$ a sub-σ-algebra of $\mathscr{F}$.*

---

[9] The problem of providing *necessary and sufficient* conditions is complex and partly still open: in this regard, see [14].

[10] A metric space is called separable if it contains a countable and dense subset.

Then there exists a family $\mu_{X|\mathscr{G}} = \big(\mu_{X|\mathscr{G}}(\cdot\,;\omega)\big)_{\omega\in\Omega}$ *of distributions on* $\mathbb{R}^d$ *such that, for every* $H \in \mathscr{B}_d$*, we have*[11]

$$\mu_{X|\mathscr{G}}(H) = E\left[\mathbb{1}_H(X)\mid\mathscr{G}\right]. \tag{4.3.3}$$

We say that $\mu_{X|\mathscr{G}}$ *is a* regular version of the conditional distribution of X given $\mathscr{G}$.

Proof See Sect. 4.4.1.   □

Remark 4.3.5 [!] *Although the existence of a regular version* $P(\cdot\mid\mathscr{G})$ *of the conditional probability given* $\mathscr{G}$ *is not guaranteed in general,* however, with a slight abuse of notation, we will write indifferently $\mu_{X|\mathscr{G}}(H)$ and $P(X \in H \mid \mathscr{G})$ to indicate a regular version of the conditional distribution of X given $\mathscr{G}$.

The proof of Theorem 4.3.4 crucially exploits the fact that X has values in $\mathbb{R}^d$ to use the density of $\mathbb{Q}^d$ in $\mathbb{R}^d$. *The result extends to the case of* X *with values in a Polish metric space,* such as the space of continuous functions $C([a, b]; \mathbb{R})$ with the maximum norm: for the general proof, see, for example, Theorem 1.1.6 in [44].

Notation 4.3.6 Hereafter, we will often omit to indicate the dependence on $\omega \in \Omega$ and write $\mu_{X|\mathscr{G}}$ instead of $\mu_{X|\mathscr{G}}(\cdot\,;\omega)$, interpreting $\mu_{X|\mathscr{G}}$ as a "random distribution". If $\mathscr{G} = \sigma(Y)$ where Y is any r.v. on $(\Omega, \mathscr{F}, P)$, we will write $\mu_{X|Y}$ instead of $\mu_{X|\sigma(Y)}$.

Example 4.3.7 [!] If $X \in m\mathscr{G}$, then $\mu_{X|\mathscr{G}} = \delta_X$. In fact, the family $(\delta_{X(\omega)})_{\omega\in\Omega}$ has the following properties:

 (i) obviously $\delta_{X(\omega)}$ is a distribution on $\mathbb{R}^d$ for every $\omega \in \Omega$;
 (ii) for every $H \in \mathscr{B}_d$ we have

$$\delta_X(H) = \mathbb{1}_H(X) =$$

(since $X \in m\mathscr{G}$ by hypothesis)

$$= E\left[\mathbb{1}_H(X) \mid \mathscr{G}\right].$$

Theorem 4.3.8 [!] *In a probability space* $(\Omega, \mathscr{F}, P)$*, let* X *be a r.v. with values in* $\mathbb{R}^d$ *and* $\mathscr{G}$ *a sub-σ-algebra of* $\mathscr{F}$*. If* $f \in m\mathscr{B}_d$ *and* $f(X) \in L^1(\Omega, P)$*, then*

$$\int_{\mathbb{R}^d} f\,d\mu_{X|\mathscr{G}} = E\left[f(X)\mid\mathscr{G}\right]. \tag{4.3.4}$$

---

[11] (4.3.3) means that, for every $H \in \mathscr{B}_d$, we have

(i) $\mu_{X|\mathscr{G}}(H)$ is a $\mathscr{G}$-measurable r.v.;
(ii) for every $W \in b\mathscr{G}$ we have

$$E\left[W\mu_{X|\mathscr{G}}(H)\right] = E\left[W\mathbb{1}_H(X)\right].$$

Proof The thesis is proved by applying the standard procedure of Remark 2.2.22, exploiting the linearity and Beppo Levi's theorem for the conditional expectation. It suffices to consider $d = 1$. Letting

$$Z(\omega) := \int_{\mathbb{R}} f(x) \mu_{X|\mathscr{G}}(dx; \omega), \qquad \omega \in \Omega,$$

we have to prove that $Z = E[f(X) \mid \mathscr{G}]$. This is true by definition (cf. (4.3.3)) if $f = \mathbb{1}_H$ with $H \in \mathscr{B}$. By linearity, (4.3.4) extends to simple functions. Moreover, if f has non-negative real values, then one considers an approximating sequence $0 \le f_n \nearrow f$ of simple functions and, applying Beppo Levi's theorem first in the classical version[12] and then for the conditional expectation, we have

$$\int_{\mathbb{R}} f d\mu_{X|\mathscr{G}} = \lim_{n \to \infty} \int_{\mathbb{R}} f_n d\mu_{X|\mathscr{G}} = \lim_{n \to \infty} E[f_n(X) \mid \mathscr{G}] = E[f(X) \mid \mathscr{G}].$$

The case of a generic f is treated as usual by separating the positive and negative parts and reusing the linearity of the conditional expectation.   □

Remark 4.3.9 [!] Theorem 4.3.8 clarifies the importance of the concept of *regular version* of the conditional distribution, as it *ensures that the integral in (4.3.4) is well defined.*

Example 4.3.10 Suppose that $X \sim \mathcal{N}_{Y,1}$ where $Y \sim \text{Exp}_\lambda$ with $\lambda > 0$ fixed. Then, by Theorem 4.3.8, we have

$$E[X \mid Y] = \int_{\mathbb{R}} x \frac{1}{\sqrt{2\pi}} e^{-\frac{(x-Y)^2}{2}} dx = Y.$$

Moreover, by (4.2.5)

$$E[X] = E[E[X \mid Y]] = E[Y] = \frac{1}{\lambda}$$

and

$$\text{cov}(X, Y) = E[XY] - E[X]E[Y]$$

$$= E[E[XY \mid Y]] - \frac{1}{\lambda^2} =$$

---

[12] Here we use the fact that $\mu_{X|\mathscr{G}} = \mu_{X|\mathscr{G}}(\cdot; \omega)$ is a distribution for every $\omega \in \Omega$.

(by (4.2.6))

$$= E\left[YE\left[X \mid Y\right]\right] - \frac{1}{\lambda^2}$$

$$= E\left[Y^2\right] - \frac{1}{\lambda^2} = \frac{1}{\lambda^2}.$$

Theorem 4.3.11 (Law of Total Probability) *[!]   In a probability space* $(\Omega, \mathscr{F}, P)$*, let X be a r.v. with values in $\mathbb{R}^d$ and $\mathscr{G}$ a sub-σ-algebra of $\mathscr{F}$. Then we have*

$$\mu_X = E\left[\mu_{X \mid \mathscr{G}}\right]. \tag{4.3.5}$$

Proof By definition, for every $H \in \mathscr{B}_d$ we have

$$E\left[\mu_{X \mid \mathscr{G}}(H)\right] = E\left[E\left[\mathbb{1}_{(X \in H)} \mid \mathscr{G}\right]\right] = E\left[\mathbb{1}_{(X \in H)}\right] = \mu_X(H).$$

$\square$

Example 4.3.12 Let us resume Example 4.3.10 again: by (4.3.5), for every $H \in \mathscr{B}$ we have

$$\mu_X(H) = E\left[\mu_{X \mid Y}(H)\right]$$

$$= E\left[\int_H \frac{1}{\sqrt{2\pi}} e^{-\frac{(x-Y)^2}{2}} dx\right] =$$

(by Fubini's theorem)

$$= \int_H \frac{1}{\sqrt{2\pi}} E\left[e^{-\frac{(x-Y)^2}{2}}\right] dx = \int_H \gamma(x) dx$$

with

$$\gamma(x) := \frac{1}{\sqrt{2\pi}} \int_0^{+\infty} e^{-\frac{(x-y)^2}{2}} \lambda e^{-\lambda y} dy$$

which is therefore the density of X.

Corollary 4.3.13 *[!] Let X, Y be random variables on $(\Omega, \mathscr{F}, P)$ with values in $\mathbb{R}^d$ and $\mathbb{R}^n$, respectively. Then we have*

$$\mu_{(X,Y)}(H \times K) = E\left[\mu_{X \mid Y}(H) \mathbb{1}_{(Y \in K)}\right], \qquad H \in \mathscr{B}_d, \ K \in \mathscr{B}_n, \tag{4.3.6}$$

$$\varphi_{(X,Y)}(\eta_1, \eta_2) = E\left[e^{i\eta_2 \cdot Y} \varphi_{X \mid Y}(\eta_1)\right], \qquad \eta_1 \in \mathbb{R}^d, \ \eta_2 \in \mathbb{R}^n. \tag{4.3.7}$$

Equation (4.3.6) *shows how to derive the joint law of* X, Y *from the conditional law* $\mu_{X|Y}$ *and the marginal law* μ_Y: in fact, the r.v. $\mu_{X|Y}(H)\mathbb{1}_{(Y \in K)}$ is a function of Y and therefore the expected value in (4.3.6) depends only on μ_Y. Similarly, Eq. (4.3.7) shows how to derive the joint CHF of X, Y from the conditional CHF $\varphi_{X|Y}$ and the marginal law μ_Y.

Proof of Corollary 4.3.13 By definition, we have

$$E\left[\mu_{X|Y}(H)\mathbb{1}_{(Y \in K)}\right] = E\left[E\left[\mathbb{1}_{(X \in H)} \mid Y\right]\mathbb{1}_{(Y \in K)}\right] =$$

(by property (ii) of Theorem 4.2.1 with $W = \mathbb{1}_{(Y \in K)}$)

$$= E\left[\mathbb{1}_{(X \in H)}\mathbb{1}_{(Y \in K)}\right] = \mu_{(X,Y)}(H \times K).$$

As for (4.3.7), we have

$$\varphi_{(X,Y)}(\eta_1, \eta_2) = E\left[e^{i\eta_1 \cdot X + i\eta_2 \cdot Y}\right]$$

$$= E\left[E\left[e^{i\eta_1 \cdot X + i\eta_2 \cdot Y} \mid Y\right]\right] =$$

(by Eq. (4.2.6))

$$= E\left[e^{i\eta_2 \cdot Y} E\left[e^{i\eta_1 \cdot X} \mid Y\right]\right]$$

$$= E\left[e^{i\eta_2 \cdot Y} \varphi_{X|Y}(\eta_1)\right].$$

$\square$

Example 4.3.14 Let us consider again Example 4.3.10: by (4.3.7) we have

$$\varphi_{(X,Y)}(\eta_1, \eta_2) = E\left[e^{i\eta_2 Y} \varphi_{X|Y}(\eta_1)\right] = E\left[e^{i\eta_2 Y} e^{i\eta_1 Y - \frac{\eta_1^2}{2}}\right]$$

$$= e^{-\frac{\eta_1^2}{2}} \frac{\lambda}{\lambda - i(\eta_1 + \eta_2)}.$$

Example 4.3.15 Given a two-dimensional r.v. (X, Y), suppose that $Y \sim \text{Unif}_{[0,1]}$ and $\mu_{X|Y} = \text{Exp}_Y$. We prove that (X, Y) is absolutely continuous and determine the joint density of X, Y and the marginal density of X. An immediate consequence of Eq. (4.3.6) is the following formula for the joint CDF: given $x \in \mathbb{R}_{\geq 0}$ and $y \in [0, 1]$, we have

$$P((X \leq x) \cap (Y \leq y)) = E\left[\text{Exp}_Y(]-\infty, x])\mathbb{1}_{(Y \leq y)}\right]$$

$$= E\left[\left(1 - e^{-xY}\right)\mathbb{1}_{(Y \leq y)}\right]$$

$$= \int_0^y \left(1 - e^{-xt}\right) dt = \frac{e^{-xy} - 1 + xy}{x}.$$

It follows that the CDF of (X, Y) is

$$F_{(X,Y)}(x, y) = \begin{cases} 0 & \text{if } (x, y) \in \mathbb{R}_{<0} \times \mathbb{R}_{<0}, \\ \frac{e^{-xy}-1+xy}{x} & \text{if } (x, y) \in \mathbb{R}_{\geq 0} \times [0, 1], \\ \frac{e^{-x}-1+x}{x} & \text{if } (x, y) \in \mathbb{R}_{\geq 0} \times [1, +\infty[. \end{cases}$$

From this, we obtain[13] the joint density

$$\gamma_{(X,Y)}(x, y) = \partial_x \partial_y F(x, y) = y e^{-xy} \mathbb{1}_{\mathbb{R}_{\geq 0} \times [0,1]}(x, y).$$

For the marginal density, we have

$$\gamma_X(x) = \partial_x P(X \leq x) = \partial_x F(x, 1) = \frac{e^{-x}(e^x - 1 - x)}{x^2} \mathbb{1}_{\mathbb{R}_{\geq 0}}(x).$$

4.3.1 Conditional Distribution Function

Theorem 4.3.16 (Regular Version of the Conditional Distribution Function)
[!] In a probability space $(\Omega, \mathscr{F}, P)$, let X be a r.v. with values in $\mathbb{R}^d$ and Y be a r.v. with values in a measurable space $(E, \mathscr{E})$. Then there exists a family $(\mu(\cdot; y))_{y \in E}$ of distributions on $\mathbb{R}^d$ such that, for each $H \in \mathscr{B}_d$,

(i) *the function $y \mapsto \mu(H; y)$ is $\mathscr{E}$-measurable;*
(ii) *$\mu(H, Y) = P(X \in H \mid Y)$, that is[14]*

$$E[W\mu(H; Y)] = E\left[W \mathbb{1}_{(X \in H)}\right], \qquad W \in b\sigma(Y).$$

We say that $(\mu(\cdot; y))_{y \in E}$ is a regular version of the conditional distribution function of X given Y *and write*

$$\mu(\cdot; y) = \mu_{X|Y=y}.$$

---

[13] Remember that

$$F(x, y) = \int_{-\infty}^x \int_{-\infty}^y \gamma_{(X,Y)}(\xi, \eta) d\xi d\eta.$$

[14] Remember the notation of Remark 4.3.5.

Proof The proof is slightly more sophisticated but essentially similar to that of Theorem 4.3.4: for this reason, we do not report it and refer to [23], Theorem 6.3, for details. □

Remark 4.3.17 If $\mu(\cdot; y) = \mu_{X|Y=y}$ then $(\mu_{X|Y}(\cdot; Y(\omega)))_{\omega \in \Omega}$ is a regular version of the conditional distribution of X given Y in the sense of Theorem 4.3.4.

Example 4.3.18 Let us take up Example 4.3.7: if Y is a real r.v. then $\mu_{Y|Y} = \delta_Y$. In other words, the random distribution δ_Y is a regular version of the conditional distribution of Y given Y.

For example, if $Y \sim \text{Unif}_{[0,1]}$ then $(\delta_y)_{y \in \mathbb{R}}$ is a regular version of the conditional distribution function of Y given Y. Actually, it would be sufficient to define the regular version only for $y \in E = [0, 1]$: the values taken outside $[0, 1]$ are irrelevant since Y takes values in $[0, 1]$ almost surely.

In Example 4.3.15 $\text{Exp}_Y = \mu_{X|Y}$, that is Exp_Y is a regular version of the conditional distribution of X given $Y \sim \text{Unif}_{[0,1]}$: equivalently $(\text{Exp}_y)_{y \in [0,1]}$ is a regular version of the conditional distribution function of X given Y.

According to notation (4.2.10), $E[X \mid Y = y]$ indicates the conditional expectation *function* of X given Y. The following result is analogous to Theorem 4.3.8.

Theorem 4.3.19 *In a probability space* $(\Omega, \mathscr{F}, P)$, *let* X *be a r.v. with values in* $\mathbb{R}^d$ *and* Y *be a r.v. with values in a measurable space* $(E, \mathscr{E})$. *For each* $f \in m\mathscr{B}_d$ *such that* $f(X) \in L^1(\Omega, P)$ *we have*

$$\int_{\mathbb{R}^d} f \, d\mu_{X|Y=y} = E[f(X) \mid Y = y].$$

4.3.2 From Joint Law to Conditional Marginals: The Absolutely Continuous Case

We have seen in Corollary 4.3.13 how to obtain the joint distribution from the conditional marginals. In this section, we consider a random vector (X, Y) in $\mathbb{R}^d \times \mathbb{R}$, that is absolutely continuous with density $\gamma_{(X,Y)}$, and derive the expression of the conditional marginal density $\gamma_{X|Y}$.

Recall that, by Fubini's theorem,

$$\gamma_Y(y) := \int_{\mathbb{R}^d} \gamma_{(X,Y)}(x, y) dx, \qquad y \in \mathbb{R}, \tag{4.3.8}$$

is a[15] density of Y and the set

$$(\gamma_Y > 0) := \{y \in \mathbb{R} \mid \gamma_Y(y) > 0\}$$

belongs to $\mathscr{B}$. The following result provides the continuous version of formula (4.1.9).

Proposition 4.3.20 *[!] Let $(X, Y) \in$ AC be a random vector with density $\gamma_{(X,Y)}$. Then the function*

$$\gamma_{X|Y}(x, y) := \frac{\gamma_{(X,Y)}(x, y)}{\gamma_Y(y)}, \qquad x \in \mathbb{R}^d, \ y \in (\gamma_Y > 0), \tag{4.3.9}$$

is a regular version of the conditional density of X given Y in the sense that the family $(\mu(\cdot\,; y))_{y \in (\gamma_Y > 0)}$ defined by

$$\mu(H; y) := \int_H \gamma_{X|Y}(x, y)dx, \qquad H \in \mathscr{B}_d, \ y \in (\gamma_Y > 0), \tag{4.3.10}$$

is a regular version of the conditional distribution function of X given Y. Consequently, for each $f \in m\mathscr{B}_d$ such that $f(X) \in L^1(\Omega, P)$ we have

$$\int_{\mathbb{R}^d} f(x)\gamma_{X|Y}(x, y)dx = E[f(X) \mid Y = y] \tag{4.3.11}$$

or equivalently

$$\int_{\mathbb{R}^d} f(x)\gamma_{X|Y}(x, Y)dx = E[f(X) \mid Y]. \tag{4.3.12}$$

Proof See Sect. 4.4.2.   □

Remark 4.3.21 **[!]** From (4.3.9), we have the formula

$$\gamma_{(X,Y)}(x, y) = \gamma_{X|Y}(x, y)\gamma_Y(y)$$

which expresses the joint density as the product of the marginal γ_Y and the conditional marginal $\gamma_{X|Y}$. This generalizes the formula

$$\gamma_{(X,Y)}(x, y) = \gamma_X(x)\gamma_Y(y)$$

only valid under the restrictive assumption that X, Y are independent.

---

[15] By Remark 1.4.19, the density of a r.v. is defined up to Borel sets with zero Lebesgue measure.

Example 4.3.22 Let (X, Y) be a random vector with uniform distribution on

$$S = \{(x, y) \in \mathbb{R}^2 \mid x > 0, \; y > 0, \; x^2 + y^2 < 1\}.$$

Determine:

(i) the conditional distribution $\mu_{X|Y}$;
(ii) $E[X \mid Y]$ and $\mathrm{var}(X \mid Y)$;
(iii) the density of the r.v. $E[X \mid Y]$.

(i) The joint density is

$$\gamma_{(X,Y)}(x, y) = \frac{4}{\pi} \mathbb{1}_S(x, y)$$

and the marginal of Y is

$$\gamma_Y(y) = \int_{\mathbb{R}} \gamma_{(X,Y)}(x, y)dx = \frac{4\sqrt{1 - y^2}}{\pi} \mathbb{1}_{]0,1[}(y).$$

Then

$$\gamma_{X|Y}(x, y) = \frac{\gamma_{(X,Y)}(x, y)}{\gamma_Y(y)} = \frac{1}{\sqrt{1 - y^2}} \mathbb{1}_{[0, \sqrt{1-y^2}]}(x), \qquad y \in \;]0, 1[,$$

from which we recognize that

$$\mu_{X|Y} = \mathrm{Unif}_{[0, \sqrt{1-Y^2}]}. \tag{4.3.13}$$

(ii) By (4.3.13) we have

$$E[X \mid Y] = \frac{\sqrt{1 - Y^2}}{2}, \qquad \mathrm{var}(X \mid Y) = \frac{1 - Y^2}{12}.$$

Alternatively, based on (4.3.11) of Proposition 4.3.20, we have, for $y \in \;]0, 1[$,

$$E[X \mid Y = y] = \int_{\mathbb{R}} x\gamma_{X|Y}(x, y)dx = \frac{\sqrt{1 - y^2}}{2},$$

$$\mathrm{var}(X \mid Y = y) = \int_{\mathbb{R}} \left(x - \frac{\sqrt{1 - y^2}}{2}\right)^2 \gamma_{X|Y}(x, y)dx = \frac{1 - y^2}{12}.$$

(iii) Finally, to determine the density of the r.v. $Z = \frac{\sqrt{1-Y^2}}{2}$ we use the CDF: we have $P(Z \leq 0) = 0$, $P(Z \leq 1/2) = 1$ and for $0 < z < 1/2$ we have

$$P(Z \leq z) = P\left(\sqrt{1-Y^2} \leq 2z\right)$$

$$= P\left(Y^2 \geq 1 - 4z^2\right)$$

$$= P\left(Y \geq \sqrt{1 - 4z^2}\right)$$

$$= 1 - \int_0^{\sqrt{1-4z^2}} \frac{4\sqrt{1-y^2}}{\pi} dy.$$

Differentiating, we obtain the density of Z:

$$\gamma_Z(z) = \frac{32z^2}{\pi\sqrt{1-4z^2}} \mathbb{1}_{]0,1/2[}(z).$$

Corollary 4.3.23 (Law of Total Probability) *Let (X, Y) be a random vector with conditional marginal density $\gamma_{X|Y}$. Then X has density*

$$\gamma_X = E\left[\gamma_{X|Y}(\cdot, Y)\right]. \tag{4.3.14}$$

Proof For every $f \in b\mathcal{B}$ we have

$$E[f(X)] = E[E[f(X) \mid Y]] =$$

(by (4.3.12))

$$= E\left[\int_{\mathbb{R}^d} f(x)\gamma_{X|Y}(x, Y)dx\right] =$$

(by the Fubini's theorem)

$$= \int_{\mathbb{R}^d} f(x)E\left[\gamma_{X|Y}(x, Y)\right]dx$$

and this proves the thesis, given the arbitrariness of f.   □

Example 4.3.24 Let X, Y be real random variables. Suppose $Y \sim \text{Exp}_\lambda$, with $\lambda > 0$, and that the conditional density of X given Y is of exponential type,

$$\gamma_{X|Y}(x, y) = ye^{-xy}\mathbb{1}_{[0,+\infty[}(x),$$

that is, $\mu_{X|Y} = \mathrm{Exp}_Y$. To determine the density of X, we use (4.3.14):

$$\gamma_X(x) = E\left[Ye^{-xY}\mathbb{1}_{[0,+\infty)}(x)\right]$$

$$= \int_0^{+\infty} ye^{-xy}\lambda e^{-\lambda y}\,dy\,\mathbb{1}_{[0,+\infty)}(x)$$

$$= \frac{\lambda}{(x+\lambda)^2}\mathbb{1}_{[0,+\infty)}(x).$$

Note that $X \notin L^1(\Omega, P)$.

Example 4.3.25 As in Example 4.2.8, we consider a two-dimensional normal random vector $(X, Y) \sim \mathcal{N}_{\mu,C}$ with

$$\mu = (\mu_1, \mu_2), \qquad C = \begin{pmatrix} \sigma_X^2 & \sigma_{XY} \\ \sigma_{XY} & \sigma_Y^2 \end{pmatrix} > 0.$$

We determine:

(i) the characteristic function $\varphi_{X|Y}$ and the conditional distribution $\mu_{X|Y}$ of X given Y;
(ii) $E[X \mid Y]$.

(i) The conditional density of X given Y is

$$\gamma_{X|Y}(x, y) = \frac{\gamma_{(X,Y)}(x, y)}{\gamma_Y(y)}, \qquad (x, y) \in \mathbb{R}^2,$$

from which, after some calculations, we find

$$\varphi_{X|Y}(\eta_1, Y) = E\left[e^{i\eta_1 X} \mid Y\right]$$

$$= \int_{\mathbb{R}} e^{i\eta_1 x}\gamma_{X|Y}(x, Y)\,dx$$

$$= e^{i\eta_1\left(\mu_1+(Y-\mu_2)\frac{\sigma_{XY}}{\sigma_Y^2}\right)-\frac{1}{2}\eta_1^2\left(\sigma_X^2-\frac{\sigma_{XY}^2}{\sigma_Y^2}\right)},$$

that is,

$$\mu_{X|Y} = \mathcal{N}_{\mu_1+(Y-\mu_2)\frac{\sigma_{XY}}{\sigma_Y^2},\,\sigma_X^2-\frac{\sigma_{XY}^2}{\sigma_Y^2}}. \qquad (4.3.15)$$

(ii) From Eq. (4.3.15), we derive

$$E[X \mid Y] = \mu_1 + (Y - \mu_2)\frac{\sigma_{XY}}{\sigma_Y^2}. \qquad (4.3.16)$$

This expression aligns with the observations made in Example 4.2.8. The same result is obtained from (4.3.11), calculating

$$E[X \mid Y = y] = \int_{\mathbb{R}} x\gamma_{X|Y}(x, y)dx = \mu_1 + (y - \mu_2)\frac{\sigma_{XY}}{\sigma_Y^2}.$$

Example 4.3.26 Let (X_1, X_2, X_3) be a random vector with normal distribution $\mathcal{N}_{\mu,C}$ where

$$\mu = (0, 1, 0), \qquad C = \begin{pmatrix} 1 & 1 & 0 \\ 1 & 2 & 1 \\ 0 & 1 & 3 \end{pmatrix}.$$

To determine

$$E[(X_1, X_2, X_3) \mid X_3],$$

let us first observe that $(X_1, X_3) \sim \mathcal{N}_{(0,0),C_2}$ and $(X_2, X_3) \sim \mathcal{N}_{(1,0),C_1}$ where

$$C_2 = \begin{pmatrix} 1 & 0 \\ 0 & 3 \end{pmatrix}, \qquad C_1 = \begin{pmatrix} 2 & 1 \\ 1 & 3 \end{pmatrix}.$$

Recalling Theorem 4.2.10-3) and observing that X_1 and X_3 are independent since $\mathrm{cov}(X_1, X_3) = 0$, we have that $E[X_1 \mid X_3] = E[X_1] = 0$. Moreover, by (4.3.16),

$$E[X_2 \mid X_3] = 1 + \frac{X_3}{3}.$$

Finally, again by Theorem 4.2.10-2), we get $E[X_3 \mid X_3] = X_3$. In conclusion

$$E[(X_1, X_2, X_3) \mid X_3] = \Big(E[X_1 \mid X_3], E[X_2 \mid X_3], E[X_3 \mid X_3] \Big)$$

$$= \Big(0, 1 + \frac{X_3}{3}, X_3 \Big).$$

Example 4.3.27 The oil received by a refinery contains a concentration of debris equal to Y Kg/barrel where $Y \sim \mathrm{Unif}_{[0,1]}$. It is estimated that the refining process brings the concentration of debris from Y to X with $X \sim \mathrm{Unif}_{[0,\alpha Y]}$ where $\alpha < 1$ is a known positive parameter. Determine:

(i) the densities $\gamma_{(X,Y)}$ and γ_X;
(ii) the expected value of the concentration of debris Y before the refining, given the concentration X after the refining.

(i) The data of the problem are:

$$\mu_Y = \mathrm{Unif}_{[0,1]}, \qquad \mu_{X|Y} = \mathrm{Unif}_{[0,\alpha Y]},$$

that is

$$\gamma_Y(y) = \mathbb{1}_{[0,1]}(y), \qquad \gamma_{X|Y}(x, y) = \frac{1}{\alpha y} \mathbb{1}_{[0,\alpha y]}(x), \qquad y \in \,]0, 1].$$

From formula (4.3.9) for the conditional density, we obtain

$$\gamma_{(X,Y)}(x, y) = \gamma_{X|Y}(x, y)\gamma_Y(y) = \frac{1}{\alpha y} \mathbb{1}_{]0,\alpha y[\times]0,1[}(x, y)$$

and

$$\gamma_X(x) = \int_{\mathbb{R}} \gamma_{(X,Y)}(x, y)dy = \int_{\frac{x}{\alpha}}^{1} \frac{1}{\alpha y} dy \, \mathbb{1}_{]0,\alpha[}(x) = \frac{\log \alpha - \log x}{\alpha} \mathbb{1}_{]0,\alpha[}(x).$$

(ii) Let us calculate $E[Y \mid X]$. We have

$$\gamma_{Y|X}(y, x) = \frac{\gamma_{(X,Y)}(x, y)}{\gamma_X(x)} \mathbb{1}_{(\gamma_X > 0)}(x) = \frac{1}{y(\log \alpha - \log x)} \mathbb{1}_{]0,\alpha y[\times]0,1[}(x, y)$$

$$(4.3.17)$$

from which

$$E[Y \mid X = x] = \int_{\mathbb{R}} y \gamma_{Y|X}(y, x)dy = \frac{1}{\log \alpha - \log x} \mathbb{1}_{]0,\alpha[}(x) \int_{\frac{x}{\alpha}}^{1} dy$$

$$= \frac{\alpha - x}{\alpha(\log \alpha - \log x)} \mathbb{1}_{]0,\alpha[}(x).$$

In conclusion, we have

$$E[Y \mid X] = \frac{\alpha - X}{\alpha(\log \alpha - \log X)}.$$

We note that in (4.3.17) we used the relation

$$\gamma_{Y|X}(y, x) = \frac{\gamma_{(X,Y)}(x, y)}{\gamma_X(x)} \mathbb{1}_{(\gamma_X > 0)}(x) = \frac{\gamma_{X|Y}(x, y)}{\gamma_X(x)} \gamma_Y(y),$$

which is a version of Bayes' formula.

Example 4.3.28 Let (X, Y) be a random vector with marginal distribution $\mu_Y = \chi^2$ and conditional distribution $\mu_{X|Y} = \mathcal{N}_{0, \frac{1}{Y}}$. We recall that the corresponding densities are

$$\gamma_Y(y) = \frac{1}{\sqrt{2\pi y}} e^{-\frac{y}{2}}, \qquad \gamma_{X|Y}(x, y) = \sqrt{\frac{y}{2\pi}} e^{-\frac{x^2 y}{2}}, \qquad y > 0.$$

Then the joint density is given by

$$\gamma_{(X,Y)}(x, y) = \gamma_{X|Y}(x, y)\gamma_Y(y) = \frac{1}{2\pi} e^{-\frac{(1+x^2)y}{2}}, \qquad y > 0,$$

and the marginal of X is

$$\gamma_X(x) = \int_0^{+\infty} \gamma_{(X,Y)}(x, y)dy = \frac{1}{\pi(1 + x^2)}, \qquad x \in \mathbb{R},$$

that is, X has a Cauchy distribution (cf. (2.5.5)).

4.4 Appendix

4.4.1 Proof of Theorem 4.3.4

We say that

$$F : \mathbb{Q} \longrightarrow [0, 1]$$

is a *cumulative distribution function (or CDF) on* $\mathbb{Q}$ if:

(i) F is monotone increasing;
(ii) F is right continuous in the sense that, for every $q \in \mathbb{Q}$, we have

$$F(q) = F(q+) := \lim_{\substack{p \downarrow q \\ p \in \mathbb{Q}}} F(p); \tag{4.4.1}$$

(iii)

$$\lim_{\substack{q \to -\infty \\ q \in \mathbb{Q}}} F(q) = 0 \qquad \text{and} \qquad \lim_{\substack{q \to +\infty \\ q \in \mathbb{Q}}} F(q) = 1. \tag{4.4.2}$$

Lemma 4.4.1 *Given a CDF F on $\mathbb{Q}$, there exists a distribution μ on $\mathbb{R}$ such that*

$$F(q) = \mu(]-\infty, q]), \qquad q \in \mathbb{Q}. \tag{4.4.3}$$

Proof It is not difficult to verify that the function defined by[16]

$$\bar{F}(x) := \lim_{\substack{y \downarrow x \\ y \in \mathbb{Q}}} F(y), \qquad x \in \mathbb{R},$$

is a CDF on $\mathbb{R}$ and $F = \bar{F}$ on $\mathbb{Q}$. Then, by Theorem 1.4.33, there exists a distribution μ that verifies (4.4.3). □

Proof of Theorem 4.3.4 It suffices to consider the case $d = 1$. For each $q \in \mathbb{Q}$, we fix a version of the conditional expectation

$$F(q) := E\left[\mathbb{1}_{(X \leq q)} \mid \mathscr{G}\right]$$

whose existence is guaranteed by Theorem 4.2.1. Actually, $F = F(q, \omega)$ also depends on $\omega \in \Omega$ but for brevity, we will write $F = F(q)$ considering $F(q)$ as a r.v. ($\mathscr{G}$-measurable, by definition). Based on the properties of the conditional expectation and since $\mathbb{Q}$ is a countable set, we have that *P-almost surely F is a CDF on* $\mathbb{Q}$: more precisely, there exists a negligible event $C \in \mathscr{G}$ such that $F = F(\cdot, \omega)$ is a CDF on $\mathbb{Q}$ for every $\omega \in \Omega \setminus C$. Indeed, if $p, q \in \mathbb{Q}$ with $p \leq q$, then $\mathbb{1}_{(X \leq p)} \leq \mathbb{1}_{(X \leq q)}$ and therefore

$$F(p) = E\left[\mathbb{1}_{(X \leq p)} \mid \mathscr{G}\right] \leq E\left[\mathbb{1}_{(X \leq q)} \mid \mathscr{G}\right] = F(q)$$

except for a $\mathscr{G}$-measurable negligible event, due to the monotonicity property of the conditional expectation. Similarly, properties (4.4.1) and (4.4.2) are proved as a consequence of the dominated convergence theorem for the conditional expectation: for example, if $(p_n)_{n \in \mathbb{N}}$ is a sequence in $\mathbb{Q}$ such that $p_n \downarrow q \in \mathbb{Q}$, then the sequence of random variables $\left(\mathbb{1}_{(X \leq p_n)}\right)_{n \in \mathbb{N}}$ is bounded and converges pointwise

$$\lim_{n \to \infty} \mathbb{1}_{(X \leq p_n)}(\omega) = \mathbb{1}_{(X \leq q)}(\omega), \qquad \omega \in \Omega,$$

from which

$$\lim_{n \to \infty} F(p_n) = \lim_{n \to \infty} E\left[\mathbb{1}_{(X \leq p_n)} \mid \mathscr{G}\right] = E\left[\mathbb{1}_{(X \leq q)} \mid \mathscr{G}\right] = F(q).$$

Based on Lemma 4.4.1, for each $\omega \in \Omega \setminus C$ there exists a distribution $\mu = \mu(\cdot, \omega)$ (but we will simply write $\mu = \mu(H)$, for $H \in \mathscr{B}$) such that

$$\mu(]-\infty, p]) = F(p), \qquad p \in \mathbb{Q}.$$

By construction, μ is a distribution on $\mathbb{R}$, except for the negligible event $C \in \mathscr{G}$: on the other hand, we can extend μ on Ω by setting, for example, $\mu(\cdot, \omega) \equiv \delta_0$ for

---

[16] The limit exists due to the monotonicity of F.

$\omega \in C$. We now prove that μ also satisfies (4.3.3): for this purpose we use Dynkin's Theorem A.0.3 and set

$$\mathcal{M} = \{H \in \mathcal{B} \mid \mu(H) = E\left[\mathbb{1}_{(X \in H)} \mid \mathcal{G}\right]\}.$$

The family

$$\mathcal{A} = \{] - \infty, p] \mid p \in \mathbb{Q}\}$$

is $\cap$-closed, $\sigma(\mathcal{A}) = \mathcal{B}$ and, by construction, $\mathcal{A} \subseteq \mathcal{M}$. If we verify that $\mathcal{M}$ is a monotone family, Dynkin's theorem will imply that $\mathcal{M} = \mathcal{B}$ from which the thesis follows. Now, we have:

(i) $\mathbb{R} \in \mathcal{M}$ since $\mathbb{1}_{\mathbb{R}}(X) \equiv 1$ is $\mathcal{G}$-measurable and therefore coincides with its own conditional expectation. On the other hand, $\mu(\mathbb{R}) = 1$ on Ω and therefore $\mu(\mathbb{R}) = E\left[\mathbb{1}_{\mathbb{R}}(X) \mid \mathcal{G}\right]$;

(ii) if $H, K \in \mathcal{M}$ and $H \subseteq K$, then

$$\mu(K \setminus H) = \mu(K) - \mu(H)$$
$$= E\left[\mathbb{1}_{K}(X) \mid \mathcal{G}\right] - E\left[\mathbb{1}_{H}(X) \mid \mathcal{G}\right] =$$

(by the linearity of conditional expectation)

$$= E\left[\mathbb{1}_{K}(X) - \mathbb{1}_{H}(X) \mid \mathcal{G}\right]$$
$$= E\left[\mathbb{1}_{K \setminus H}(X) \mid \mathcal{G}\right];$$

(iii) let $(H_n)_{n \in \mathbb{N}}$ be an increasing sequence of elements of $\mathcal{M}$. By the continuity from below of distributions, we have

$$\mu(H) = \lim_{n \to \infty} \mu(H_n), \qquad H := \bigcup_{n \geq 1} H_n.$$

On the other hand, by Beppo Levi's theorem for conditional expectation, we have

$$\lim_{n \to \infty} \mu(H_n) = \lim_{n \to \infty} E\left[\mathbb{1}_{H_n}(X) \mid \mathcal{G}\right] = E\left[\mathbb{1}_{H}(X) \mid \mathcal{G}\right].$$

$\square$

4.4.2   Proof of Proposition 4.3.20

Consider an absolutely continuous random vector (X, Y) in $\mathbb{R}^d \times \mathbb{R}$, with density $\gamma_{(X,Y)}$.

Lemma 4.4.2 *For each* $g \in b\mathscr{B}_{d+1}$, *we have*

$$\int_{(\gamma_Y=0)} \int_{\mathbb{R}^d} g(x, y)\gamma_{(X,Y)}(x, y)dxdy = 0. \tag{4.4.4}$$

Proof Let γ_Y be the density of Y in (4.3.8). Since $\gamma_{(X,Y)} \geq 0$, by Corollary 2.2.15 we have

$$\gamma_Y(y) = 0 \quad \Longrightarrow \quad \gamma_{(X,Y)}(\cdot, y) = 0 \quad \text{a.s.}$$

Then, for each $g \in b\mathscr{B}_{d+1}$ and for each y such that $\gamma_Y(y) = 0$, we have

$$\int_{\mathbb{R}^d} g(x, y)\gamma_{(X,Y)}(x, y)dx = 0,$$

from which follows (4.4.4).   □

Proof of Proposition 4.3.20 We have to prove that the family $(\mu(\cdot; y))_{y \in (\gamma_Y > 0)}$ defined in (4.3.10)–(4.3.9) is a regular version of the conditional distribution *function* of X given Y according to the definition of Theorem 4.3.16.

First of all, $\mu(\cdot; y)$ is a distribution: in fact, $\gamma_{X|Y}(\cdot, y)$ in (4.3.9) is a density since it is a measurable, non-negative function such that, by (4.3.8),

$$\int_{\mathbb{R}^d} \gamma_{X|Y}(x, y)dx = \frac{1}{\gamma_Y(y)} \int_{\mathbb{R}^d} \gamma_{(X,Y)}(x, y)dx = 1.$$

Fix $H \in \mathscr{B}_d$. Regarding i) of Theorem 4.3.16, the fact that $y \mapsto \mu(H; y) \in m\mathscr{B}$ follows from Fubini's theorem and the fact that $\gamma_{X|Y}$ is a Borel-measurable function. As for ii) of Theorem 4.3.16, consider $W \in b\sigma(Y)$: by Doob's theorem, $W = g(Y)$ with $g \in b\mathscr{B}$ and thus we have

$$E[W\mu(H; Y)] = \int_{\mathbb{R}} g(y)\mu(H; y)\gamma_Y(y)dy =$$

(by Fubini's theorem)

$$= \int_{(\gamma_Y > 0)} g(y) \left(\int_H \gamma_{X|Y}(x, y)dx \right) \gamma_Y(y)dy$$

$$= \int_{(\gamma_Y > 0)} \int_H g(y)\gamma_{(X,Y)}(x, y)dx\, dy =$$

(by (4.4.4))

$$= \iint_{\mathbb{R}^d \times \mathbb{R}} g(y)\mathbb{1}_H(x)\gamma_{(X,Y)}(x, y)dx\, dy = E\left[W\mathbb{1}_{(X \in H)}\right].$$

□

Chapter 5
Summary Exercises

5.1 Measures and Probability Spaces

Exercise 5.1.1 Let A, B, C be independent events on the probability space $(\Omega, \mathscr{F}, P)$. Determine if:

  (i) A and B^c are independent;
 (ii) A and $B \cup C$ are independent;
(iii) $A \cup C$ and $B \cup C$ are independent.

Solution

  (i) This is the content of Proposition 1.3.25, according to which $A, B \in \mathscr{F}$ are independent if and only if A^c, B or A, B^c or A^c, B^c are independent;
 (ii) based on point (i), to prove that A and $B \cup C$ are independent, it is sufficient to verify that A and $(B \cup C)^c = B^c \cap C^c$ are independent or that A and $B \cap C$ are independent: by the hypothesis of independence of A, B, C we have

$$P(A \cap (B \cap C)) = P(A)P(B)P(C) = P(A)P(B \cap C)$$

which proves the thesis.
(iii) in general $A \cup C$ and $B \cup C$ are not independent; to show this, we still use Proposition 1.3.25 and verify that $A \cap C$ and $B \cap C$ are not, in general, independent: in fact we have

$$P((A \cap C) \cap (B \cap C)) = P(A \cap B \cap C) = P(A)P(B)P(C),$$

but

$$P(A \cap C)P(B \cap C) = P(A)P(B)P(C)^2.$$

© The Editor(s) (if applicable) and The Author(s), under exclusive license to Springer Nature Switzerland AG 2024
A. Pascucci, *Probability Theory I*, La Matematica per il 3+2 165,
https://doi.org/10.1007/978-3-031-63190-0_5

Exercise 5.1.2 Let A, B, C be independent events on the probability space $(\Omega, \mathscr{F}, P)$, with $P(A) = P(B) = P(C) = \frac{1}{2}$. Determine:

(i) $P(A \cup B)$;
(ii) $P(A \cup B \cup C)$.

Solution

(i) We have

$$P(A \cup B) = 1 - P(A^c \cap B^c) = 1 - P(A^c)P(B^c) = 1 - \frac{1}{4} = \frac{3}{4}.$$

Alternatively, remembering that the symbol $\uplus$ indicates disjoint union, we have

$$P(A \cup B) = P\left(A \uplus (B \cap A^c)\right) = P(A) + P(B \cap A^c) =$$

(by the independence of B and A^c)

$$= \frac{1}{2} + \frac{1}{2} \cdot \frac{1}{2} = \frac{3}{4}.$$

(ii) Similarly, we have

$$P(A \cup B \cup C) = 1 - P(A^c \cap B^c \cap C^c) = 1 - P(A^c)P(B^c)P(C^c)$$

$$= 1 - \frac{1}{8} = \frac{7}{8},$$

or

$$P(A \cup B \cup C) = P(A \cup B) + P\left(C \cap (A \cup B)^c\right) =$$

(by point (i))

$$= \frac{3}{4} + P\left(C \cap A^c \cap B^c\right) =$$

(by the hypothesis of independence)

$$= \frac{3}{4} + P(C)P(A^c)P(B^c) = \frac{3}{4} + \frac{1}{8} = \frac{7}{8}.$$

Exercise 5.1.3 Tests show that a vaccine is effective against virus α in 55 cases out of 100, against virus β in 65 cases out of 100 and against at least one of the two viruses in 80 cases out of 100. Determine the probability that the vaccine is effective against both viruses.

Solution Consider the events $A=$"the vaccine is effective against virus α" and $B=$"the vaccine is effective against virus β". We know that $P(A) = 55\%$, $P(B) = 65\%$ and $P(A \cup B) = 80\%$. Then

$$P(A \cap B) = P(A) + P(B) - P(A \cup B) = 40\%.$$

Exercise 5.1.4 For $n \geq 2$, consider Ω as the set of permutations of $I_n :=$ $\{1, 2, \dots, n\}$, i.e., the collection of bijective functions from I_n to itself, endowed with the uniform probability distribution P. A permutation ω has $i \in I_n$ as a fixed point if and only if $\omega(i) = i$. Define the event A_i as the event "the permutation has i as a fixed point". Determine:

(i) $P(A_i)$ for $i = 1, \dots, 10$;
(ii) whether these events are independent or not;
(iii) the expected value of the number of fixed points.

Solution

(i)   A permutation with i as a fixed point is equivalent to a permutation of the remaining $(n-1)$ elements, so there are $(n-1)!$ such permutations (regardless of i), hence $P(A_i) = \frac{(n-1)!}{n!} = \frac{1}{n}$.
(ii) Proceeding as in the previous point, for $i \neq j$ we have

$$P(A_i \cap A_j) = \frac{(n-2)!}{n!} = \frac{1}{n(n-1)} \neq \frac{1}{n^2} = P(A_i)P(A_j)$$

and therefore the events are not independent.
(iii) We need to determine the expected value of the r.v.

$$\mathbb{1}_{A_1} + \mathbb{1}_{A_2} + \cdots + \mathbb{1}_{A_n}.$$

By linearity of expectation, this is equal to $n \cdot \frac{1}{n} = 1$.

Exercise 5.1.5 Three draws are made without replacement from an urn containing 3 white balls, 2 black balls, and 2 red balls. Let X and Y be the number of white balls and black balls drawn, respectively. Determine:

(i) $P((X = 1) \cap (Y = 0))$;
(ii) $P(X = 1 \mid Y = 0)$.

Solution

(i)   We have

$$P((X = 1) \cap (Y = 0)) = \frac{3}{\binom{7}{3}} = \frac{3}{35}.$$

(ii) Since

$$P(Y = 0) = \frac{\binom{5}{3}}{\binom{7}{3}} = \frac{2}{7}$$

we have

$$P(X = 1 \mid Y = 0) = \frac{P((X = 1) \cap (Y = 0))}{P(Y = 0)} = \frac{3}{10}.$$

Exercise 5.1.6 Let $X, Y \sim \text{Be}_p$, with $0 < p < 1$, be independent. Let $Z = \mathbb{1}_{(X+Y=0)}$ and determine:

(i) the distribution of Z;
(ii) whether X and Z are independent.

Solution

(i) Z can only take the values 0, 1 and is equal to

$$P(Z = 1) = P((X = 0) \cap (Y = 0)) = (1 - p)^2$$

so that

$$Z \sim (1 - p)^2 \delta_0 + (1 - (1 - p)^2)\delta_1.$$

(ii) X and Z are not independent because, for example, we have

$$P((X = 0) \cap (Z = 1)) = P(Y = 0) = 1 - p$$

and

$$P(X = 0)P(Z = 1) = (1 - p)^3.$$

Exercise 5.1.7 Let X and Y be the values (natural numbers from 1 to 10) of two cards drawn in sequence from a deck of 40 cards, without replacement. Determine:

(i) the joint distribution function of X and Y;
(ii) $P(X < Y)$;
(iii) the distribution function of Y. Are the random variables X and Y independent?

Solution

(i) For $h, k \in I_{10}$, we have $P(X = h) = \frac{1}{10}$, i.e., $X \sim \text{Unif}_{10}$ and

$$P(Y = k \mid X = h) = \begin{cases} \frac{3}{39} & \text{if } h = k, \\ \frac{4}{39} & \text{if } h \neq k. \end{cases}$$

Then the distribution function of (X, Y) is given by

$$\bar{\mu}_{(X,Y)}(h, k) = P((X = h) \cap (Y = k)) = P(Y = k \mid X = h) P(X = h)$$

$$= \begin{cases} \frac{1}{130} & \text{if } h = k, \\ \frac{2}{195} & \text{if } h \neq k. \end{cases}$$

(ii) We have

$$P(X < Y) = \sum_{1 \leq h < k \leq 10} \bar{\mu}_{(X,Y)}(h, k) = \frac{2}{195} \sum_{k=2}^{10} (k - 1) = \frac{2}{195} \cdot 45 = \frac{6}{13}.$$

(iii) The distribution function of Y is obtained from

$$\bar{\mu}_Y(k) = \sum_{h=1}^{10} \bar{\mu}_{(X,Y)}(h, k) = \frac{1}{10} \sum_{h=1}^{10} P(Y = k \mid X = h)$$

$$= \frac{1}{10} \left(\frac{3}{39} + 9 \cdot \frac{4}{39} \right) = \frac{1}{10}$$

i.e., also $Y \sim \text{Unif}_{10}$. It also follows that X, Y are not independent since the joint distribution function is not the product of the marginals (cf. Theorem 2.3.23).

Exercise 5.1.8 From a deck of 40 cards, three cards are drawn in sequence and without replacement, whose values (integers from 1 to 10) are indicated respectively with X_1, X_2, and X_3. Determine:

(i) the distribution of X_2;
(ii) the probabilities of the events

$A = (X_1 \leq 4) \cap (X_2 \geq 5) \cap (X_3 \geq 5)$;
$B = $ "at most one card drawn has a value less than or equal to 4";

(iii) if A, B are independent and $P(A \mid B)$;
(iv) consider the r.v.

$N = $ "number of cards drawn whose value is less than or equal to 4".

Are the random variables X_2 and N independent?

Solution

(i) X_2 has a uniform distribution on $I_{10} = \{n \in \mathbb{N} \mid n \leq 10\}$, i.e., $X_2 \sim \text{Unif}_{I_{10}}$: to verify this rigorously, one can proceed as in Example 2.3.24 or using the Law of total probability:

$$P(X_2 = n) = P(X_2 = n \mid X_1 = n)P(X_1 = n)$$

$$+ P(X_2 = n \mid X_1 \neq n) P(X_1 \neq n)$$

$$= \frac{3}{39} \cdot \frac{1}{10} + \frac{4}{39} \cdot \frac{9}{10} = \frac{1}{10}, \qquad n \in I_{10}.$$

(ii) We solve the question in two ways: using conditional probability and in particular formula (1.3.5), we have

$$P(A) = P(X_1 \leq 4) P(X_2 \geq 5 \mid X_1 \leq 4) P(X_3 \geq 5 \mid (X_1 \leq 4) \cap (X_2 \geq 5))$$

$$= \frac{4}{10} \cdot \frac{24}{39} \cdot \frac{23}{38}.$$

The same result is obtained with the method of successive choices: we observe that we have to use arrangements because we are interested in the order of extraction of the cards. Therefore,

$$P(A) = \frac{16 \cdot |\mathbf{D}_{24,2}|}{|\mathbf{D}_{40,3}|}.$$

Then $B = B_0 \uplus B_1$ where B_0 is the event "no card drawn has a value less than or equal to 4" and B_1 is the event "exactly one card drawn has a value less than or equal to 4". We have $P(B) = P(B_0) + P(B_1)$ and

$$P(B_0) = \frac{|\mathbf{C}_{24,3}|}{|\mathbf{C}_{40,3}|} = \frac{|\mathbf{D}_{24,3}|}{|\mathbf{D}_{40,3}|}$$

$$P(B_1) = \frac{16 \cdot |\mathbf{C}_{24,2}|}{|\mathbf{C}_{40,3}|} = \frac{3 \cdot 16 \cdot |\mathbf{D}_{24,2}|}{|\mathbf{D}_{40,3}|}.$$

The factor "3" that appears in the last expression is due to the fact that, if we use the arrangements, then we must take into account the order and therefore we must also make the choice of the position (among the three possible) of the card that has a value less than or equal to 4.

(iii) $A \subseteq B$ and therefore $A \cap B = A$. But $P(A \cap B) = P(A) \neq P(A)P(B)$ and therefore they are not independent events. Moreover, we have $P(A \mid B) = \frac{P(A)}{P(B)}$.

(iv) X_2 and N are not independent because, for example, $(X_2 = 4) \cap (N = 0) = \emptyset$ but

$$P(X_2 = 4) P(N = 0) \neq 0.$$

Exercise 5.1.9 Two urns each contain 1 white ball and 4 black balls.

(i) Draw 3 balls from the first urn and 3 balls from the second urn, and calculate the probability that at least one of them is white;

(ii) put all the balls in the same urn (which then contains 2 white balls and 8 black balls) and draw 6 balls. Calculate the probability that at least one of them is white;

(iii) as in point (ii) assuming that the extraction takes place with replacement, i.e. drawing a ball at a time and putting it back in the urn. Calculate the probability that the color of at least one of the six balls drawn is white.

Solution

(i) The probability of drawing a white ball from the first urn (event A) is equal to $\frac{3}{5}$ and the same for the second urn (event B). Moreover, A and B are independent. Then

$$P(A \cup B) = P(A) + P(B) - P(A \cap B)$$

$$= P(A) + P(B) - P(A)P(B) = \frac{21}{25} = 0.84.$$

(ii) Number the two white balls (ball 1 and ball 2) and indicate with A_i, $i = 1, 2$, the event according to which among the 6 balls drawn there is ball i. Then we have $P(A_1) = P(A_2) = \frac{6}{10}$, $P(A_1 \mid A_2) = \frac{5}{9}$ and

$$P(A_1 \cup A_2) = P(A_1) + P(A_2) - P(A_1 \cap A_2)$$

$$= P(A_1) + P(A_2) - P(A_1 \mid A_2)P(A_2) = \frac{13}{15} \approx 0.87.$$

Alternatively, we can consider the r.v. $X \sim \mathrm{Hyper}_{n,b,N}$ with hypergeometric distribution, according to formula (2.1.9) with $b = 2$, $N = 10$ and $n = 6$. Then X indicates the number of white balls drawn. Thus we have

$$P(X = 1) + P(X = 2) = \frac{13}{15}.$$

(iii) In this case, we can consider the r.v. $S \sim \mathrm{Bin}_{n,p}$ with binomial distribution, according to the formula (2.1.5) with $n = 6$ and $p = \frac{2}{10}$. Then S indicates the number of white balls drawn. Then we have

$$\sum_{i=1}^{6} P(S = i) \approx 0.74.$$

Exercise 5.1.10 An urn contains 3 white balls, 6 red balls, and 6 black balls. Two balls are drawn: if they have the same color they are thrown away, while if they have different colors they are put back in the urn. Then two more balls are drawn. Determine the probability of the following events:

(i) $A_1 = $ "the two balls of the first draw are white";

(ii) $A_2 =$ "the two balls of the first draw have the same color";
(iii) $A_3 =$ "all four drawn balls are white";
(iv) $A_4 =$ "all four drawn balls are red".

Solution

(i) $P(A_1) = \dfrac{|C_{3,2}|}{|C_{15,2}|} = \dfrac{\binom{3}{2}}{\binom{15}{2}} = \dfrac{1}{35}.$

(ii) $P(A_2) = \dfrac{|C_{3,2}|+|C_{6,2}|+|C_{6,2}|}{|C_{15,2}|} = \dfrac{\binom{3}{2}+2\binom{6}{2}}{\binom{15}{2}} = \dfrac{11}{35}.$

(iii) if $B =$ "the two balls of the second draw are white" then

$$P(A_3) = P(B \mid A_1)P(A_1) = 0.$$

(iv) if $C_i =$ "the two balls of the i-th draw are red" then

$$P(A_4) = P(C_1 \cap C_2) = P(C_2 \mid C_1)P(C_1)$$

$$= \frac{|C_{4,2}|}{|C_{13,2}|}\frac{|C_{6,2}|}{|C_{15,2}|} = \frac{\binom{4}{2}}{\binom{13}{2}}\frac{\binom{6}{2}}{\binom{15}{2}} = \frac{1}{91}.$$

Exercise 5.1.11 Nine students randomly and independently choose a professor, among three available, to take the exam with. Consider the events:

$A =$ "exactly three students choose the first professor";
$B =$ "each professor is chosen by three students";
$C =$ "one professor is chosen by two students, another by three students, and the remaining one by four students".

Determine:

(i) $P(A)$;
(ii) $P(B)$;
(iii) $P(A \mid B)$ and $P(B \mid A)$;
(iv) $P(C)$.

Solution The sample space of all possible choices of the students is $\Omega = \mathbf{DR}_{3,9}$, from which $|\Omega| = 3^9$. Remember that Ω is the space of functions from I_9 to I_3 and each function corresponds to a possible choice of the nine students.

(i) There are $|C_{9,3}|$ possible ways to determine the three students who choose the first professor and consequently

$$P(A) = \frac{|C_{9,3}|\,|\mathbf{DR}_{2,6}|}{|\mathbf{DR}_{3,9}|} = \frac{\binom{9}{3}2^6}{3^9} \approx 27\%.$$

We have equivalently $P(A) = \mathrm{Bin}_{9,\frac{1}{3}}(\{3\})$.

(ii) There are $|C_{9,3}|$ possible ways to determine the three students who choose the first professor and $|C_{6,3}|$ possible ways to determine the three students who choose the second professor: consequently

$$P(B) = \frac{|C_{9,3}|\,|C_{6,3}|}{|DR_{3,9}|} = \frac{\binom{9}{3}\binom{6}{3}}{3^9} \approx 8.5\%.$$

(iii) Since $B \subseteq A$ we have

$$P(A \mid B) = 1, \qquad P(B \mid A) = \frac{P(B)}{P(A)} \approx 31\%.$$

(iv) Proceed in a similar way to point (ii) but with the difference that a factor 3! must be added because the order of choice of the professors is not specified. We get

$$P(C) = 3!\frac{|C_{9,2}|\,|C_{7,3}|}{|DR_{3,9}|} = 6\frac{\binom{9}{3}\binom{6}{3}}{3^9} \approx 38\%.$$

Exercise 5.1.12 An urn contains 3 red balls, 3 white balls, and 4 black balls. Two coins are tossed: if two heads are obtained, a red ball is added to the urn; if two tails are obtained, a white ball is added; in other cases, nothing is added. Two balls are drawn in sequence and without replacement from the urn. Determine the probability:

(i) that the first ball drawn is black;
(ii) of having obtained at least one tail, knowing that the first ball drawn is black;
(iii) that both balls drawn are black, knowing that no balls were added.

Solution

(i) Consider the following events: $N_1 =$ "the first ball drawn is black", $TT =$ "the result of the two coin tosses is two heads", $CT =$ "the result of the first coin toss is tails and the second is heads" and similarly we define CC and TC. Using the Law of total probability, we have

$$P(N_1) = P(N_1 \mid TT)P(TT) + P(N_1 \mid CC)P(CC)$$
$$+ P(N_1 \mid CT \cup TC)P(CT \cup TC)$$
$$= \frac{4}{11} \cdot \frac{1}{4} + \frac{4}{11} \cdot \frac{1}{4} + \frac{4}{10} \cdot \frac{2}{4} = \frac{21}{55}.$$

(ii) Using Bayes' formula, we have

$$P(CT \cup TC \cup CC \mid N_1) = 1 - P(TT \mid N_1)$$
$$= 1 - \frac{P(N_1 \mid TT)P(TT)}{P(N_1)} = \frac{16}{21}.$$

(iii) Let $\bar{P} = P(\cdot \mid CT \cup TC)$. By the multiplication rule (1.3.5), we have

$$\bar{P}(N_1 \cap N_2) = \bar{P}(N_1)\bar{P}(N_2 \mid N_1) = \frac{4}{10} \cdot \frac{3}{9} = \frac{2}{15}.$$

Exercise 5.1.13 Six coins are randomly and independently placed in three boxes. Consider the events:

$A =$ "the first box contains two coins";
$B =$ "each box contains two coins".

Determine:

(i) $P(A)$;
(ii) $P(B)$;
(iii) $P(A \mid B)$ and $P(B \mid A)$.

Solution The sample space of all possible coin arrangements is $\Omega = \mathbf{DR}_{3,6}$, from which $|\Omega| = 3^6$. Remember that Ω is the space of functions from I_6 to I_3 and each function corresponds to a possible arrangement of the six coins.

(i) There are $|\mathbf{C}_{6,2}|$ possible ways to place the two coins in the first box and consequently

$$P(A) = \frac{|\mathbf{C}_{6,2}| \, |\mathbf{DR}_{2,4}|}{|\mathbf{DR}_{3,6}|} = \frac{\binom{6}{2}2^4}{3^6} \approx 33\%.$$

Alternatively, $P(A) = \mathrm{Bin}_{6,\frac{1}{3}}(\{2\})$.

(ii) There are $|\mathbf{C}_{6,2}|$ possible ways to place the two coins in the first box and $|\mathbf{C}_{4,2}|$ possible ways to place the two coins in the second box: consequently

$$P(B) = \frac{|\mathbf{C}_{6,2}| \, |\mathbf{C}_{4,2}|}{|\mathbf{DR}_{3,6}|} = \frac{\binom{6}{2}\binom{4}{2}}{3^6} \approx 12\%.$$

(iii) Since $B \subseteq A$, we have

$$P(A \mid B) = 1, \qquad P(B \mid A) = \frac{P(B)}{P(A)} \approx 37.5\%.$$

Exercise 5.1.14 A LED light bulb has, each day, independently of the other days, a probability $p = 0.1\%$ of burning out. Determine:

(i) the average lifespan (in days) of the light bulb;
(ii) the probability that the light bulb lasts at least one year.

In a city, there are 10,000 streetlights that use this light bulb. Write a formula to determine (no need to calculate it) the minimum number of spare light bulbs needed so that, with a 99% probability, all the light bulbs, among the 10,000 installed, that burn out in one day can be replaced.

Solution

(i)–(ii) Let T be the r.v. that indicates the day the light bulb burns out. Then $T \sim$ Geom$_p$ (cf. Example 2.1.25). Therefore, the average lifespan (in days) of the light bulb is

$$E[T] = \frac{1}{p} = 1000.$$

Moreover, the probability that the light bulb lasts at least one year is (cf. Theorem 2.1.26)

$$P(T > 365) = (1-p)^{365} \approx 69.4\%$$

(iii) Let X be the number of light bulbs, among the 10,000 installed, that burn out in one day. Then $X \sim$ Bin$_{10000,p}$ (cf. Example 2.1.21). We need to determine the minimum N such that

$$P(X \le N) \ge 99\%.$$

Now we have (we could also use the Poisson approximation, cf. Example 2.1.23):

$$P(X \le N) = \sum_{k=0}^{N} \binom{10000}{k} p^k (1-p)^{n-k}.$$

An explicit calculation shows that

$$P(X \le 17) = 98.57\%, \qquad P(X \le 18) = 99.28\%,$$

so $N = 18$.

Exercise 5.1.15 In a portion of the sky, N stars are counted, positioned uniformly and independently of each other. Suppose that the portion of the sky is divided into two parts A and B whose area is one double the other, $|A| = 2|B|$, and let N_A be the number of stars in A.

(i) Determine $P(N_A = k)$;
(ii) the number N depends on the power of the telescope used. Then suppose that N is a Poisson random variable, $N \sim$ Poisson$_\lambda$ with $\lambda > 0$: determine the probability that there is only one star in A.

Solution

(i) Since the position distribution is uniform, each star has a probability $p = \frac{2}{3}$ of being in A independently of the others. Then

$$P(N_A = k) = \text{Bin}_{N,p}(k) = \binom{N}{k}\frac{2^k}{3^N}.$$

(ii) By the Law of total probability, we have

$$\sum_{N=0}^{\infty} P(N_A = 1)\frac{e^{-\lambda}\lambda^N}{N!} = e^{-\lambda}\sum_{N=1}^{\infty}\frac{2N}{3^N}\frac{\lambda^N}{N!} = \frac{2\lambda}{3}e^{-\frac{2\lambda}{3}}.$$

Exercise 5.1.16 Let $X \sim$ Poisson$_\lambda$ with $\lambda > 0$. Give an example of $f \in m\mathscr{B}$ such that $f(X)$ is not absolutely integrable.

Solution Just consider any measurable function (for instance, piecewise constant) such that $f(k) = \frac{k!}{\lambda^k}$ for $k \in \mathbb{N}$.

Exercise 5.1.17 Two consecutive extractions without replacement are performed from an urn containing 90 numbered balls. Let p_1 and p_2 be the numbers of the two balls drawn. Determine:

(i) the probability of event $A = (p_2 > p_1)$;
(ii) the distribution of the r.v. $\mathbb{1}_A$;
(iii) the probability that $p_1 \geq 45$ knowing that $p_2 > p_1$.

Solution

(i) By the Law of total probability, we have

$$P(A) = \sum_{k=1}^{90} P(A \mid p_1 = k)P(p_1 = k) = \sum_{k=1}^{90}\frac{90-k}{89}\cdot\frac{1}{90} = \frac{1}{2}.$$

(ii) $\mathbb{1}_A$ has a Bernoulli distribution, $\mathbb{1}_A \sim \text{Be}_{\frac{1}{2}}$.

(iii)

$$P(p_1 \geq 45 \mid A) = \frac{P\left((p_1 \geq 45)\cap A\right)}{P(A)} = 2\sum_{k=45}^{90}\frac{90-k}{89}\cdot\frac{1}{90} \approx 25.8\%.$$

Exercise 5.1.18 In a supermarket, a certain number, N, of customers evenly distribute themselves among the 5 available cash registers upon exiting. Let N_1 denote the count of customers heading to the first cash register.

(i) Assuming $N = 100$, determine (or explain how it is possible to determine) the maximum value $\bar{n} \in \mathbb{N}$ such that

$$P(N_1 \geq \bar{n}) \geq 90\%;$$

(ii) assuming that $N \sim \text{Poisson}_{100}$, write a formula to calculate $P(N_1 \geq 15)$.

Solution

(i) Each customer has a probability of $\frac{1}{5}$ of going to the first cash register, independently of the others, and therefore $N_1 \sim \text{Bin}_{100,\frac{1}{5}}$. Then we need to determine the maximum value of n such that

$$90\% \leq P(N_1 \geq n) = \sum_{k=n}^{100} \binom{100}{k} \left(\frac{1}{5}\right)^k \left(\frac{4}{5}\right)^{100-k}.$$

We find that $P(N_1 \geq 16) \approx 87.1\%$ and $P(N_1 \geq 15) \approx 91.9\%$, so $\bar{n} = 15$.

(ii) We have

$$P(N_1 \geq 15) = \sum_{h=0}^{\infty} P(N_1 \geq 15 \mid N = h) P(N = h)$$

$$= \sum_{h=15}^{\infty} \sum_{k=15}^{h} \binom{h}{k} \left(\frac{1}{5}\right)^k \left(\frac{4}{5}\right)^{h-k} \frac{e^{-100} 100^h}{h!} \approx 89.5\%.$$

Exercise 5.1.19 Two friends, A and B, play by each rolling a die: A's die is normal while B's die has the numbers from 2 to 7 on its faces. The one who gets the strictly higher number wins: in case of a tie, they roll the dice again. Determine:

(i) the probability that, rolling the dice once, A wins;
(ii) the probability that A wins within the first ten rolls (≤ 10);
(iii) the probability that there are no winners in the first ten rolls;
(iv) the expected number of A's wins within the first ten rolls (≤ 10).

Solution

(i) Let N_A and N_B be the numbers obtained in the first roll of the dice: then

$$P(N_A > N_B) = \sum_{k=2}^{7} P(N_A > k \mid N_B = k) P(N_B = k)$$

$$= \frac{1}{6}\left(\frac{4}{6} + \frac{3}{6} + \frac{2}{6} + \frac{1}{6}\right) = \frac{10}{36} =: p.$$

(ii) The r.v. T indicating the first time A wins has a geometric distribution with parameter p: therefore

$$P(T \leq 10) = 1 - P(T > 10) = 1 - (1 - p)^{10} \approx 96\%.$$

(iii) As in point (i), we calculate

$$P(N_A = N_B) = \frac{5}{36}$$

and therefore the probability sought is $\left(\frac{5}{36}\right)^{10}$.

(iv) If N represents the number of wins of A in the first ten throws, then $N \sim Bin_{10,p}$ and therefore $E[N] = \frac{100}{36}$.

Exercise 5.1.20 A switchboard randomly distributes incoming calls among 10 operators. Let Y_n be the uniform r.v. on $\{1, 2, 3, \ldots, 10\}$ that indicates the operator chosen by the switchboard for the n-th call. When the i-th operator receives the n-th call (event $Y_n = i$), there is a probability p_i in $]0, 1[$ that the operator is on break and therefore the call is lost. Let X_n be the r.v. that indicates whether the n-th call is lost ($X_n = 1$) or received ($X_n = 0$). We assume that the random variables X_n are independent.

(i) Determine the distribution of X_n;
(ii) let N be the sequence number of the first lost call. Determine the distribution and the mean of N;
(iii) calculate the probability that none of the first 100 calls are lost.

Solution

(i) X_n is a Bernoulli r.v. and, by the Law of total probability, we have

$$P(X_n = 1) = \sum_{i=1}^{10} P(X_n = 1 \mid Y_n = i)P(Y_n = i) = \frac{1}{10}\sum_{i=1}^{10} p_i =: p.$$

Hence $X_n \sim Be_p$.

(ii) $N \sim Geom_p$ and therefore $E[N] = \frac{1}{p}$.

(iii) We have (cf. Theorem 2.1.26)

$$P(N > 100) = (1 - p)^{100}.$$

Exercise 5.1.21 In a 100-m race, T_1 and T_2 are respectively the times (in seconds) obtained by two runners. We assume that T_1, T_2 are independent random variables with $T_i \sim \text{Exp}_{\lambda_i}$, $\lambda_i > 0$ for $i = 1, 2$. Let $T_{\max} = T_1 \vee T_2$ and $T_{\min} = T_1 \wedge T_2$, determine:

(i) the CDF of $T_{\max}$ and $T_{\min}$;
(ii) the probability that at least one of the two runners obtains a time less than 10 seconds, assuming $\lambda_1 = \lambda_2 = \frac{1}{10}$;
(iii) the probability that both runners obtain a time less than 10 seconds, assuming $\lambda_1 = \lambda_2 = \frac{1}{10}$;
(iv) $E[t \vee T_2]$ for every $t > 0$ and, using the freezing lemma, $E[T_{\max} \mid T_1]$.

Solution

(i) By Proposition 2.6.9 on the maximum and minimum of independent variables, we have the following relationship between the distribution functions

$$F_{T_{\max}}(t) = F_{T_1}(t) F_{T_2}(t) = \left(1 - e^{-\lambda_1 t}\right)\left(1 - e^{-\lambda_2 t}\right), \qquad 0 t \geq 0,$$

$$F_{T_{\min}}(t) = 1 - \left(1 - F_{T_1}(t)\right)\left(1 - F_{T_2}(t)\right) = 1 - e^{-(\lambda_1 + \lambda_2)t}, \qquad t \geq 0.$$

(ii) $F_{T_{\min}}(10) \approx 86\%$;
(iii) $F_{T_{\max}}(10) \approx 40\%$;
(iv)

$$E[t \vee T_2] = \int_0^{+\infty} (t \vee s)\lambda_2 e^{-\lambda_2 s} ds$$

$$= \int_0^t t\lambda_2 e^{-\lambda_2 s} ds + \int_t^{+\infty} s\lambda_2 e^{-\lambda_2 s} ds = t + \frac{e^{-\lambda_2 t}}{\lambda_2}.$$

By the freezing lemma (cf. Theorem 4.2.10), we have

$$E[T_{\max} \mid T_1] = T_1 + \frac{e^{-\lambda_2 T_1}}{\lambda_2}.$$

Exercise 5.1.22 Beginning at 8 in the morning, Mr. Smith typically receives an average of two phone calls per hour. We assume that, for each hour, the number of calls received is a Poisson random variable, and that these random variables are independent. Determine:

(i) the distribution of the number of calls received between 8 and 10;
(ii) the probability of receiving at least 4 calls between 8 and 10;
(iii) the probability of receiving at least 2 calls per hour between 8 and 10;
(iv) the probability of receiving at least 4 calls between 8 and 10, knowing that at least 2 are received between 8 and 10;
(v) the probability of receiving at least 4 calls between 8 and 10, knowing that at least 2 are received between 8 and 9.

Solution Let N_{n-m} be the number of calls received from hour n to hour m. Then $N_{8-9} \sim$ Poisson$_2$.

(i) $N_{8-10} = N_{8-9} + N_{9-10} \sim$ Poisson$_4$ by independence assumption (Example 2.6.5);

(ii)

$$P(N_{8-10} \geq 4) = 1 - P(N_{8-10} \leq 3) = 1 - e^{-4} \sum_{k=0}^{3} \frac{4^k}{k!};$$

(iii) due to independence

$$P((N_{8-9} \geq 2) \cap (N_{9-10} \geq 2)) = \left(1 - e^{-2} \sum_{k=0}^{1} \frac{2^k}{k!}\right)^2;$$

(iv)

$$P\left(N_{8-10} \geq 4 \mid N_{8-10} \geq 2\right) = \frac{P(N_{8-10} \geq 4)}{P(N_{8-10} \geq 2)}$$

and the probabilities are calculated as in point (ii);

(v)

$$P\left(N_{8-10} \geq 4 \mid N_{8-9} \geq 2\right) = \frac{P\left((N_{8-10} \geq 4) \cap (N_{8-9} \geq 2)\right)}{P(N_{8-9} \geq 2)}$$

$$= \frac{1}{P(N_{8-9} \geq 2)} \sum_{k \geq 2}$$

$$P(N_{9-10} \geq 4 - k) P(N_{8-9} = k).$$

Exercise 5.1.23 Suppose that countries can be divided into three groups based on their financial stability: A (excellent stability), B (good), or C (sufficient). For a generic country, the probability of being in group A, B, or C is considered equal to $\frac{1}{3}$. To determine which group a particular country belongs to, an economic analysis is carried out, the outcome of which can only be positive or negative. It is known that the economic analysis of countries in group A has a positive outcome with a probability of 99%; moreover, for countries in groups B and C, the outcome is positive with a probability of 80 and 30%, respectively.

(i) Determine the probability that the economic analysis of Probabiland has a positive outcome;

(ii) knowing that the economic analysis of Probabiland had a negative outcome, what is the probability of being in group C?

Solution

(i) Let E be the event "the economic analysis of Probabiland has a positive outcome". By the Law of total probability, we have

$$P(E) = P(E \mid A)P(A) + P(E \mid B)P(B) + P(E \mid C)P(C)$$

$$= \frac{1}{3}(99\% + 80\% + 30\%) \approx 70\%.$$

(ii) Knowing that $P(C) = \frac{1}{3}$ and

$$P(E^c \mid C) = 1 - P(E \mid C) = 70\%,$$

by Bayes' formula, we have

$$P(C \mid E^c) = \frac{P(E^c \mid C)P(C)}{P(E^c)} \approx 77\%.$$

Exercise 5.1.24 (Paradox of the Three Cards) Three cards are given, of which the first is red on both sides, the second is red on one side and white on the other, and the third is white on both sides. A card is randomly chosen, and only one side is observed, which is red. What is the probability that the other side of the drawn card is also red?

Solution We use Bayes' theorem: let A denote the event where the first card is chosen (red on both sides) and R denote the event where the visible side of the chosen card is red. We have

$$P(A \mid R) = \frac{P(R \mid A)P(A)}{P(R)} = \frac{1 \cdot \frac{1}{3}}{\frac{3}{6}} = \frac{2}{3}.$$

Exercise 5.1.25 A company has two production lines A and B that produce 30 and 70% of the products, respectively. The percentage of defective products from lines A and B is equal to 0.5 and 0.1%, respectively. Determine:

(i) the probability that there is exactly one defective product in a box containing 10 products all coming from the same line;

(ii) the probability that a box containing exactly one defective product comes from line A;

(iii) the probability that there is exactly one defective product in a box containing 10 products assuming that the products are boxed without distinguishing the production line.

Solution

(i) Let D be the event for which we need to calculate the probability. The probability that a box produced by A has exactly one defective product is $p_A = \text{Bin}_{10,0.5\%}(\{1\}) \approx 4.78\%$. Similarly, $p_B = \text{Bin}_{10,0.1\%}(\{1\}) \approx 0.99\%$. Then, with notations whose meaning should be evident, the probability we are looking for is

$$P(D) = P(D \mid A)P(A) + P(D \mid B)P(B) = p_A \cdot 30\% + p_B \cdot 70\% \approx 2.13\%.$$

(ii) By Bayes' formula, we have

$$P(A \mid D) = \frac{P(D \mid A)P(A)}{P(D)} = \frac{p_A \cdot 30\%}{2.13\%} \approx 67.39\%.$$

(iii) The probability that a single product is defective is equal to

$$p_D = 0.5\% \cdot 30\% + 0.1\% \cdot 70\% \approx 0.22\%.$$

Then the probability we are looking for is equal to $\text{Bin}_{10,p_D}(\{1\}) \approx 2.15\%$.

Exercise 5.1.26 An antispam algorithm classifies as "suspicious" emails that contain certain keywords. To train the antispam algorithm, data is used concerning a set of 100 emails, of which 60 are spam, 90% of spam emails are suspicious, and only 1% of emails that are not spam are suspicious. Based on this data, estimate the probability that a suspicious email is actually spam.

Solution Let X be the event "an email is spam" and S be the event "an email is suspicious". By assumption, we have

$$P(X) = 60\%, \qquad P(S \mid X) = 90\%, \qquad P(S \mid X^c) = 1\%.$$

Then, by Bayes' formula, we obtain

$$P(X \mid S) = \frac{P(S \mid X)P(X)}{P(S)} =$$

(by the Law of total probability)

$$= \frac{P(S \mid X)P(X)}{P(S \mid X)P(X) + P(S \mid X^c)P(X^c)} \approx 99.26\%.$$

Exercise 5.1.27 Every year, the probability of contracting an infectious disease is 1% if vaccinated and 80% if not vaccinated.

(i) Knowing that in a year 10% of the population contracts the disease, estimate the percentage of vaccinated people;

(ii) calculate the probability that a sick person is vaccinated.

Solution

(i) If M is the event "contracting the disease" and V is the event "being vaccinated", we have

$$P(M) = P(M \mid V)P(V) + P(M \mid V^c)(1 - P(V))$$

from which

$$P(V) = \frac{P(M) - P(M \mid V^c)}{P(M \mid V) - P(M \mid V^c)} \approx 89\%$$

(ii) By Bayes' theorem, we have

$$P(V \mid M) = \frac{P(M \mid V)P(V)}{P(M)} \approx 0.09\%$$

Exercise 5.1.28 A bag contains two coins: a gold one that is balanced and a silver one for which the probability of getting heads is equal to $p \in]0, 1[$. We randomly draw one of the two coins and toss it n times: let X be the r.v. that indicates the number of heads obtained. Given $k \in \mathbb{N}_0$, determine:

(i) the probability that X is equal to k, knowing that the silver coin has been drawn;
(ii) $P(X = k)$;
(iii) the probability that the silver coin has been drawn, knowing that $X = n$;
(iv) the expectation of X.

Solution

(i) Let $A = $ "the silver coin is drawn". Then for $k = 0, 1, \ldots, n$ we have

$$P(X = k \mid A) = \mathrm{Bin}_{n,p}(\{k\}) = \binom{n}{k} p^k (1 - p)^{n-k}.$$

(ii) By the Law of total probability, we have

$$P(X = k) = \frac{1}{2} \left(P(X = k \mid A^c) + P(X = k \mid A) \right)$$

$$= \frac{1}{2} \left(\mathrm{Bin}_{n,\frac{1}{2}}(\{k\}) + \mathrm{Bin}_{n,p}(\{k\}) \right) \qquad (5.1.1)$$

(iii) First of all,

$$P(X = n) = \frac{1}{2} \left(\frac{1}{2^n} + p^n \right).$$

By Bayes' theorem, we have

$$P(A \mid X = n) = \frac{P(X = n \mid A)P(A)}{P(X = n)} = \frac{p^n}{\frac{1}{2^n} + p^n}.$$

(iv) Recalling that the expectation of a r.v. with distribution $\text{Bin}_{n,p}$ is equal to np, by (5.1.1) we have

$$E[X] = \frac{1}{2}\left(\frac{n}{2} + np\right).$$

Exercise 5.1.29 Urn A contains one red ball and one green ball. Urn B instead contains two red balls and four green balls. We randomly draw a ball from urn A and put it in urn B, then we draw a ball from urn B. Determine the probability that:

(i) the ball drawn from urn B is red;
(ii) the ball drawn from urn A is red, knowing that also the ball drawn from urn B is red;
(iii) the two balls drawn are of the same color.

Solution Introduce the events:

$$R_A = \text{"the ball drawn from urn } A \text{ is red"},$$
$$V_A = \text{"the ball drawn from urn } A \text{ is green"} = R_A^c,$$
$$R_B = \text{"the ball drawn from urn } B \text{ is red"},$$
$$V_B = \text{"the ball drawn from urn } B \text{ is green"} = R_B^c.$$

(i) By the Law of total probability, we have

$$P(R_B) = P(R_B \mid R_A)P(R_A) + P(R_B \mid V_A)P(V_A) = \frac{3}{7} \cdot \frac{1}{2} + \frac{2}{7} \cdot \frac{1}{2} = \frac{5}{14}.$$

(ii) By Bayes' formula, we have

$$P(R_A|R_B) = \frac{P(R_B \mid R_A)P(R_A)}{P(R_B)} = \frac{\frac{3}{7} \cdot \frac{1}{2}}{\frac{5}{14}} = \frac{3}{5}.$$

(iii) Again, by the Law of total probability, if E denotes the for which we want to calculate the probability,

$$P(E) = P(E \mid R_A)P(R_A) + P(E \mid V_A)P(V_A) = \frac{3}{7} \cdot \frac{1}{2} + \frac{5}{7} \cdot \frac{1}{2} = \frac{4}{7}.$$

Exercise 5.1.30 A winery produces a numbered series of wine bottles. In a quality control check, each bottle must pass three tests to be deemed suitable: the probability of passing the first test is 90%; if the first test is passed, the probability of passing the second test is 95%; if the second test is also passed, the probability of passing the third test is 99%. We assume that the outcomes of the checks on different bottles are independent of each other.

 (i) Determine the probability that a bottle is suitable;
 (ii) determine the probability that an unsuitable bottle did not pass the first test;
 (iii) let X_n be the r.v. that takes the value 0 or 1 depending on whether the n-th bottle is suitable. Determine the distribution of X_n and of (X_n, X_{n+1});
 (iv) let N be the number corresponding to the first unsuitable bottle. Determine the distribution and the mean of N;
 (v) calculate the probability that all of the first 100 bottles are suitable.

Solution

 (i) Let T_i, $i = 1, 2, 3$, be the event "the i-th test is passed", and $T = T_1 \cap T_2 \cap T_3$. By the multiplication rule, we have

$$P(T) = P(T_1)P(T_2 \mid T_1)P(T_3 \mid T_1 \cap T_2) = \frac{90 \cdot 95 \cdot 99}{100^3} \approx 85\%;$$

 (ii) by Bayes' formula, we have

$$P(T_1^c \mid T^c) = \frac{P(T^c \mid T_1^c)P(T_1^c)}{P(T^c)} = \frac{1 \cdot 10\%}{1 - P(T)} \approx 65\%;$$

 (iii) $X_n \sim Be_p$ with $p = P(T)$. Due to independence, $(X_1, X_2) \sim Be_p \otimes Be_p$;
 (iv) $N \sim Geom_{1-p}$ and $E[N] = \frac{1}{1-p}$;
 (v) we have (cf. Theorem 2.1.26)

$$P(N > 100) = (1 - (1 - p))^{100} = p^{100}.$$

Exercise 5.1.31 An urn contains 4 white balls, 4 red balls, and 4 black balls. A series of draws is performed as follows: a ball is drawn and then replaced in the urn along with another ball of the same color as the one drawn. Determine the probability:

 (i) of drawing a white ball on the second draw;
 (ii) of drawing a red ball on the first draw knowing that a white ball is drawn on the second draw;
 (iii) after three draws, having drawn all white balls;
 (iv) after three draws, not having drawn balls that are all the same color.

Solution Let us indicate with B_n the event "the ball drawn at the n-th draw is white", with $n \in \mathbb{N}$. Similarly, let N_n and R_n be defined.

(i) By the Law of total probability, we have

$$P(B_2) = P(B_2 \mid B_1)P(B_1) + P(B_2 \mid R_1)P(R_1) + P(B_2 \mid N_1)P(N_1)$$

$$= \frac{5}{13} \cdot \frac{1}{3} + \frac{4}{13} \cdot \frac{1}{3} + \frac{4}{13} \cdot \frac{1}{3} = \frac{1}{3};$$

(ii) by Bayes' formula, we have

$$P(R_1 \mid B_2) = \frac{P(B_2 \mid R_1)}{P(B_2)} P(R_1) = \frac{\frac{4}{13} \cdot \frac{1}{3}}{\frac{1}{3}} = \frac{4}{13};$$

(iii) by the multiplication rule, we have

$$P(B_1 \cap B_2 \cap B_3) = P(B_1)P(B_2 \mid B_1)P(B_3 \mid B_1 \cap B_2)$$

$$= \frac{1}{3} \cdot \frac{5}{13} \cdot \frac{6}{14} = \frac{5}{91};$$

(iv) from point (iii), the probability that all balls have the same color is $\frac{15}{91}$. The probability we are looking for is therefore $1 - \frac{15}{91}$.

Exercise 5.1.32 Based on a recent analysis, individuals who participate in sports activities have a 90% probability of achieving good academic performance, compared to 70% for those who do not engage in sports activities.

(i) Knowing that in a year the percentage of students with good academic performance is 85%, estimate the percentage of students who engage in sports activities;

(ii) calculate the probability that those who have good academic performance engage in sports activities.

Solution

(i) If B is the event "having good academic performance" and S is the event "engaging in sports activities", we have

$$P(B) = P(B \mid S)P(S) + P(B \mid S^c)(1 - P(S))$$

from which

$$P(S) = \frac{P(B) - P(B \mid S^c)}{P(B \mid S) - P(B \mid S^c)} = 75\%;$$

(ii) by Bayes' theorem, we have

$$P(S \mid B) = \frac{P(B \mid S)P(S)}{P(B)} \approx 79\%.$$

Exercise 5.1.33 In a production chain, a bolt is suitable if it passes two quality tests: the probability of passing the first test is 90%; if the first test is passed, the probability of passing the second test is 95%. Assume that the outcomes of the checks on different bolts are independent of each other. Determine:

  (i)   the probability that a bolt is suitable;
  (ii) the probability that an unsuitable bolt has passed the first test;
  (iii) the distribution of the number N of suitable bolts among the first 100 produced;
  (iv) the distribution and mean of M, where M is the number corresponding to the first unsuitable bolt.

Solution

  (i) Let T_i, $i = 1, 2$, be the event "the i-th test is passed" and $T = T_1 \cap T_2$. By the multiplication rule, we have

$$p := P(T) = P(T_1)P(T_2 \mid T_1) = \frac{90 \cdot 95}{100^2} = 85.5\%;$$

  (ii) using Bayes' formula and since $P(T^c \mid T_1) = P(T_2^c \mid T_1) = 5\%$, we have

$$P(T_1 \mid T^c) = \frac{P(T^c \mid T_1)P(T_1)}{P(T^c)} = \frac{5\% \cdot 90\%}{14.5\%} \approx 31\%;$$

  (iii) $N \sim \mathrm{Bin}_{100,p}$;
  (iv) $M \sim \mathrm{Geom}_{1-p}$ and $E[M] = \frac{1}{1-p}$.

5.2 Random Variables

Exercise 5.2.1 Two dice (not rigged) with three faces each, numbered from 1 to 3, are rolled. On the sample space $\Omega = \{(m, n) \mid 1 \leq m, n \leq 3\}$, let X_1 and X_2 be the random variables that indicate the outcomes of the rolls of the first and second die, respectively. Let $X = X_1 + X_2$, determine $\sigma(X)$ and whether X_1 is $\sigma(X)$-measurable.

Solution $\sigma(X)$ is the σ-algebra whose elements are $\emptyset$ and the unions of

$$(X = 2) = \{(1, 1)\},$$
$$(X = 3) = \{(1, 2), (2, 1)\},$$
$$(X = 4) = \{(1, 3), (3, 1), (2, 2)\},$$
$$(X = 5) = \{(2, 3), (3, 2)\},$$
$$(X = 6) = \{(3, 3)\}.$$

The event $(X_1 = 1) \notin \sigma(X)$: intuitively, we cannot infer the outcome of the first throw knowing the sum of the two rolls.

Exercise 5.2.2 Let $B \sim \text{Unif}_{[-2,2]}$. Determine the probability that the quadratic equation

$$x^2 + 2Bx + 1 = 0$$

have real solutions. What is the probability that such solutions are coincident?

Solution We have $\Delta = 4B^2 - 4$. The solutions are real if and only if $\Delta \geq 0$ that is $|B| \geq 1$: now we simply have $P(|B| \geq 1) = \frac{1}{2}$. Moreover, the solutions are coincident if and only if $|B| = 1$, so with zero probability.

Exercise 5.2.3 Given a random point Q in $[0, 1]$, let X be the length of the interval with the greater amplitude between the two in which $[0, 1]$ is divided by Q. Determine the distribution and the expected value of X.

Solution We observe that $X = \max\{Q, 1 - Q\}$ and $\frac{1}{2} \leq X \leq 1$. We determine the CDF of X: for $\frac{1}{2} \leq x \leq 1$ we have

$$P(X \leq x) = P\left((Q \leq x) \cap (Q \geq \tfrac{1}{2})\right) + P\left((1 - Q \leq x) \cap (Q \leq \tfrac{1}{2})\right)$$

$$= P(\tfrac{1}{2} \leq Q \leq x) + P(1 - x \leq Q \leq \tfrac{1}{2}) = 2x - 1.$$

From this, we find that $X \in AC$ and precisely $X \sim \text{Unif}_{\left[\frac{1}{2},1\right]}$. In particular $E[X] = \frac{3}{4}$.

Exercise 5.2.4 Let $Y = Y(t)$ be the solution of the Cauchy problem

$$\begin{cases} Y'(t) = AY(t), \\ Y(0) = y_0, \end{cases}$$

where $A \sim \mathcal{N}_{\mu,\sigma^2}$ and $y_0 > 0$.

(i) For each $t > 0$ determine the distribution and density of the r.v. $Y(t)$;
(ii) write the expression of the CHF φ_A of the r.v. A and from it derive

$$E\left[e^A\right] = \varphi_A(-i),$$

and then calculate $E[Y(t)]$;
(iii) are the random variables $Y(1)$ and $Y(2)$ independent?

Solution

(i) We have

$$Y(t) = y_0 e^{tA}$$

and therefore $Y(t)$ has a log-normal distribution. More precisely, for each $y > 0$ we have

$$P(Y(t) \le y) = P\left(A \le \frac{1}{t} \log \frac{y}{y_0}\right) = F_A\left(\frac{1}{t} \log \frac{y}{y_0}\right)$$

where F_A is the CDF of A. By differentiating, we obtain the density of $Y(t)$ which is zero for $y \le 0$ and equals

$$\gamma(y) = \frac{d}{dy} P(Y(t) \le y) = \frac{1}{ty} F_A'\left(\frac{1}{t} \log \frac{y}{y_0}\right)$$

$$= \frac{1}{ty\sqrt{2\pi\sigma^2}} e^{-\frac{\left(\frac{1}{t}\log\frac{y}{y_0}-\mu\right)^2}{2\sigma^2}},$$

for $y > 0$.

(ii) Recalling (2.5.7) we have

$$E\left[e^A\right] = \varphi_A(-i) = e^{\mu + \frac{\sigma^2}{2}}.$$

Since $tA \sim \mathcal{N}_{t\mu, t^2\sigma^2}$ we have

$$E[Y(t)] = E\left[y_0 e^{tA}\right] = y_0 e^{t\mu + \frac{t^2\sigma^2}{2}}.$$

(iii) We observe that

$$E[Y(1)Y(2)] = y_0^2 E\left[e^{3A}\right] = y_0^2 e^{3\mu + \frac{9\sigma^2}{2}}$$

is different from

$$E[Y(1)] E[Y(2)] = y_0^2 E\left[e^A\right] E\left[e^{2A}\right] = y_0^2 e^{\mu + \frac{\sigma^2}{2}} e^{2\mu + \frac{4\sigma^2}{2}}$$

except when $\sigma = 0$ (when clearly $Y(1)$, $Y(2)$ are independent).

Exercise 5.2.5 Consider a chessboard of infinite size, with alternating black and white squares extending indefinitely in all directions. As the first move, a pawn is placed on a random square; at each subsequent move, the pawn is randomly moved to one of the 8 adjacent squares (horizontally, vertically, or diagonally). Let X be the number of the first move in which the pawn is in a white square, Y the number of the first move in which the pawn is in a black square, and finally Z the number of the first move in which the pawn has been on both a white and a black square (that is, Z represents the minimum number of moves required for the pawn to visit at least one white square and one black square). Determine:

  (i) the laws of X, Y and Z;
 (ii) the joint law of X, Y e Z;
(iii) $\operatorname{cov}(X, Z)$.

Solution

  (i) X and Y have both geometric distribution $\operatorname{Geom}_{\frac{1}{2}}$. Moreover, $Z = \max\{X, Y\} = X\mathbb{1}_{(Y=1)} + Y\mathbb{1}_{(X=1)}$ and, for $n \geq 2$, we have

$$P(Z = n) = P((X = n) \cap (Y = 1)) + P((X = 1) \cap (Y = n))$$
$$= P(X = n \mid Y = 1)P(Y = 1) + P(Y = n \mid X = 1)P(X = 1)$$
$$= \operatorname{Geom}_{\frac{1}{2}}(n - 1) = \frac{1}{2^{n-1}};$$

 (ii) the random vector (X, Y, Z) can only take the values $(1, n, n)$ and $(n, 1, n)$ with $n \geq 2$. By the multiplication rule, we have

$$P((X, Y, Z) = (1, n, n)) = P(X = 1)P(Y = n \mid X = 1)$$
$$P(Z = n \mid (X, Y) = (1, n))$$
$$= \frac{1}{2}\operatorname{Geom}_{\frac{1}{2}}(\{n - 1\}) = \frac{1}{2^n},$$

which is also equal to $P((X, Y, Z) = (n, 1, n))$;
(iii) since $X \sim \operatorname{Geom}_{\frac{1}{2}}$, we have

$$E[X] = \sum_{n=1}^{\infty} n \operatorname{Geom}_{\frac{1}{2}}(\{n\}) = 2,$$

and

$$E[Z] = \sum_{n=2}^{\infty} n \operatorname{Geom}_{\frac{1}{2}}(\{n - 1\}) = \sum_{n=1}^{\infty}(n + 1)\operatorname{Geom}_{\frac{1}{2}}(\{n\})) = 3.$$

Moreover,

$$E[XZ] = \sum_{n=2}^{\infty} \left(nP((X, Y, Z) = (1, n, n)) + n^2 P((X, Y, Z) = (n, 1, n)) \right)$$

$$= \sum_{n=2}^{\infty} \frac{n}{2^n} + \sum_{n=2}^{\infty} \frac{n^2}{2^n} = \frac{3}{2} + \frac{11}{2} = 7.$$

In conclusion, we have

$$\mathrm{cov}(X, Z) = E[XZ] - E[X]E[Z] = 1.$$

Exercise 5.2.6 Given $\gamma \in \mathbb{R}$, consider the function

$$\mu_\gamma(n) = (1 - \gamma)\gamma^n, \qquad n \in \mathbb{N}_0 := \mathbb{N} \cup \{0\}.$$

(i) Determine the values of γ for which μ_γ is a discrete distribution function. It may be useful to remember that

$$\sum_{n=0}^{\infty} x^n = \frac{1}{1-x}, \qquad |x| < 1;$$

(ii) let γ be such that μ_γ is a distribution function and consider the r.v. X with distribution function μ_γ. Given $m \in \mathbb{N}$, calculate the probability that X is divisible by m;

(iii) find a function $f : \mathbb{R} \to \mathbb{R}$ such that $Y = f(X)$ has distribution Geom_p and determine p as a function of γ;

(iv) calculate $E[X]$.

Solution

(i) The values $\mu_\gamma(n)$ must be non-negative, from which $0 < \gamma < 1$. For such values of γ, we have that μ_γ is a distribution function since

$$\sum_{n=0}^{\infty} \mu_\gamma(n) = (1 - \gamma) \sum_{n=0}^{\infty} \gamma^n = 1.$$

(ii) X is divisible by m if there exists $k \in \mathbb{N}_0$ such that $X = km$. Since $P(X = km) = (1 - \gamma)\gamma^{km}$, then we have

$$\sum_{k=0}^{\infty} P(X = km) = (1 - \gamma) \sum_{k=0}^{\infty} \gamma^{km} = \frac{1 - \gamma}{1 - \gamma^m}.$$

(iii) The r.v. $Y = X + 1$ is such that

$$P(Y = n) = P(X = n - 1) = (1 - \gamma)\gamma^{n-1}, \qquad n \in \mathbb{N}.$$

Therefore, $Y \sim \text{Geom}_{1-\gamma}$.

(iv) By point (iii), we have

$$E[X] = E[Y] - 1 = \frac{1}{1 - \gamma} - 1 = \frac{\gamma}{1 - \gamma}.$$

Exercise 5.2.7 Let X, Y be independent random variables with distribution Exp_λ. Determine:

(i) the densities of $X + Y$ and $X - Y$;
(ii) the characteristic functions of $X + Y$ and $X - Y$;
(iii) are $X + Y$ and $X - Y$ independent?

Solution

(i) We know (cf. Example 2.6.7) that if $X, Y \sim \text{Exp}_\lambda \equiv \text{Gamma}_{1,\lambda}$ are independent r.v., then

$$X + Y \sim \text{Gamma}_{2,\lambda}$$

with density

$$\gamma_{X+Y}(z) = \lambda^2 z e^{-\lambda z} \mathbb{1}_{\mathbb{R}_{>0}}(z).$$

Now let us compute the density of $X - Y$ as the convolution of the densities of X and $-Y$. To do this, first, let us compute the density of $-Y$: we have $P(-Y \le y) = 1$ if $y \ge 0$ and, for $y < 0$,

$$P(-Y \le y) = P(Y \ge -y) = \int_{-y}^{\infty} \lambda e^{-\lambda x} dx = \int_{-\infty}^{y} \lambda e^{\lambda z} dt$$

from which

$$\gamma_{-Y}(y) = \lambda e^{\lambda y} \mathbb{1}_{\mathbb{R}_{<0}}(y).$$

Now

$$\gamma_{X-Y}(w) = (\gamma_X * \gamma_{-Y})(w) = \int_{\mathbb{R}} \gamma_X(x) \gamma_{-Y}(w - x) dx = \frac{\lambda}{2} e^{-\lambda|w|}, \quad w \in \mathbb{R}.$$

(ii) Recalling that $\varphi_X(\eta) = \frac{\lambda}{\lambda - i\eta}$, by the independence of X and Y we have

$$\varphi_{X+Y}(\eta) = E\left[e^{i\eta(X+Y)}\right] = E\left[e^{i\eta X}\right] E\left[e^{i\eta Y}\right] = \frac{\lambda^2}{(\lambda - i\eta)^2},$$

and similarly

$$\varphi_{X-Y}(\eta) = E\left[e^{i\eta(X-Y)}\right] = \frac{\lambda^2}{(\lambda - i\eta)(\lambda + i\eta)} = \frac{\lambda^2}{\lambda^2 + \eta^2}.$$

(iii) $X + Y$ and $X - Y$ are independent if and only if

$$\varphi_{(X+Y, X-Y)}(\eta_1, \eta_2) = \varphi_{X+Y}(\eta_1)\varphi_{X-Y}(\eta_2).$$

We already have the expression of φ_{X+Y} and φ_{X-Y} from point (ii). Let us compute

$$\varphi_{(X+Y, X-Y)}(\eta_1, \eta_2) = E\left[e^{i\eta_1(X+Y) + i\eta_2(X-Y)}\right]$$
$$= E\left[e^{iX(\eta_1 + \eta_2) + iY(\eta_1 - \eta_2)}\right] =$$

(by independence of X and Y)

$$= E\left[e^{iX(\eta_1 + \eta_2)}\right] E\left[e^{iY(\eta_1 - \eta_2)}\right] = \frac{\lambda}{\lambda - i(\eta_1 + \eta_2)} \frac{\lambda}{\lambda - i(\eta_1 - \eta_2)}.$$

Hence, $X + Y$ and $X - Y$ are not independent.

Exercise 5.2.8 Let $X \sim \mathrm{Exp}_\lambda$ and $Y \sim \mathrm{Be}_p$ be independent random variables with $\lambda > 0$ and $0 < p < 1$. Determine:

(i)   the CDF of $X + Y$ and XY;
(ii) whether $X + Y$ and XY are absolutely continuous and, if so, find their density;
(iii) the CHF of $X + Y$ and XY.

Solution

(i) We have

$$P(X + Y \leq z) = P\left((X + Y \leq z) \cap (Y = 0)\right) + P\left((X + Y \leq z) \cap (Y = 1)\right)$$

(by independence of X and Y)

$$= P(X \leq z)P(Y = 0) + P(X \leq z - 1)P(Y = 1)$$
$$= (1 - p)P(X \leq z) + pP(X \leq z - 1),$$

and also remember that $P(X \leq z) = 1 - e^{-\lambda z}$. Then we have

$$F_{X+Y}(z) := P(X + Y \leq z)$$

$$= \begin{cases} 0 & \text{if } z < 0, \\ (1-p)\left(1 - e^{-\lambda z}\right) & \text{if } 0 \leq z \leq 1, \\ (1-p)\left(1 - e^{-\lambda z}\right) + p\left(1 - e^{-\lambda(z-1)}\right) & \text{if } z > 1. \end{cases}$$

Similarly, we have

$$F_{XY}(z) := P(XY \leq z) = P\left((XY \leq z) \cap (Y = 0)\right)$$
$$+ P\left((XY \leq z) \cap (Y = 1)\right)$$

(by independence of X and Y)

$$= P(0 \leq z)P(Y = 0) + P(X \leq z)P(Y = 1)$$

$$= \begin{cases} 0 & \text{if } z < 0, \\ (1-p) + p\left(1 - e^{-\lambda z}\right) & \text{if } z \geq 0. \end{cases}$$

(ii) The function F_{X+Y} is absolutely continuous and the density of $X + Y$ is obtained simply by differentiating (cf. Theorem 1.4.33):

$$\frac{d}{dz}F_{X+Y}(z) = \begin{cases} 0 & \text{if } z < 0, \\ (1-p)\lambda e^{-\lambda z} & \text{if } 0 \leq z \leq 1, \\ (1-p)\lambda e^{-\lambda z} + p\lambda e^{-\lambda(z-1)} & \text{if } z > 1. \end{cases}$$

The function F_{XY} is discontinuous at 0 and therefore the r.v. XY is not absolutely continuous: indeed, we have (cf. (1.4.9))

$$P(XY = 0) = F_{XY}(0) - F_{XY}(0-) = 1 - p.$$

(iii) By independence (cf. Proposition 2.5.11), we have

$$\varphi_{X+Y}(\eta) = \varphi_X(\eta)\varphi_Y(\eta) = \frac{\lambda}{\lambda - i\eta}(1 + p(e^{i\eta} - 1)).$$

Furthermore,

$$\varphi_{XY}(\eta) = E\left[e^{i\eta XY}\right] = \iint_{\mathbb{R}^2} e^{i\eta xy} \left(\text{Exp}_\lambda \otimes \text{Be}_p\right)(dx, dy) =$$

(by Fubini's theorem)

$$= \int_{\mathbb{R}} \left(\int_{\mathbb{R}} e^{i\eta xy} \operatorname{Be}_p(dy) \right) \operatorname{Exp}_\lambda(dx)$$

$$= \int_{\mathbb{R}} \left(1 - p + p e^{i\eta x} \right) \operatorname{Exp}_\lambda(dx)$$

$$= 1 - p + p \frac{\lambda}{\lambda - i\eta}.$$

Exercise 5.2.9 Let X, Y be independent random variables with distribution $\mu = \frac{1}{2}(\delta_{-1} + \delta_1)$. Determine:

(i) the joint CHF $\varphi_{(X,Y)}$;
(ii) the CHF φ_{X+Y} of the sum $X + Y$;
(iii) the CHF φ_{XY} and the distribution of the product XY;
(iv) if X and XY are independent.

Solution

(i) Since the random variables are independent, the joint CHF is the product of the marginal CHFs:

$$\varphi_{(X,Y)}(\eta_1, \eta_2) = E\left[e^{i(\eta_1 X + \eta_2 Y)} \right] = E\left[e^{i\eta_1 X} \right] E\left[e^{i\eta_2 Y} \right] = \cos(\eta_1) \cos(\eta_2),$$

since

$$\varphi_Y(\eta) = \varphi_X(\eta) = E\left[e^{i\eta X} \right] = \frac{1}{2}\left(e^{i\eta} + e^{-i\eta} \right) = \cos \eta.$$

(ii) Again, due to independence, the CHF of the sum is

$$\varphi_{X+Y}(\eta) = E\left[e^{i\eta(X+Y)} \right] = E\left[e^{i\eta X} \right] E\left[e^{i\eta Y} \right] = (\cos \eta)^2 .$$

(iii) We have

$$\varphi_{XY}(\eta) = E\left[e^{i\eta XY} \right] = \iint_{\mathbb{R}^2} e^{i\eta xy} (\mu \otimes \mu)(dx, dy) =$$

(by Fubini's theorem)

$$= \int_{\mathbb{R}} \left(\int_{\mathbb{R}} e^{i\eta xy} \mu(dx) \right) \mu(dy)$$

$$= \int_{\mathbb{R}} \cos(\eta y) \mu(dy)$$

$$= \frac{1}{2} (\cos \eta + \cos(-\eta)) = \cos \eta.$$

Thus, XY has the same CHF as X and therefore also the same distribution μ.

(iv) To prove that X and XY are independent, we determine the CHF of X and XY, and verify that it is equal to the product of the marginal CHFs:

$$\varphi_{(X,XY)}(\eta_1, \eta_2) = E\left[e^{i(\eta_1 X + \eta_2 XY)}\right] = \iint_{\mathbb{R}^2} e^{ix(\eta_1 + \eta_2 y)} (\mu \otimes \mu)(dx, dy) =$$

(by Fubini's theorem)

$$= \int_{\mathbb{R}} \left(\int_{\mathbb{R}} e^{ix(\eta_1 + \eta_2 y)} \mu(dx) \right) \mu(dy)$$

$$= \frac{1}{2} \int_{\mathbb{R}} \left(e^{-i(\eta_1 + \eta_2 y)} + e^{-i(\eta_1 + \eta_2 y)} \right) \mu(dy)$$

$$= \frac{1}{4} \left(e^{-i(\eta_1 - \eta_2)} + e^{-i(\eta_1 + \eta_2)} + e^{i(\eta_1 - \eta_2)} + e^{i(\eta_1 + \eta_2)} \right)$$

$$= \cos(\eta_1) \cos(\eta_2) = \varphi_X(\eta_1) \varphi_{XY}(\eta_2).$$

Exercise 5.2.10 Verify that the function

$$\gamma(x, y) = (x + y) \mathbb{1}_{[0,1] \times [0,1]}(x, y), \qquad (x, y) \in \mathbb{R}^2,$$

is a density and consider a random vector (X, Y) with density γ. Determine:

(i)   if X, Y are independent;
(ii) the expected value $E[XY]$;
(iii) the density of the sum $X + Y$.

Solution The function γ is non-negative and we have

$$\iint_{\mathbb{R}^2} \gamma(x, y) dxdy = \left[\frac{x^2 + y^2}{2} \right]_{x=y=0}^{x=y=1} = 1.$$

Therefore γ is a density. Moreover:

(i) the density of X is

$$\gamma_X(x) := \int_{\mathbb{R}} \gamma(x, y) dy = \left(x + \frac{1}{2} \right) \mathbb{1}_{[0,1]}(x), \qquad x \in \mathbb{R}.$$

Similarly, we compute γ_Y and verify that X, Y are not independent since $\gamma \neq \gamma_X \gamma_Y$;

(ii) we have

$$E[XY] = \int_0^1 \int_0^1 xy(x+y)dxdy = \frac{1}{3};$$

(iii) by Theorem 2.6.1, the density of $X + Y$ is

$$\gamma_{X+Y}(z) = \int_{\mathbb{R}} \gamma(x, z-x)dx, \qquad z \in [0, 2].$$

Imposing the condition $(x, z-x) \in [0, 1] \times [0, 1]$, we have

$$\int_{\mathbb{R}} \gamma(x, z-x)dx = \begin{cases} z^2 & \text{if } z \in [0, 1], \\ z(2-z) & \text{if } z \in [1, 2]. \end{cases}$$

Exercise 5.2.11 The delivery time of a courier is described by a r.v. $T \sim \text{Exp}_\lambda$ with $\lambda > 0$. We assume that the unit of time is the day, i.e., $T = 1$ is equivalent to one day, and we denote by N the r.v. that indicates the day of delivery, defined by $N = n$ if $T \in [n-1, n[$ for $n \in \mathbb{N}$. Determine:

(i) the law and the CDF of N;
(ii) $E[N]$ and $E[N \mid T > 1]$;
(iii) $E[N \mid T]$.

Solution

(i) N is a discrete r.v. that takes only values in $\mathbb{N}$: we have

$$P(N = n) = P(n-1 \leq T < n)$$

$$= \int_{n-1}^n \lambda e^{-\lambda t} dt = e^{-\lambda n}(e^\lambda - 1) =: p_n, \quad n \in \mathbb{N}.$$

Then

$$N \sim \sum_{n=1}^\infty p_n \delta_n$$

and the CDF of N is

$$F_N(x) = \begin{cases} 0 & \text{if } x < 0, \\ \sum_{k=1}^n p_k & \text{if } n-1 \leq x < n; \end{cases}$$

(ii) we have

$$E\,[N] = \sum_{n=1}^{\infty} np_n = \frac{e^\lambda}{e^\lambda - 1},$$

$$E\,[N \mid T > 1] = \frac{E\left[N\mathbb{1}_{(T>1)}\right]}{P(T > 1)} = e^\lambda \sum_{n=2}^{\infty} np_n = \frac{2e^\lambda - 1}{e^\lambda - 1};$$

(iii) we observe that N is $\sigma(T)$-measurable because it is a (measurable) function of T: precisely $N = 1 + [T]$ where $[x]$ denotes the *integer part of $x \in \mathbb{R}$*. Consequently,

$$E\,[N \mid T] = N.$$

Exercise 5.2.12 Given a r.v. $C \sim \mathrm{Unif}_{[0,\lambda]}$, where $\lambda > 0$, determine the maximum value of λ such that the equation

$$x^2 - 2x + C = 0$$

has, with probability one, real solutions. For this value of λ, determine the density of one of the solutions of the equation.

Solution The equation has real solutions if it has a non-negative discriminant:

$$\Delta = 4 - 4C \geq 0$$

that is, $C \leq 1$. Therefore, if $\lambda \leq 1$, the equation has real solutions with probability one, while if $\lambda > 1$ then the probability that the equation does not have real solutions is equal to $\mathrm{Unif}_\lambda(]1, \lambda]) = \frac{\lambda-1}{\lambda} > 0$. Hence, the maximum value sought is $\lambda = 1$.

We consider the solution $X = 1 + \sqrt{1 - C}$ and calculate its distribution function. First, if $C \sim \mathrm{Unif}_{[0,1]}$ then X takes values in $[1, 2]$: thus for $x \in [1, 2]$ we have

$$P\,(X \leq x) = P\left(\sqrt{1 - C} \leq x - 1\right)$$

$$= P\left(C \geq 1 - (x - 1)^2\right)$$

$$= \int_{1-(x-1)^2}^{1} dy = (x - 1)^2.$$

By differentiating, we obtain the density of X:

$$\gamma_X(x) = 2(x - 1)\mathbb{1}_{[1,2]}(x), \qquad x \in \mathbb{R}.$$

Exercise 5.2.13 Determine the values of $a, b \in \mathbb{R}$ such that the function

$$F(x) = a \arctan x + b$$

is a CDF. For these values, let X be a r.v. with CDF equal to F: determine the density of X and establish whether X is absolutely integrable.

Solution In order for the properties of a CDF to be satisfied, we have $a = \frac{1}{\pi}$ and $b = \frac{1}{2}$. The density is determined simply by differentiating F:

$$\gamma(x) = F'(x) = \frac{1}{\pi(1 + x^2)}.$$

The r.v. X is not absolutely integrable because the function $\frac{|x|}{\pi(1+x^2)} \notin L^1(\mathbb{R})$.

Exercise 5.2.14 Let $(X, Y) \sim \mathcal{N}_{0,C}$ with

$$C = \begin{pmatrix} 1 & \rho \\ \rho & 1 \end{pmatrix}, \qquad |\rho| \le 1.$$

Determine:

(i) the values of ρ such that the random variables $X+Y$ and $X-Y$ are independent;
(ii) the distribution of $X + Y$, the values of ρ for which it is absolutely continuous and, for these values, the density γ_{X+Y}.

Solution

(i) We have

$$\begin{pmatrix} X + Y \\ X - Y \end{pmatrix} = \alpha \begin{pmatrix} X \\ Y \end{pmatrix}, \qquad \alpha = \begin{pmatrix} 1 & 1 \\ 1 & -1 \end{pmatrix},$$

and so $(X + Y, X - Y) \sim \mathcal{N}_{0,\alpha C \alpha^*}$ with

$$\alpha C \alpha^* = \begin{pmatrix} 2(1 + \rho) & 0 \\ 0 & 2(1 - \rho) \end{pmatrix}.$$

Then, $X + Y$ and $X - Y$ are independent for every $\rho \in [-1, 1]$ because they have joint normal distribution and are uncorrelated;

(ii) From (i) it also follows that $X + Y \sim \mathcal{N}_{0,2(1+\rho)}$ and therefore $X + Y \in AC$ for $\varrho \in]-1, 1]$ with normal density

$$\gamma_{X+Y}(z) = \frac{1}{2\sqrt{\pi(1 + \varrho)}} e^{-\frac{z^2}{4(1+\varrho)}}, \qquad z \in \mathbb{R}.$$

Exercise 5.2.15 Let X be a real r.v. with density γ_X.

(i) Prove that

$$\gamma(x) := \frac{\gamma_X(x) + \gamma_X(-x)}{2}$$

is a density;

(ii) let Y be a r.v. with density γ: is there a relationship between the CHFs φ_X and φ_Y?

(iii) determine a r.v. Z such that $\varphi_Z(\eta) = \varphi_X^2(\eta)$.

Solution

(i) Clearly $\gamma \geq 0$ and

$$\int_{\mathbb{R}} \gamma(x)dx = \frac{1}{2}\left(\int_{\mathbb{R}} \gamma_X(x)dx + \int_{\mathbb{R}} \gamma_X(-x)dx\right) = \int_{\mathbb{R}} \gamma_X(x)dx = 1.$$

(ii) We have

$$\varphi_Y(\eta) = E\left[e^{i\eta Y}\right]$$

$$= \frac{1}{2}\int_{\mathbb{R}} e^{i\eta x}(\gamma_X(x) + \gamma_X(-x))dx$$

$$= \frac{1}{2}(\varphi_X(\eta) + \varphi_X(-\eta)) = \text{Re}\,(\varphi_X(\eta)).$$

(iii) Let X_1 and X_2 be independent random variables, equal in law to X. Then

$$\varphi_{X_1+X_2}(\eta) = \varphi_{X_1}(\eta)\varphi_{X_2}(\eta) = \varphi_X(\eta)^2.$$

Exercise 5.2.16 Let $X = (X_1, X_2, X_3) \sim \mathcal{N}_{0,C}$ with

$$C = \begin{pmatrix} 1 & 0 & 0 \\ 0 & 1 & -1 \\ 0 & -1 & 1 \end{pmatrix}.$$

Given the random vectors $Y := (X_1, X_2)$ and $Z := (X_2, X_3)$, determine:

(i) the distribution of Y and Z, specifying if they are absolutely continuous;

(ii) if Y and Z are independent;

(iii) the CHFs φ_Y and φ_Z.

Solution

(i) Since

$$Y = \begin{pmatrix} 1 & 0 & 0 \\ 0 & 1 & 0 \end{pmatrix} X, \qquad Z = \begin{pmatrix} 0 & 1 & 0 \\ 0 & 0 & 1 \end{pmatrix} X,$$

we have $Y \sim \mathcal{N}_{0,C_Y}$ and $Z \sim \mathcal{N}_{0,C_Z}$ where

$$C_Y = \begin{pmatrix} 1 & 0 \\ 0 & 1 \end{pmatrix}, \qquad C_Z = \begin{pmatrix} 1 & -1 \\ -1 & 1 \end{pmatrix}.$$

This implies that Y is absolutely continuous, while Z is not because C_Z is singular.

(ii) To see that Y and Z are not independent, it is enough to observe that for every $H \in \mathscr{B}_1$, we have

$$P\left((Y \in \mathbb{R} \times H) \cap (Z \in H \times \mathbb{R})\right) = P(X_2 \in H),$$

and

$$P(Y \in \mathbb{R} \times H) = P(X_2 \in H) = P(Z \in H \times \mathbb{R}).$$

(iii) We have

$$\varphi_Y(\eta_1, \eta_2) = e^{-\frac{1}{2}(\eta_1^2 + \eta_1^2)}, \qquad \varphi_Z(\eta_1, \eta_2) = e^{-\frac{1}{2}(\eta_1^2 + \eta_1^2 - 2\eta_1\eta_2)}.$$

Exercise 5.2.17 Let $X \sim \mathcal{N}_{\mu,1}$ with $\mu \in \mathbb{R}$, and let φ_X be the CHF of X.

(i) For $c \in \mathbb{R}$, calculate $E\left[e^{cX}\right]$: for this purpose, choose an appropriate complex value η_c such that $E\left[e^{cX}\right] = \varphi_X(\eta_c)$;

(ii) let $Y \sim \mathrm{Unif}_n$, with $n \in \mathbb{N}$, be independent of X. Find the joint distribution of X and Y, and determine $E\left[e^{\frac{X}{Y}}\right]$;

(iii) determine the CDF of $Z := \frac{X}{Y}$. If $Z \in \mathrm{AC}$, find its density.

Solution

(i) Taking $\eta_c = -ic$, we have

$$E\left[e^{cX}\right] = \varphi_X(-ic) = e^{c\mu + \frac{c^2}{2}}.$$

(ii) By independence, we have $\mu_{(X,Y)} = \mathcal{N}_{\mu,1} \otimes \mathrm{Unif}_n$ and

$$E\left[e^{\frac{X}{Y}}\right] = \iint_{\mathbb{R}^2} e^{\frac{x}{y}} \mathcal{N}_{\mu,1} \otimes \mathrm{Unif}_n(dx, dy) =$$

(by Fubini's theorem)

$$= \frac{1}{n} \sum_{k=1}^{n} \int_{\mathbb{R}} e^{\frac{x}{k}} \mathcal{N}_{\mu,1}(dx) =$$

(as seen in point (i) with $c = \frac{1}{k}$)

$$= \frac{1}{n} \sum_{k=1}^{n} e^{\frac{\mu}{k} + \frac{1}{2k^2}}.$$

(iii) By the Law of total probability, we have

$$F_Z(z) = P(Z \le z) = \sum_{k=1}^{n} P(Z \le z \mid Y = k) P(Y = k)$$

$$= \frac{1}{n} \sum_{k=1}^{n} P(X \le kz) = \frac{1}{n} \sum_{k=1}^{n} \int_{-\infty}^{kz} \Gamma(x - \mu)dx$$

where $\Gamma(x) = \frac{1}{\sqrt{2\pi}} e^{-\frac{x^2}{2}}$ is the standard normal density. Next, $Z \in AC$ since $F_Z \in C^{\infty}(\mathbb{R})$ and we have

$$F_Z'(z) = \frac{1}{n} \sum_{k=1}^{n} k \Gamma(kz - \mu).$$

Exercise 5.2.18 Let F be a CDF and $\alpha > 0$.

(i) Prove that F^{α} is still a CDF;
(ii) suppose F is the CDF of the exponential distribution Exp_{λ}. Determine the density of a random variable with CDF F^{α};
(iii) suppose F is the CDF of the discrete uniform distribution Unif_n, where $n \in \mathbb{N}$ is fixed. As α approaches $+\infty$, does F^{α} converge to a CDF? If so, which distribution does it correspond to? What if F is the CDF of the standard normal distribution?

Solution

(i) For every $\alpha > 0$ the function $f(x) = x^{\alpha}$ is continuous, monotone increasing on $[0, 1]$, $f(0) = 0$ and $f(1) = 1$. It follows that the properties of monotonicity, right continuity and the limits at $\pm\infty$ are preserved by composing f with a CDF F.

(ii) The function $F^\alpha(t) = \left(1 - e^{-\lambda t}\right)^\alpha \mathbb{1}_{\mathbb{R}_{\geq 0}}(t)$ is absolutely continuous and by differentiating we obtain the density

$$\gamma(t) = \alpha\lambda e^{-\lambda t}(1 - e^{-\lambda t})^{\alpha-1}\mathbb{1}_{\mathbb{R}_{\geq 0}}(t).$$

(iii) Since $F(x) < 1$ for $x < n$ and $F(x) = 1$ for $x \geq n$, we have

$$G(x) = \lim_{\alpha \to +\infty} F^\alpha(x) = \begin{cases} 0 & \text{if } x < n, \\ 1 & \text{if } x \geq n, \end{cases}$$

that is, G is the CDF of the Dirac delta centered at n. If F is the CDF of the standard normal distribution then we have $0 < F(x) < 1$ for every $x \in \mathbb{R}$ and therefore, for $\alpha \to +\infty$, F^α converges pointwise to the identically null function which is not a CDF.

Exercise 5.2.19 Given a real r.v. X, are there implications between the following properties?

 (i) X is absolutely continuous;
(ii) the CHF $\varphi_X \in L^1(\mathbb{R})$.

Solution (i) does not imply (ii): for example, $X \sim \text{Unif}_{[-1,1]}$ is absolutely continuous but $\varphi_X(\eta) = \frac{\sin\eta}{\eta}$ is not absolutely integrable as it can be verified directly or by means of the inversion theorem, see also Remark 2.5.7. Instead, (ii) implies (i) by the inversion theorem.

Exercise 5.2.20 Let (X, Y) be a two-dimensional r.v. with density

$$f(x, y) = \begin{cases} 2xy & \text{if } 0 < x < 1,\ 0 < y < \frac{1}{\sqrt{x}}, \\ 0 & \text{otherwise.} \end{cases}$$

 (i) Find the marginal densities of X, Y and determine if X, Y are independent;
(ii) do the random variables X and Y have finite mean and variance?

Solution

 (i) We have

$$f_X(x) = \begin{cases} \int_0^{\frac{1}{\sqrt{x}}} 2xy\, dy = 1 & \text{if } 0 < x < 1, \\ 0 & \text{otherwise,} \end{cases}$$

$$f_Y(y) = \begin{cases} \int_0^{\frac{1}{y^2}} 2xy\, dx = \frac{1}{y^3} & \text{if } y > 1, \\ \int_0^1 2xy\, dx = y & \text{if } 0 < y < 1, \\ 0 & \text{if } y < 0. \end{cases}$$

X, Y are not independent because the joint density is not the product of the marginals.

(ii) $X \sim \mathrm{Unif}_{[0,1]}$ and therefore has finite mean and variance. The density of Y is bounded on compact sets and is equal to y^{-3} for $y > 1$. It follows that Y has finite mean and infinite variance.

Exercise 5.2.21 Given three independent random variables X, Y, α with $X, Y \sim \mathcal{N}_{0,1}$ and $\alpha \sim \mathrm{Unif}_{[0,2\pi]}$, let

$$Z = X \cos \alpha + Y \sin \alpha.$$

Determine:

(i) the CHF and the distribution of Z;
(ii) $\mathrm{cov}(X, Z)$;
(iii) the value of the joint CHF $\varphi_{(X,Z)}(1, 1)$. Utilizing the approximation $\int_0^{2\pi} e^{-\cos t} dt \approx 8$, determine if X and Z are independent.

Solution

(i) Let us determine the distribution of Z by calculating its CHF:

$$\varphi_Z(\eta) = E\left[e^{i\eta(X \cos \alpha + Y \sin \alpha)} \right] =$$

(by the independence assumption)

$$= \frac{1}{2\pi} \int_0^{2\pi} \int_{\mathbb{R}} \int_{\mathbb{R}} e^{i\eta(x \cos t + y \sin t)} \mathcal{N}_{0,1}(dx) \mathcal{N}_{0,1}(dy) dt$$

$$= \frac{1}{2\pi} \int_0^{2\pi} e^{-\frac{1}{2}\eta^2 (\cos^2 t + \sin^2 t)} dt = e^{-\frac{\eta^2}{2}}$$

and therefore $Z \sim \mathcal{N}_{0,1}$.

(ii)

$$\mathrm{cov}(X, Z) = E[XZ] = E\left[X^2 \cos \alpha + XY \sin \alpha \right] =$$

(by independence)

$$= E\left[X^2 \right] E[\cos \alpha] = 0$$

since $E\left[X^2 \right] = \mathrm{var}(X) = 1$ and

$$E[\cos \alpha] = \frac{1}{2\pi} \int_0^{2\pi} \cos t \, dt = 0.$$

(iii) We have

$$\varphi_{(X,Z)}(1, 1) = E\left[e^{i(X+Z)}\right] = E\left[e^{iX(1+\cos\alpha)+iY\sin\alpha}\right] =$$

(by the independence assumption)

$$= \frac{1}{2\pi}\int_0^{2\pi}\int_{\mathbb{R}}\int_{\mathbb{R}} e^{ix(1+\cos t)+iy\sin t}\,\mathcal{N}_{0,1}(dx)\mathcal{N}_{0,1}(dy)dt$$

$$= \frac{1}{2\pi}\int_0^{2\pi} e^{-\frac{1}{2}(1+\cos t)^2-\frac{1}{2}\sin^2 t}\,dt$$

$$= \frac{e^{-1}}{2\pi}\int_0^{2\pi} e^{-\cos t}\,dt.$$

Then X and Z are not independent because otherwise it should be

$$\varphi_{(X,Z)}(1, 1) = \varphi_X(1)\varphi_Z(1) = e^{-1}.$$

Exercise 5.2.22 Let $X \sim \mathrm{Unif}_{[-1,1]}$. Give an example of $f \in m\mathcal{B}$ such that $f(X)$ is absolutely integrable and has infinite variance.

Solution For example

$$f(x) = \begin{cases} \frac{\mathrm{sgn}(x)}{\sqrt{|x|}} & \text{if } x \neq 0, \\ 0 & \text{if } x = 0. \end{cases}$$

We have

$$E[f(X)] = \frac{1}{2}\int_{-1}^{1} f(x)dx = 0$$

and

$$\mathrm{var}(f(X)) = E\left[f(X)^2\right] = \int_{-1}^{1}\frac{1}{|x|}dx = +\infty.$$

Exercise 5.2.23 Let X and Y be random variables with joint density

$$\gamma_{(X,Y)}(x, y) = \frac{1}{y}\mathbb{1}_{]0,\lambda y[\,\times\,]0,\frac{1}{\lambda}[}(x, y), \qquad \lambda > 0.$$

(i) Determine the marginal densities;
(ii) are the random variables $Z := e^X$ and $W := e^Y$ independent?

Solution

(i) We have

$$
\gamma_X(x) = \int_{\mathbb{R}} \gamma_{(X,Y)}(x, y)dy = \int_{\frac{x}{\lambda}}^{\frac{1}{\lambda}} \frac{1}{y}dy = -\log x, \qquad x \in {]0, 1[},
$$

$$
\gamma_Y(y) = \int_{\mathbb{R}} \gamma_{(X,Y)}(x, y)dx = \int_0^{\lambda y} \frac{1}{y}dx = \lambda, \qquad y \in \left]0, \frac{1}{\lambda}\right[.
$$

Thus, $\gamma_X(x) = \log x \cdot \mathbb{1}_{]0,1[}(x)$ and $\gamma_Y(y) = \lambda \mathbb{1}_{\left]0, \frac{1}{\lambda}\right[}(y)$.

(ii) If Z and W were independent, then $X = \log Z$ and $Y = \log W$ would also be independent. However, X and Y are not independent since the joint density is not equal to the product of the marginal densities.

Exercise 5.2.24 Let $X \sim \mathrm{Exp}_{\lambda_1}$ and $Y \sim \mathrm{Exp}_{\lambda_2}$ be independent random variables with $\lambda_1, \lambda_2 > 0$. Determine:

(i) the density of X^2;
ii) the joint CHF $\varphi_{(X,Y)}$;
(iii) the CHF of the sum φ_{X+Y}.

Solution

(i) The CDF of X^2 is given by

$$
F_{X^2}(z) = P(X^2 \le z) = P(X \le \sqrt{z}) = \int_0^{\sqrt{z}} \lambda_1 e^{-\lambda_1 t}dt = 1 - e^{-\lambda_1 \sqrt{z}}
$$

if $z \ge 0$ and $F_{X^2} \equiv 0$ on $] -\infty, 0]$. Since it is an AC function, we obtain the density of X^2 by differentiating

$$
\gamma_{X^2}(z) = \frac{d}{dz}F_{X^2}(z) = \frac{\lambda_1 e^{-\lambda_1 \sqrt{z}}}{2\sqrt{z}} \mathbb{1}_{\mathbb{R}_{\ge 0}}(z).
$$

(ii) By independence, we have

$$
\varphi_{(X,Y)}(\eta_1, \eta_2) = \varphi_X(\eta_1)\varphi_Y(\eta_2) = \frac{\lambda_1 \lambda_2}{(\lambda_1 - i\eta_1)(\lambda_2 - i\eta_2)}.
$$

(iii) Similarly,

$$
\varphi_{X+Y}(\eta) = \varphi_X(\eta)\varphi_Y(\eta) = \frac{\lambda_1 \lambda_2}{(\lambda_1 - i\eta)(\lambda_2 - i\eta)}.
$$

Exercise 5.2.25 Consider the function

$$F(x) = \begin{cases} \beta - e^{-x^\alpha} & \text{if } x \geq 0, \\ 0 & \text{if } x < 0. \end{cases}$$

(i) Are there values of α and β such that F is the CDF of the Dirac delta distribution? Determine all values of α and β for which F is a CDF;
(ii) for such values, consider a r.v. X that has F as its CDF. Calculate $P(X \leq 0)$ and $P(X \geq 1)$;
(iii) for the values of α, β for which $X \in AC$, determine a density of X;
(iv) for $\alpha = 2$, determine $E\left[X^{-1}\right]$ and the density of $Z := X^2 + 1$.

Solution

(i) If $\alpha = 0$ and $\beta = 1 + \frac{1}{e}$, then F is the CDF of the Dirac delta distribution centered at 0. The other values for which F is a CDF are $\alpha > 0$ and $\beta = 1$;
(ii) if $\alpha > 0$ and $\beta = 1$, then

$$P(X \leq 0) = F(0) = 0, \qquad P(X \geq 1) = 1 - F(1) = \frac{1}{e}.$$

If $\alpha = 0$ and $\beta = 1 + \frac{1}{e}$, then $P(X \leq 0) = 1$ and $P(X \geq 1) = 0$.
(iii) $X \in AC$ if $\alpha > 0$ and $\beta = 1$, and in this case, a density is determined by differentiating F:

$$\gamma(x) = F'(x) = \begin{cases} \alpha x^{\alpha-1} e^{-x^\alpha} & \text{if } x > 0, \\ 0 & \text{if } x < 0. \end{cases}$$

(iv) If $\alpha = 2$ we have

$$E\left[X^{-1}\right] = 2\int_0^{+\infty} e^{-x^2} dx = \sqrt{\pi}.$$

Let us determine the CDF of Z: first, $P(Z \leq 1) = 0$ and for $z > 1$ we have

$$P(X^2 + 1 \leq z) = P(-\sqrt{z-1} \leq X \leq \sqrt{z-1})$$
$$= P(X \leq \sqrt{z-1}) = 1 - e^{1-z}.$$

Then the density of Z is

$$\gamma_Z(z) = e^{1-z} \mathbb{1}_{[1,+\infty[}(z).$$

Exercise 5.2.26 Let X, Y be random variables with standard normal distribution, i.e., $X, Y \sim \mathcal{N}_{0,1}$, and T a r.v. with Bernoulli distribution, $T \sim \mathrm{Be}_{\frac{1}{2}}$. We assume that X, Y and T are independent.

(i) Prove that the random variables

$$Z := X - Y, \qquad W := TX + (1 - T)Y,$$

have normal distribution;

(ii) calculate $\mathrm{cov}(Z, W)$;

(iii) determine the joint CHF $\varphi_{(Z,W)}$;

(iv) are the random variables Z and W independent?

Solution

(i) The random vector (X, Y) has a standard bivariate normal distribution (since, by hypothesis, X, Y are independent). Moreover, we have

$$Z = \alpha \begin{pmatrix} X \\ Y \end{pmatrix}, \qquad \alpha = \begin{pmatrix} 1 & -1 \end{pmatrix}$$

and therefore, denoting by I the 2×2 identity matrix, we have $Z \sim \mathcal{N}_{0,\alpha I \alpha^*} = \mathcal{N}_{0,2}$.

Under the independence assumption, the joint distribution of X, Y and T is the product distribution

$$\mathcal{N}_{0,1} \otimes \mathcal{N}_{0,1} \otimes \mathrm{Be}_{\frac{1}{2}}$$

and therefore for every bounded $f \in m\mathcal{B}$ we have

$$E[f(W)] = \int_{\mathbb{R}^3} f(tx + (1 - t)y) \left(\mathcal{N}_{0,1} \otimes \mathcal{N}_{0,1} \otimes \mathrm{Be}_{\frac{1}{2}} \right)(dx, dy, dt) =$$

(by Fubini's theorem)

$$= \int_{\mathbb{R}} \left(\int_{\mathbb{R}} \left(\int_{\mathbb{R}} f(tx + (1 - t)y) \mathcal{N}_{0,1}(dx) \right) \mathcal{N}_{0,1}(dy) \right) \mathrm{Be}_{\frac{1}{2}}(dt)$$

$$= \frac{1}{2} \int_{\mathbb{R}} \left(\int_{\mathbb{R}} f(x) \mathcal{N}_{0,1}(dx) \right) \mathcal{N}_{0,1}(dy)$$

$$+ \frac{1}{2} \int_{\mathbb{R}} \left(\int_{\mathbb{R}} f(y) \mathcal{N}_{0,1}(dx) \right) \mathcal{N}_{0,1}(dy)$$

$$= \frac{1}{2} \int_{\mathbb{R}} f(x) \mathcal{N}_{0,1}(dx) + \frac{1}{2} \int_{\mathbb{R}} f(y) \mathcal{N}_{0,1}(dy)$$

$$= \int_{\mathbb{R}} f(x) \mathcal{N}_{0,1}(dx).$$

Thus, $W \sim \mathcal{N}_{0,1}$.

(ii) We have

$$\text{cov}(Z, W) = E\left[(X - Y)(TX + (1 - T)Y)\right]$$

$$= E\left[TX^2\right] + E\left[(1 - 2T)XY\right] - E\left[(1 - T)Y^2\right] =$$

(by independence of X, Y, T)

$$= E\left[T\right] E\left[X^2\right] - E\left[1 - T\right] E\left[Y^2\right] = 0.$$

(iii) The joint CHF is given by

$$\varphi_{(Z,W)}(\eta_1, \eta_2) = E\left[e^{i(\eta_1(X-Y)+\eta_2(TX+(1-T)Y))}\right]$$

$$= E\left[e^{i(\eta_1(X-Y)+\eta_2 X)}\mathbb{1}_{(T=1)}\right]$$

$$+ E\left[e^{i(\eta_1(X-Y)+\eta_2 Y)}\mathbb{1}_{(T=0)}\right] =$$

(by the independence of X, Y, T)

$$= \frac{1}{2} E\left[e^{i(\eta_1+\eta_2)X}\right] E\left[e^{-i\eta_1 Y}\right] + \frac{1}{2} E\left[e^{i\eta_1 X}\right] E\left[e^{i(\eta_2-\eta_1)Y}\right] =$$

(since $X, Y \sim \mathcal{N}_{0,1}$)

$$= \frac{e^{-\frac{\eta_1^2}{2}}}{2}\left(e^{-\frac{(\eta_1+\eta_2)^2}{2}} + e^{-\frac{(y_1-\eta_2)^2}{2}}\right),$$

which is not the CHF of a two-dimensional normal distribution. This also proves that

$$\varphi_{(Z,W)}(\eta_1, \eta_2) \neq \varphi_Z(\eta_1)\varphi_W(\eta_2)$$

and therefore Z, W are not independent.

Exercise 5.2.27 Let X be a r.v. with CDF

$$F(x) = \begin{cases} 0 & x < 0, \\ \lambda x & 0 \leq x < 1, \\ 1 & x \geq 1, \end{cases}$$

where λ is a fixed parameter such that $0 < \lambda < 1$. Let $Y \sim \text{Unif}_{[0,1]}$ be independent of X. Determine:

(i) if X is absolutely continuous;
(ii) the distribution of

$$Z := X\mathbb{1}_{(X<1)} + Y\mathbb{1}_{(X\geq 1)}.$$

Solution

(i) X is not absolutely continuous because $P(X = 1) = F(1) - F(1-) = 1 - \lambda > 0$.

(ii) Let us calculate the CDF of Z. For $z \in [0, 1]$ we have

$$P(Z \leq z) = P\left((Z \leq z) \cap (X < 1)\right) + P\left((Z \leq z) \cap (X \geq 1)\right)$$
$$= P\left(X \leq z\right) + P\left((Y \leq z) \cap (X \geq 1)\right) =$$

(by independence)

$$= \lambda z + P(Y \leq z)P(X \geq 1) = \lambda z + z(1 - \lambda) = z.$$

As a result, $Z \sim \text{Unif}_{[0,1]}$.

Exercise 5.2.28 Let (X, Y) be a two-dimensional r.v. with a uniform distribution on the triangle T with vertices $(0, 0)$, $(2, 0)$, and $(0, 2)$. Determine:

(i) the density of X;
(ii) if X and Y are independent;
(iii) the density and expectation of $Z := X + Y$.

Solution

(i) The density of (X, Y) is

$$\gamma_{(X,Y)}(x, y) = \frac{1}{2}\mathbb{1}_T(x, y), \qquad T = \{x, y \in \mathbb{R} \mid x, y \geq 0, \ x + y \leq 2\}.$$

We have

$$\gamma_X(x) = \int_{\mathbb{R}} \gamma_{(X,Y)}(x, y)dy = \int_0^{2-x} \frac{1}{2}\mathbb{1}_{[0,2]}(x)dy = \frac{2-x}{2}\mathbb{1}_{[0,2]}(x).$$

The calculation of γ_Y is analogous.

(ii) X, Y are not independent because the joint density is not the product of the marginals.

(iii) We have

$$\gamma_Z(z) = \int_{\mathbb{R}} \gamma_{(X,Y)}(x, z - x)dx = \frac{1}{2}\int_{\mathbb{R}} \mathbb{1}_T(x, z - x)dx$$

$$= \frac{z}{2}\mathbb{1}_{[0,2]}(z).$$

Therefore

$$E[Z] = \int_0^2 \frac{z^2}{2}dz = \frac{4}{3}.$$

Exercise 5.2.29 Verify that the function

$$\gamma(x, y) = \begin{cases} 4y & \text{if } x > 0 \text{ and } 0 < y < e^{-x}, \\ 0 & \text{otherwise,} \end{cases}$$

is a density. Let X, Y be random variables with joint density γ. Determine:

(i) the marginal densities γ_X and γ_Y;
(ii) if X and Y are independent;
(iii) the conditional density $\gamma_{X|Y}$ and identify which known density it corresponds to;
(iv) $E[X \mid Y]$ and $\mathrm{Var}(X \mid Y)$.

Solution The function γ is non-negative and measurable with

$$\int_{\mathbb{R}^2} \gamma(x, y)dxdy = \int_0^{+\infty} \int_0^{e^{-x}} 4ydydx = \int_0^{+\infty} 2e^{-2x}dx = 1.$$

(i) We have just calculated

$$\gamma_X(x) = \int_{\mathbb{R}} \gamma(x, y)dy = \int_0^{e^{-x}} 4ydy = 2e^{-2x}\mathbb{1}_{]0,+\infty[}(x)$$

from which we recognize that $X \sim \mathrm{Exp}_2$. Then we observe that

$$\gamma(x, y) = 4y\mathbb{1}_{]0,-\log y[}(x)\mathbb{1}_{]0,1[}(y)$$

so that

$$\gamma_Y(y) = \int_{\mathbb{R}} \gamma(x, y)dx = \int_0^{-\log y} 4y\mathbb{1}_{]0,1[}(y)dx = -4y\log y\,\mathbb{1}_{]0,1[}(y).$$

(ii) X and Y are not independent because the joint density is not the product of the marginals.
(iii) We have

$$\gamma_{X|Y}(x,y) = \frac{\gamma(x,y)}{\gamma_Y(y)}\mathbb{1}_{(\gamma_Y>0)}(y) = -\frac{1}{\log y}\mathbb{1}_{]0,-\log y[}(x)\mathbb{1}_{]0,1[}(y)$$

and therefore X has a uniform conditional density on $]0,-\log Y[$.
(iv) By point (iii), we have

$$E[X\mid Y] = \frac{-\log Y}{2}, \qquad \mathrm{var}(X\mid Y) = \frac{(\log Y)^2}{12}.$$

Exercise 5.2.30 Given the function

$$\gamma(x) = (ax+b)\mathbb{1}_{[-1,1]}(x), \qquad x \in \mathbb{R},$$

determine the values of a and b in $\mathbb{R}$ such that:

(i) γ is a density;
(ii) the corresponding CHF is real-valued.

Solution

(i) By imposing

$$1 = \int_{\mathbb{R}} \gamma(x)dx = 2b$$

we get $b = \frac{1}{2}$. Furthermore, $\gamma \geq 0$ if and only if $ax \geq -\frac{1}{2}$ for every $x \in [-1,1]$, from which we obtain the condition $-\frac{1}{2} \leq a \leq \frac{1}{2}$.
(ii) The CHF is given by

$$\int_{-1}^{1} e^{i\eta x}\left(ax+\frac{1}{2}\right)dx = \frac{\sin\eta}{\eta} + 2ia\frac{\sin\eta - \eta\cos\eta}{\eta^2}$$

and it takes real values if $a = 0$.

Exercise 5.2.31 Let (X,Y) be a random vector with uniform distribution on the unit disk C centered at the origin in $\mathbb{R}^2$.

(i) Write the density of (X,Y) and calculate $E[X]$;
(ii) are X and $X-Y$ independent?

Let

$$Z_\alpha = \left(X^2 + Y^2\right)^\alpha, \qquad \alpha > 0.$$

(iii) Write the CDF of Z_α and plot its graph;
(iv) determine if $Z_\alpha \in AC$, and if so, determine its density;
 (v) find the values of $\alpha > 0$ for which $\frac{1}{Z_\alpha}$ is absolutely integrable, and calculate its expected value for those values of α.

Solution

 (i) $\gamma_{(X,Y)} = \frac{1}{\pi} \mathbb{1}_C$ and $E[X] = 0$.
(ii) If X and $X - Y$ were independent, then we would have

$$0 = E[X]E[X - Y] = E[X(X - Y)] = E\left[X^2\right] - E[XY] = \frac{1}{4},$$

where the expected values are determined by a simple calculation as in Example 2.3.34.
(iii) We have

$$F(t) := P(Z_\alpha \le t) = \begin{cases} 0 & \text{if } t \le 0, \\ 1 & \text{if } t \ge 1 \end{cases}$$

and, for $0 < t < 1$,

$$P(Z_\alpha \le t) = P\left(X^2 + Y^2 \le t^{\frac{1}{\alpha}}\right) = t^{\frac{1}{\alpha}}$$

where the probability is calculated as the ratio between the area of the circle with radius $t^{\frac{1}{2\alpha}}$ and that of radius one: see Fig. 5.1.
(iv) F is absolutely continuous because it is a.e. differentiable and $F(t) = \int_0^t F'(s)ds$ (cf. Definition 1.4.30). A density of Z_α is given by

$$F'(t) = \frac{1}{\alpha} t^{\frac{1}{\alpha}-1} \mathbb{1}_{]0,1[}(t).$$

 (v) We have

$$E\left[Z_\alpha^{-1}\right] = \int_0^1 \frac{F'(t)}{t} dt < \infty$$

if $2 - \frac{1}{\alpha} < 1$ that is $0 < \alpha < 1$. In this case $E\left[Z_\alpha^{-1}\right] = \frac{1}{1-\alpha}$.

Exercise 5.2.32 Let $(X, Y, Z) \sim \mathcal{N}_{\mu,C}$ with

$$\mu = \begin{pmatrix} 0 \\ 1 \\ 2 \end{pmatrix}, \qquad C = \begin{pmatrix} 1 & 0 & -1 \\ 0 & 2 & 2 \\ -1 & 2 & 3 \end{pmatrix}.$$

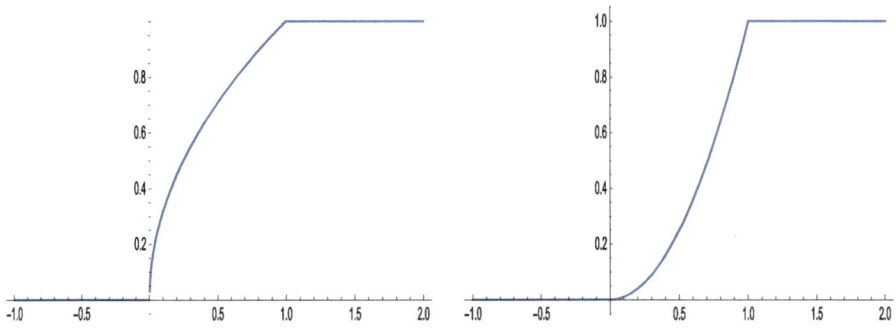

Fig. 5.1 On the left: graph of F for $\alpha > 1$. **On the right:** graph of F for $0 < \alpha < 1$

Determine:

(i) the CHF of (X, Y);
(ii) if the random variables $X + Y$ and Z are independent.

Solution

(i) We have $(X, Y) \sim \mathcal{N}_{\bar{\mu}, \bar{C}}$ with $\bar{\mu} = \begin{pmatrix} 0 \\ 1 \end{pmatrix}$ and $\bar{C} = \begin{pmatrix} 1 & 0 \\ 0 & 2 \end{pmatrix}$ and therefore

$$\varphi_{(X,Y)}(\eta_1, \eta_2) = e^{i\eta_2 - \frac{1}{2}(\eta_1^2 + 2\eta_2^2)}.$$

(ii) $(X + Y, Z)$ has a bivariate normal distribution since it is a linear combination of (X, Y, Z). Consequently, $X + Y$ and Z are independent if and only if they are uncorrelated: since

$$\mathrm{cov}(X + Y, Z) = \mathrm{cov}(X, Z) + \mathrm{cov}(Y, Z) = -1 + 2,$$

then $X + Y$ and Z are not independent.

Exercise 5.2.33 A stopwatch is started and stops automatically at a random time $T \sim \mathrm{Exp}_1$. We wait until time 3 and at that moment we observe the value X displayed on the stopwatch. Determine:

(i) the CDF of X, calculating $F_X(x)$ separately for $x < 3$ and $x \geq 3$;
(ii) if X is absolutely continuous;
(iii) $E[X]$;
(iv) $E[X \mid T]$;
(v) if X is a discrete random variable.

Solution Observe that

$$X = \min\{T, 3\} = T\mathbb{1}_{(T \leq 3)} + 3\mathbb{1}_{(T > 3)}.$$

(i) We have $P(X \leq 0) = 0$ and

$$P(X \leq x) = P\left((X \leq x) \cap (T \leq 3)\right) + P\left((X \leq x) \cap (T > 3)\right)$$

$$= \begin{cases} P(T \leq x) = 1 - e^{-x} & \text{if } 0 \leq x < 3, \\ 1 & \text{if } x \geq 3. \end{cases}$$

(ii) X is not absolutely continuous because the CDF is discontinuous at point 3.

(iii) We have

$$E[X] = E\left[T\mathbb{1}_{(T \leq 3)} + 3\mathbb{1}_{(T > 3)}\right] = \int_0^3 te^{-t}dt + 3P(T > 3) = 1 - e^{-3}.$$

(iv) X is $\sigma(T)$-measurable because it is a (measurable) function of T. Consequently

$$E[X \mid T] = X = \min\{T, 3\}.$$

(v) X is not discrete because $P(X = 3) = P(T \geq 3)$ is positive and strictly less than 1, and $P(X = x) = 0$ for every $x \neq 3$.

Exercise 5.2.34 Verify that the function

$$\gamma(x, y) = \frac{e^{-x}}{e - 1}\mathbb{1}_A(x, y), \qquad A = \{(x, y) \in \mathbb{R}^2 \mid x + y > 0, \ 0 < y < 1\},$$

is a density and consider (X, Y) with density $\gamma_{(X,Y)} = \gamma$. Justify the validity of the formula (without performing the calculations)

$$\gamma_X(x) = \begin{cases} 0 & \text{if } x \leq -1, \\ \frac{(1+x)e^{-x}}{e-1} & \text{if } -1 < x < 0, \\ \frac{e^{-x}}{e-1} & \text{if } x \geq 0, \end{cases}$$

and determine:

(i) if X and Y are independent;
(ii) the density of Y^2;
(iii) the conditional density $\gamma_{X|Y}$.

Solution The function γ is measurable, non-negative, and has an integral equal to one. The expression of γ_X can be obtained from the formula

$$\gamma_X(x) = \int_{\mathbb{R}} \gamma_{(X,Y)}(x, y)dy.$$

(i) Since

$$\gamma_Y(y) = \int_{\mathbb{R}} \gamma_{(X,Y)}(x, y)dx = \frac{e^y}{e-1}\mathbb{1}_{[0,1]}(y),$$

we recognize that X, Y are not independent because the joint density is not the product of the marginals.

(ii) First, we calculate the CDF for $0 < z < 1$:

$$F_{Y^2}(z) = P(Y^2 \leq z) = P(Y \leq \sqrt{z}) = \int_0^{\sqrt{z}} \frac{e^y}{e-1}dy = \frac{e^{\sqrt{z}}-1}{e-1}.$$

By differentiating, we obtain

$$\gamma_{Y^2}(z) = \frac{e^{\sqrt{z}}}{2(e-1)\sqrt{z}}\mathbb{1}_{[0,1]}(z).$$

(iii) We have

$$\gamma_{X|Y}(x, y) = \frac{\gamma_{(X,Y)}(x, y)}{\gamma_Y(y)}\mathbb{1}_{(\gamma_Y>0)}(y) = e^{-(x+y)}\mathbb{1}_A(x, y).$$

Exercise 5.2.35 Let $X = (X_1, X_2, X_3) \sim \mathcal{N}_{0,C}$ with

$$C = \begin{pmatrix} 2 & 1 & -1 \\ 1 & 1 & -1 \\ -1 & -1 & 1 \end{pmatrix}$$

and consider the random vectors $Y := (X_1, X_3)$ and $Z := (X_2, 2X_3)$. Determine:

(i) the distributions of Y and Z, specifying whether they are absolutely continuous;
(ii) if Y and Z are independent;
(iii) the CHF φ_Z specifying whether it is absolutely integrable on $\mathbb{R}^2$.

Solution

(i) Since

$$Y = \alpha X, \qquad \alpha = \begin{pmatrix} 1 & 0 & 0 \\ 0 & 0 & 1 \end{pmatrix},$$

$$Z = \beta X, \qquad \beta = \begin{pmatrix} 0 & 1 & 0 \\ 0 & 0 & 2 \end{pmatrix},$$

we have $Y \sim \mathcal{N}_{0,\alpha C\alpha^*}$ and $Z \sim \mathcal{N}_{0,\beta C\beta^*}$ with

$$\alpha C\alpha^* = \begin{pmatrix} 2 & -1 \\ -1 & 2 \end{pmatrix}, \qquad \beta C\beta^* = \begin{pmatrix} 1 & -2 \\ -2 & 4 \end{pmatrix}.$$

It follows that Y is absolutely continuous, while Z is not because $\beta C\beta^*$ is singular.

(ii) Y and Z are not independent. Observing that they have the second component proportional, we let $f(x_1, x_2) = x_2$: then, we have

$$E[f(Y)f(Z)] = 2E\left[X_3^2\right] = 2$$

but $E[f(Y)] = E[f(Z)] = 0$.

(iii) Since $Z \sim \mathcal{N}_{0,\beta C\beta^*}$ we have

$$\varphi_Z(\eta_1, \eta_2) = e^{-\frac{1}{2}\left(\eta_1^2 + 4\eta_2^2 - 4\eta_1\eta_2\right)}.$$

φ_Z is not absolutely integrable otherwise, by the inversion theorem, Z would be absolutely continuous.

Exercise 5.2.36 Let X, Y be random variables with standard normal distribution, i.e., $X, Y \sim \mathcal{N}_{0,1}$, and $T \sim \mu := \frac{1}{2}(\delta_{-1} + \delta_1)$. We assume that X, Y and T are independent.

(i) Prove that the random variables

$$Z := X + Y, \qquad W := X + TY,$$

have the same law;

(ii) are Z and W independent?

(iii) determine the joint CHF $\varphi_{(Z,W)}$.

Solution

(i) The random vector (X, Y) has a standard bivariate normal distribution (since, by hypothesis, X, Y are independent). Moreover, we have

$$Z = \alpha\begin{pmatrix} X \\ Y \end{pmatrix}, \qquad \alpha = (1 \ 1)$$

and therefore, denoting by I the 2×2 identity matrix, we have $Z \sim \mathcal{N}_{0,\alpha I\alpha^*} = \mathcal{N}_{0,2}$.

By independence, the joint law of X, Y and T is the product distribution

$$\mathcal{N}_{0,1} \otimes \mathcal{N}_{0,1} \otimes \mu$$

and therefore for every bounded $f \in m\mathscr{B}$ we have

$$E[f(W)] = \int_{\mathbb{R}^3} f(x + ty) \left(\mathcal{N}_{0,1} \otimes \mathcal{N}_{0,1} \otimes \mu\right)(dx, dy, dt) =$$

(by Fubini's theorem)

$$= \int_{\mathbb{R}} \left(\int_{\mathbb{R}} \left(\int_{\mathbb{R}} f(x + ty)\mathcal{N}_{0,1}(dx) \right) \mathcal{N}_{0,1}(dy) \right) \mu(dt)$$

$$= \frac{1}{2} \int_{\mathbb{R}} \left(\int_{\mathbb{R}} f(x + y)\mathcal{N}_{0,1}(dx) \right) \mathcal{N}_{0,1}(dy)$$

$$+ \frac{1}{2} \int_{\mathbb{R}} \left(\int_{\mathbb{R}} f(x - y)\mathcal{N}_{0,1}(dx) \right) \mathcal{N}_{0,1}(dy)$$

(by the change of variables $z = -y$ in the second integral)

$$= \int_{\mathbb{R}^2} f(x + y)\mathcal{N}_{0,1}(dx)\mathcal{N}_{0,1}(dy) = E[f(Z)].$$

It follows that Z and W both have distribution $\mathcal{N}_{0,2}$.
(ii) Z and W are not independent because

$$\mathrm{cov}(Z, W) = E[(X + Y)(X + TY)]$$

$$= E\left[X^2\right] + E[(1 + T)XY] + E\left[TY^2\right] = 1$$

by the independence of X, Y, T.
(iii) The joint CHF is given by

$$\varphi_{(Z,W)}(\eta_1, \eta_2) = E\left[e^{i(\eta_1(X+Y)+\eta_2(X+TY))}\right]$$

$$= E\left[e^{i(\eta_1+\eta_2)(X+Y)}\mathbb{1}_{(T=1)}\right]$$

$$+ E\left[e^{i(\eta_1+\eta_2)X+i(\eta_1-\eta_2)Y}\mathbb{1}_{(T=-1)}\right] =$$

(by independence of X, Y, T)

$$= \frac{1}{2}\left(E\left[e^{i(\eta_1+\eta_2)(X+Y)}\right] + E\left[e^{i(\eta_1+\eta_2)X}\right] E\left[e^{i(\eta_1-\eta_2)Y}\right]\right) =$$

(since $X, Y \sim \mathcal{N}_{0,1}$ and $X + Y \sim \mathcal{N}_{0,2}$)

$$= \frac{1}{2}\left(e^{-(\eta_1+\eta_2)^2} + e^{-\eta_1^2-\eta_2^2}\right).$$

Exercise 5.2.37 Consider the function

$$\gamma(x, y) = \frac{1}{4}(ax + by + 1)\mathbb{1}_{[-1,1]\times[-1,1]}(x, y), \qquad (x, y) \in \mathbb{R}^2.$$

Determine:

(i) for which $a, b \geq 0$, the function γ is a density;
(ii) the density of X and Y assuming that γ is the density of (X, Y);
(iii) for which $a, b \geq 0$ the random variables X and Y are independent.

Solution

(i) γ is a measurable function with

$$\iint_{\mathbb{R}^2} \gamma(x, y)dxdy = 1$$

for every $a, b \geq 0$. Moreover, since $a, b \geq 0$, we have

$$\gamma(x, y) \geq \gamma(-1, -1) = -a - b + 1, \qquad (x, y) \in [-1, 1] \times [-1, 1]$$

and therefore $\gamma \geq 0$ if $a + b \leq 1$.

(ii)

$$\gamma_X(x) = \int_{-1}^{1} \gamma(x, y)dy = \frac{ax + 1}{2}\mathbb{1}_{[-1,1]}(x),$$

$$\gamma_Y(y) = \int_{-1}^{1} \gamma(x, y)dx = \frac{by + 1}{2}\mathbb{1}_{[-1,1]}(y).$$

(iii) X and Y are independent if and only if $\gamma(x, y) = \gamma_X(x)\gamma_Y(y)$, that is

$$(ax + 1)(by + 1) = ax + by + 1$$

which means $abxy = 0$ i.e. $a = 0$ or $b = 0$.

Exercise 5.2.38 Let $X = (X_1, X_2, X_3) \sim \mathcal{N}_{0,C}$ with

$$C = \begin{pmatrix} 2 & 1 & -1 \\ 1 & 1 & 0 \\ -1 & 0 & 1 \end{pmatrix}.$$

Determine for which values of $a \in \mathbb{R}$:

(i) $Y := (aX_1 + X_2, X_3)$ is an absolutely continuous random variable;
(ii) $aX_1 + X_2$ e X_3 are independent;
(iii) the CHF φ_Y is absolutely integrable on $\mathbb{R}^2$.

Solution

(i) Since

$$Y = \alpha X, \qquad \alpha = \begin{pmatrix} a & 1 & 0 \\ 0 & 0 & 1 \end{pmatrix},$$

We have $Y \sim \mathcal{N}_{0,\alpha C \alpha^*}$ with

$$\alpha C \alpha^* = \begin{pmatrix} 1 + 2a + 2a^2 & -a \\ -a & 1 \end{pmatrix}, \qquad \det(\alpha C \alpha^*) = (1 + a)^2.$$

Only for $a = -1$, the matrix $\alpha C \alpha^*$ is singular, and for that value of a, the random variable Y is not absolutely continuous.

(ii) Given the expression of the covariance matrix $\alpha C \alpha^*$, we have that $aX_1 + X_2$ and X_3 are uncorrelated (and therefore independent) if $a = 0$.

(iii) Since $Y \sim \mathcal{N}_{0,\alpha C \alpha^*}$ we have

$$\varphi_Y(\eta) = e^{-\frac{1}{2}\langle C \alpha^* \eta, \alpha^* \eta \rangle}.$$

φ_Y is not absolutely integrable if $a = -1$. Otherwise, by the inversion theorem, Y would be absolutely continuous.

Exercise 5.2.39 Let $X \sim \mathcal{N}_{\mu,\sigma^2}$ and $Y \sim Be_p$, with $0 < p < 1$, be independent random variables. Let $Z = X^Y$. Determine:

(i) $E[Z]$;
(ii) the CDF of Z and whether Z is absolutely continuous;
(iii) the CHF of Z and use it to calculate $E[Z^2]$.

Solution

(i) By independence, we have

$$E[Z] = \iint_{\mathbb{R}^2} x^y \mathcal{N}_{\mu,\sigma^2} \otimes Be_p(dx, dy) =$$

(by Fubini's theorem)

$$= p \int_{\mathbb{R}} x \mathcal{N}_{\mu,\sigma^2}(dx) + (1-p) \int_{\mathbb{R}} \mathcal{N}_{\mu,\sigma^2}(dx) = p\mu + (1-p).$$

(ii) We have

$$F_Z(z) = P(Z \le z) = P((Z \le z) \cap (Y = 1)) + P((Z \le z) \cap (Y = 0)) =$$

(by independence of X and Y)

$$= P(X \leq z)P(Y = 1) + P(1 \leq z)P(Y = 0)$$
$$= pF_X(z) + (1 - p)\mathbb{1}_{[1,+\infty[}(z).$$

Since F_Z has a jump at $z = 1$ with size $1 - p$, the random variable Z is not absolutely continuous.

(iii) We have

$$\varphi_Z(\eta) = E\left[e^{i\eta Z}\right] = pE\left[e^{i\eta X}\right] + (1 - p)E\left[e^{i\eta}\right]$$

$$= p\varphi_X(\eta) + (1 - p)e^{i\eta}, \quad \varphi_X(\eta) = e^{i\mu\eta - \frac{\sigma^2\eta^2}{2}}.$$

By Theorem 2.5.20 we have

$$E\left[Z^2\right] = -\partial_\eta^2\varphi_Z(\eta)|_{\eta=0} = p(\mu^2 + \sigma^2) + (1 - p).$$

Exercise 5.2.40 In pharmacology, the half-life is the time required (expressed in days) to reduce the amount of a drug in the body by 50%. Let $T \sim$ Gamma$_{2,1}$ be the half-life of an antibiotic upon taking the first dose and let $S \sim$ Unif$_{[T,2T]}$ be the half-life at the assumption of the second dose. Determine:

  (i) the joint density $\gamma_{(S,T)}$ and marginal γ_S;
 (ii) the conditional expectation of T given $(S < 2)$.
(iii) the conditional expectation of T given S (it is sufficient to write the formulas without performing all the calculations).

Solution

  (i) By assumption, $\gamma_T(t) = te^{-t}\mathbb{1}_{\mathbb{R}_{\geq 0}}(t)$ and $\gamma_{S|T}(s,t) = \frac{1}{t}\mathbb{1}_{[t,2t]}(s)$. From formula (4.3.9) for the conditional density, we obtain

$$\gamma_{(S,T)}(s,t) = \gamma_{S|T}(s,t)\gamma_T(t) = e^{-t}\mathbb{1}_{[t,2t]\times\mathbb{R}_{\geq 0}}(s,t) = e^{-t}\mathbb{1}_{\mathbb{R}_{\geq 0}\times[s/2,s]}(s,t)$$

and

$$\gamma_S(s) = \int_{\mathbb{R}} \gamma_{(S,T)}(s,t)dt = \int_{s/2}^{s} e^{-t}dt\,\mathbb{1}_{\mathbb{R}_{\geq 0}}(s) = \left(e^{-\frac{s}{2}} - e^{-s}\right)\mathbb{1}_{\mathbb{R}_{\geq 0}}(s).$$

 (ii) We have

$$P(S < 2) = \int_0^2 \gamma_S(s)ds = \left(1 - \frac{1}{e}\right)^2 \approx 40\%,$$

$$E\left[T \mid S < 2\right] = \frac{1}{P(S < 2)} \int_0^2 \int_0^{+\infty} t\gamma_{(S,T)}(s,t)\,dt\,ds = \frac{2(e-2)}{e-1} \approx 0.84.$$

(iii) First of all,

$$\gamma_{T \mid S}(t,s) = \frac{\gamma_{(S,T)}(s,t)}{\gamma_S(s)}\mathbb{1}_{(\gamma_S > 0)}(s) = \frac{e^{-t}}{e^{-\frac{s}{2}} - e^{-s}}\mathbb{1}_{\mathbb{R}_{\geq 0} \times [s/2,s]}(s,t).$$

Then we have

$$E\left[T \mid S\right] = \int_0^{+\infty} t\,\gamma_{T \mid S}(t,S)\,dt = \frac{1}{2}\left(-\frac{S}{e^{S/2}-1} + S + 2\right).$$

Exercise 5.2.41 Let X and Y be random variables with joint density

$$\gamma_{(X,Y)}(x,y) = \frac{e^{-y|x|}}{\log 4}\mathbb{1}_{[1,2]}(y), \qquad (x,y) \in \mathbb{R}^2.$$

Determine:

  (i) the marginal densities;
  (ii) if the random variables $Z := e^X$ and $W := e^Y$ are independent;
  (iii) $E\left[Y \mid X > 0\right]$.

Solution

  (i) We have

$$\gamma_X(x) = \int_{\mathbb{R}} \gamma_{(X,Y)}(x,y)\,dy = \frac{e^{-|x|} - e^{-2|x|}}{|x|\log 4},$$

$$\gamma_Y(y) = \int_{\mathbb{R}} \gamma_{(X,Y)}(x,y)\,dx = \frac{1}{y\log 2}\mathbb{1}_{]1,2]}(y).$$

  (ii) If Z and W were independent, then $X = \log Z$ and $Y = \log W$ would also be independent. However, X and Y are not independent because the joint density is not equal to the product of the marginals.
  (iii) By symmetry, $P(X > 0) = \frac{1}{2}$ and we have

$$E\left[Y \mid X > 0\right] = \frac{1}{P(X > 0)} \int_{(X>0)} Y\,dP$$

$$= 2\int_1^2 \frac{y}{\log 4} \int_0^{+\infty} e^{-y|x|}\,dx\,dy = \frac{1}{\log 2}.$$

Exercise 5.2.42 Let $(X, Y) \sim \mathcal{N}_{\mu,C}$ with $\mu = (0, 0)$ and $C = \begin{pmatrix} 1 & 0 \\ 0 & 2 \end{pmatrix}$.
Determine:

(i) the law of (Y, X);
(ii) the law and the CHF of (X, X). Is it an absolutely continuous random variable? Is it true that

$$\lim_{|(\eta_1,\eta_2)| \to +\infty} \varphi_{(X,X)}(\eta_1, \eta_2) = 0?$$

(iii) if (Y, X) and (X, X) are independent.

Solution

(i) Since $\begin{pmatrix} Y \\ X \end{pmatrix} = \alpha \begin{pmatrix} X \\ Y \end{pmatrix}$ with $\alpha = \begin{pmatrix} 0 & 1 \\ 1 & 0 \end{pmatrix}$, we have $(X, Y) \in \mathcal{N}_{(0,0),C_1}$ with $C_1 = \alpha C \alpha^* = \begin{pmatrix} 2 & 0 \\ 0 & 1 \end{pmatrix}$.

(ii) Similarly, we can show that $(X, X) \in \mathcal{N}_{(0,0),C_2}$ with $C_2 = \begin{pmatrix} 1 & 1 \\ 1 & 1 \end{pmatrix}$. In this case, the covariance matrix is degenerate and (X, X) is not absolutely continuous. We have

$$\varphi_{(X,X)}(\eta_1, \eta_2) = e^{-\frac{1}{2}(\eta_1^2 + 2\eta_1\eta_2 + \eta_2^2)}$$

and $\varphi_{(X,X)}(\eta_1, -\eta_1) = 1$ for every $\eta_1 \in \mathbb{R}$; in particular, $\varphi_{(X,X)}$ does not converge to 0 at infinity.

(iii) If (Y, X) and (X, X) were independent, then their second components, which are both equal to X, would also be independent.

Exercise 5.2.43 Let

$$\Gamma(y) = \frac{1}{\sqrt{2\pi}} e^{-\frac{y^2}{2}}, \qquad y \in \mathbb{R},$$

be the standard Gaussian function.

(i) Verify that the function

$$\gamma(x, y) = \mathbb{1}_H(x, y), \qquad H := \{(x, y) \in \mathbb{R}^2 \mid 0 \le x \le \Gamma(y)\}$$

is a density;
(ii) let X, Y be random variables with joint density γ. Determine the marginal densities γ_X and γ_Y. Are X and Y independent?

(iii) recalling formula (4.3.9) for the conditional density

$$\gamma_{X|Y}(x, y) := \frac{\gamma(x, y)}{\gamma_Y(y)}, \qquad x \in \mathbb{R}, \ y \in (\gamma_Y > 0),$$

determine $\gamma_{X|Y}$ and the conditional expected value $E[X^n \mid Y]$ for $n \in \mathbb{N}$.

Solution

(i) γ is a measurable, non-negative function and

$$\iint_{\mathbb{R}^2} \gamma(x, y) dx dy = \int_{\mathbb{R}} \int_0^{\Gamma(y)} dx dy = \int_{\mathbb{R}} \Gamma(y) dy = 1.$$

(ii) We have

$$\gamma_X(x) = \int_{\mathbb{R}} \gamma(x, y) dy = 2\sqrt{-2 \log\left(x\sqrt{2\pi}\right)} \mathbb{1}_{]0, \frac{1}{\sqrt{2\pi}}]}(x),$$

$$\gamma_Y(y) = \int_{\mathbb{R}} \gamma(x, y) dx = \Gamma(y).$$

X and Y are not independent since the joint density is not the product of the marginals.

(iii) We have

$$\gamma_{X|Y}(x, y) = \frac{1}{\Gamma(y)} \mathbb{1}_H(x, y)$$

and

$$E[X^n \mid Y] = \int_{\mathbb{R}} x^n \gamma_{X|Y}(x, y) = \frac{1}{\Gamma(y)} \int_0^{\Gamma(y)} x^n dx = \frac{1}{n+1} \Gamma^n(y).$$

5.3 Sequences of Random Variables

Exercise 5.3.1

(i) Determine the values of $a, b \in \mathbb{R}$ such that the function

$$\gamma(x) = (2ax + b)\mathbb{1}_{[0,1]}(x), \qquad x \in \mathbb{R},$$

is a density;

(ii) consider a sequence of random variables $(X_n)_{n\in\mathbb{N}}$ i.i.d. with density γ with $b = 0$. Determine the CDF of $\sqrt{n}X_1$ and of

$$Y_n = \min\{\sqrt{n}X_1, \ldots, \sqrt{n}X_n\};$$

(iii) prove that $(Y_n)_{n\in\mathbb{N}}$ converges weakly and determine the density of the limiting random variable.

Solution

(i) Imposing

$$1 = \int_{\mathbb{R}} \gamma(x)dx = \int_0^1 (2ax + b)dx = a + b$$

we get $b = 1 - a$. Moreover, γ must be non-negative: if $a \geq 0$ then the minimum of γ is assumed for $x = 0$ and we have the condition $1 - a \geq 0$; if $a < 0$ then the minimum of γ is assumed for $x = 1$ and we have the condition $a + 1 \geq 0$. In conclusion, for $|a| \leq 1$ and $b = 1 - a$, γ is a density.

(ii) We have

$$P(\sqrt{n}X_1 \leq x) = \begin{cases} 0 & \text{if } x < 0, \\ \int_0^{\frac{x}{\sqrt{n}}} 2ydy = \frac{x^2}{n} & \text{if } 0 \leq x < \sqrt{n}, \\ 1 & \text{if } x \geq \sqrt{n}. \end{cases}$$

By Proposition 2.6.9, we have

$$F_{Y_n}(x) = 1 - (1 - F_{\sqrt{n}X_1}(x))^n = \begin{cases} 0 & \text{if } x < 0, \\ 1 - \left(1 - \frac{x^2}{n}\right)^n & \text{if } 0 \leq x < \sqrt{n}, \\ 1 & \text{if } x \geq \sqrt{n}. \end{cases}$$

(iii) We have

$$\lim_{n\to\infty} F_{Y_n}(x) = F_Y(x) := \begin{cases} 0 & \text{if } x < 0, \\ 1 - e^{-x^2} & \text{if } x \geq 0, \end{cases}$$

and thus by Theorem 3.3.3 $Y_n \xrightarrow{d} Y$ as $n \to \infty$ with Y having density $\gamma_Y(x) = F_Y'(x) = 2xe^{-x^2}\mathbb{1}_{[0,+\infty[}(x)$.

Exercise 5.3.2 Let $(X_n)_{n\in\mathbb{N}}$ be a sequence of random variables with distribution $X_n \sim \left(1 - \frac{1}{n}\right)\delta_0 + \frac{1}{n}\delta_n$. Determine:

(i) the mean, variance, and CHF of X_n;

(ii) the CHF of $Z_n := \frac{X_n - 1}{\sqrt{n-1}}$ and deduce that $Z_n \xrightarrow{d} 0$ by Lévy's continuity theorem;

(iii) if $Z_n \xrightarrow{L^2} 0$;

(iv) if $Z_n \xrightarrow{P} 0$.

Solution

(i) We have

$$E[X_n] = 0 \cdot \left(1 - \frac{1}{n}\right) + n \cdot \frac{1}{n} = 1, \quad \mathrm{var}(X_n) = E\left[(X_n - 1)^2\right] = n - 1.$$

Moreover,

$$\varphi_{X_n}(\eta) = E\left[e^{i\eta X_n}\right] = 1 - \frac{1}{n} + \frac{1}{n}e^{i\eta n}.$$

(ii) We have

$$\varphi_{Z_n}(\eta) = e^{-i\frac{\eta}{\sqrt{n-1}}} E\left[e^{i\frac{\eta}{\sqrt{n-1}}X_n}\right]$$

$$= e^{-i\frac{\eta}{\sqrt{n-1}}} \varphi_{X_n}\left(\frac{\eta}{\sqrt{n-1}}\right)$$

$$= e^{-i\frac{\eta}{\sqrt{n-1}}}\left(1 - \frac{1}{n} + \frac{1}{n}e^{in\frac{\eta}{\sqrt{n-1}}}\right) \xrightarrow[n\to\infty]{} 1.$$

Now the constant function 1 is the CHF of the Dirac delta centered at zero, hence the result.

(iii) We have

$$\|Z_n\|_2^2 = E\left[Z_n^2\right] = \frac{1}{n-1}\mathrm{var}(X_n) = 1$$

and therefore there is no convergence in $L^2(\Omega, P)$.

(iv) Convergence in probability holds due to point vi) of Theorem 3.1.9.

Exercise 5.3.3 Let $(X_n)_{n\in\mathbb{N}}$ be a sequence of i.i.d. random variables with distribution $\mathrm{Unif}_{[0,\lambda]}$, with $\lambda > 0$. Determine:

(i) the CDF of the r.v. nX_1 for $n \in \mathbb{N}$;

(ii) the CDF of the r.v.

$$Y_n := \min\{nX_1, \ldots, nX_n\},$$

for $n \in \mathbb{N}$;

(iii) the limit in law of $(Y_n)_{n \in \mathbb{N}}$, recognizing which notable distribution it is.

Solution

(i) We have

$$F_{nX_1}(x) = P\left(X_1 \le \frac{x}{n}\right) = \begin{cases} 0 & \text{if } x \le 0, \\ \frac{x}{\lambda n} & \text{if } 0 < x < \lambda n, \\ 1 & \text{if } x \ge \lambda n. \end{cases}$$

(ii) By Proposition 2.6.9, we have

$$F_{Y_n}(x) = 1 - (1 - F_{nX_1}(x))^n = \begin{cases} 0 & \text{if } x \le 0, \\ 1 - \left(1 - \frac{x}{\lambda n}\right)^n & \text{if } 0 < x < \lambda n, \\ 1 & \text{if } x \ge \lambda n. \end{cases}$$

(iii) We have

$$\lim_{n \to \infty} F_{Y_n}(x) = \begin{cases} 0 & \text{if } x \le 0, \\ 1 - e^{-\frac{x}{\lambda}} & \text{if } x > 0, \end{cases}$$

and thus by Theorem 3.3.3 $Y_n \xrightarrow{d} Y \sim \text{Exp}_{\frac{1}{\lambda}}$ as $n \to \infty$.

Exercise 5.3.4 Let X and $(X_n)_{n \in \mathbb{N}}$ be a r.v. and a sequence of random variables, respectively, defined on a probability space $(\Omega, \mathscr{F}, P)$ such that $(X, X_n) \sim \text{Unif}_{[-1,1] \times \left[-1-\frac{1}{n}, 1+\frac{1}{n}\right]}$ for every $n \in \mathbb{N}$. Determine:

(i) the distribution of X_n, for each $n \in \mathbb{N}$. Are the random variables X and X_n independent?
(ii) $E[X]$, $E[X_n]$, var(X) and var(X_n);
(iii) if X_n converges to X in $L^2(\Omega, P)$;
(iv) if $X_n \xrightarrow{d} X$;
(v) if $X_n \xrightarrow{P} X$.

Solution

(i) Integrating the joint density, we see that $X_n \sim \text{Unif}_{\left[-1-\frac{1}{n}, 1+\frac{1}{n}\right]}$. The joint density is the product of the marginal densities, and therefore X and X_n are independent.

(ii) It is known that $E[X] = E[X_n] = 0$, var$(X) = \frac{1}{3}$ and var$(X_n) = \frac{1}{3}\left(1 + \frac{1}{n}\right)^2$.

(iii) We have

$$E\left[(X - X_n)^2\right] = E\left[X^2\right] + E\left[X_n^2\right] - 2E\left[XX_n\right] =$$

(due to independence)

$$= \mathrm{var}(X) + \mathrm{var}(X_n) = \frac{1}{3} + \frac{1}{3}\left(1 + \frac{1}{n}\right)^2$$

and therefore there is no convergence in $L^2(\Omega, P)$.

(iv) Given the expression of the uniform CHF, we have that

$$\varphi_{X_n}(\omega) = \frac{e^{i\omega\left(1+\frac{1}{n}\right)} - e^{-i\omega\left(1+\frac{1}{n}\right)}}{2i\eta\left(1 + \frac{1}{n}\right)}$$

converges pointwise to φ_X as $n \to \infty$. Alternatively, without using the explicit expression of CHF, it suffices to simply note that

$$\lim_{n\to\infty}\varphi_{X_n}(\eta) = \lim_{n\to\infty}\int_{-1}^{1} e^{i\eta y}\gamma_{X_n}(y)dy = \frac{1}{2}\int_{-1}^{1} e^{i\eta y}dy = \varphi_X(\eta).$$

by the dominated convergence theorem. In any case, by Lévy's continuity theorem, we have that $X_n \xrightarrow{d} X$.

(v) X_n does not converge in probability to X, since for every $0 < \varepsilon < 1$

$$P(|X - X_n| \geq \varepsilon) = \iint_{|x-y|>\varepsilon} \gamma_{(X,X_n)}(x, y)dxdy$$

does not converge to zero as $n \to \infty$.

Exercise 5.3.5 Let $(X_n)_{n\in\mathbb{N}}$ be a sequence of random variables such that $X_n \sim \mathrm{Exp}_{\frac{1}{n^\alpha}}$ with $0 < \alpha \leq 1$.

(i) For every $0 < \alpha < 1$, determine if $Y_n := \frac{X_n - 1}{n}$ converges in L^2;
(ii) for $\alpha = 1$, does the sequence $(Y_n)_{n\in\mathbb{N}}$ converge in distribution? If so, determine the limit.

Solution

(i) We have

$$E\left[Y_n^2\right] = \frac{1}{n^2}\int_0^{+\infty} (t - 1)^2 e^{-\frac{t}{n^\alpha}}\frac{dt}{n^\alpha} =$$

(by the change of variables $\tau = \frac{t}{n^\alpha}$)

$$= \frac{n^{2\alpha}}{n^2} \int_0^{+\infty} (\tau - n^{-\alpha})^2 e^{-\tau} d\tau = \frac{2n^{2a} - 2n^\alpha + 1}{n^2}$$

which converges to zero as $n \to \infty$. Alternatively and more simply, without explicitly calculating the integral, we have

$$0 \le \frac{n^{2\alpha}}{n^2} \int_0^{+\infty} (\tau - n^{-\alpha})^2 e^{-\tau} d\tau \le \frac{c}{n^{2-2\alpha}} \longrightarrow 0, \quad c = \int_0^{+\infty} (\tau+1)^2 e^{-\tau} d\tau.$$

(ii) We have

$$\varphi_{X_n}(\eta) = \frac{1}{1 - i\eta n^\alpha}$$

and, for $\alpha = 1$,

$$\varphi_{Y_n}(\eta) = e^{-\frac{i\eta}{n}} \varphi_{X_n}\left(\frac{\eta}{n}\right) = \frac{e^{-\frac{i\eta}{n}}}{1 - i\eta} \xrightarrow[n\to\infty]{} \frac{1}{1 - i\eta}.$$

Thus, for $\alpha = 1$, we have $Y_n \xrightarrow{d} Y \sim \mathrm{Exp}_1$.

Exercise 5.3.6 Given $X \in \mathcal{N}_{0,1}$, consider the sequence

$$X_n = \frac{1}{n} - \sqrt{1 + \frac{1}{n}}X, \quad n \in \mathbb{N}.$$

Determine if:

(i) $X_n \xrightarrow[n\to\infty]{d} X$;

(ii) $X_n \xrightarrow[n\to\infty]{L^2} X$;

(iii) $X_n \xrightarrow[n\to\infty]{a.s.} X$.

Solution

(i) We have $X_n \sim \mathcal{N}_{\frac{1}{n}, 1+\frac{1}{n}}$. Since

$$\varphi_{X_n}(\eta) = e^{i\frac{\eta}{n} - \frac{\eta^2}{2}\left(1+\frac{1}{n}\right)} \xrightarrow[n\to\infty]{} e^{-\frac{\eta^2}{2}} = \varphi_X(\eta),$$

by Lévy's continuity theorem, we have $X_n \xrightarrow{d} X$.

(ii) We have

$$E\left[(X_n - X)^2\right] = E\left[\left(\frac{1}{n} - \left(\sqrt{1 + \frac{1}{n}} + 1\right)X\right)^2\right]$$

$$= \frac{1}{n^2} + \left(\sqrt{1 + \frac{1}{n}} + 1\right)^2 E\left[X^2\right] \xrightarrow[n \to \infty]{} 4$$

and therefore there is no convergence in L^2.

(iii) For every $\omega \in \Omega$ we have

$$X_n(\omega) \xrightarrow[n \to \infty]{} -X(\omega)$$

and therefore there is no almost sure convergence: X_n converges to X only on the negligible event $(X = 0)$.

Exercise 5.3.7 Let $(X_n)_{n \in \mathbb{N}}$ be a sequence of random variables with $X_n \sim \text{Unif}_{[0,n]}$.

(i) Study the pointwise convergence of the sequence of CHFs φ_{X_n} and determine if $(X_n)_{n \in \mathbb{N}}$ converges weakly;

(ii) does $(X_n)_{n \in \mathbb{N}}$ converge almost surely?

Solution

(i) We have

$$\varphi_{X_n}(\eta) = E\left[e^{i\eta X_n}\right] = \begin{cases} 1 & \text{if } \eta = 0, \\ \frac{e^{i\eta n} - 1}{i\eta n} & \text{otherwise.} \end{cases}$$

Note that φ_{X_n} is a continuous function since, for every $n \in \mathbb{N}$, we have

$$\lim_{\eta \to 0} \frac{e^{i\eta n} - 1}{i\eta n} = 1.$$

Then

$$\lim_{n \to \infty} \varphi_{X_n}(\eta) = \begin{cases} 1 & \text{if } \eta = 0, \\ 0 & \text{otherwise,} \end{cases}$$

which is not continuous at $\eta = 0$. Therefore, by Lévy's continuity theorem, the sequence $(X_n)_{n \in \mathbb{N}}$ does not converge weakly.

(ii) Since $(X_n)_{n \in \mathbb{N}}$ does not converge weakly, by Theorem 3.1.9, it does not converge almost surely either.

Exercise 5.3.8 Consider the function

$$F_p(x) := \left(1 - \frac{p}{p - 1 + e^x} \right) \mathbb{1}_{\mathbb{R} \geq 0}(x), \qquad x \in \mathbb{R}.$$

(i) Prove that F_p is a cumulative distribution function for every $p \geq 0$ and not for $p < 0$;
(ii) let μ_p be the distribution with CDF F_p: for which p, is μ_p absolutely continuous?
(iii) study the weak convergence of μ_{p_n} as $p_n \longrightarrow 0^+$ and as $p_n \longrightarrow 1$ and recognize the limiting distributions.

Solution By differentiating

$$F'_p(x) = \frac{pe^x}{(p - 1 + e^x)^2} \mathbb{1}_{\mathbb{R} \geq 0}(x),$$

we see that F_p is monotone increasing for $p \geq 0$ and decreasing for $p < 0$. For $p = 0$, F_p is the CDF of the Dirac delta centered at zero. If $p > 0$ then F_p is an absolutely continuous function on $\mathbb{R}$:

$$F_p(x) = \int_0^x F'_p(y)dy, \qquad x \in \mathbb{R}.$$

Moreover, $F_p(x) \equiv 0$ for $x < 0$ and

$$\lim_{x \to +\infty} F_p(x) = 1.$$

We have

$$\lim_{p_n \to 0^+} F_{p_n}(x) = F_0(x), \qquad x \in \mathbb{R} \setminus \{0\}$$

with 0 being the unique point of discontinuity of F_0: thus, by Theorem 3.3.3, μ_{p_n} weakly converges to the Dirac delta centered at zero. We have

$$\lim_{p_n \to 1} F_{p_n}(x) = F_1(x) = 1 - e^{-x}, \qquad x \in \mathbb{R}$$

and therefore μ_{p_n} weakly converges to Exp_1.

Exercise 5.3.9 Let $(X_n)_{n \in \mathbb{N}}$ be a sequence of random variables with distribution

$$X_n \sim \mu_n := \frac{1}{2n} \left(\delta_{-\sqrt{n}} + \delta_{\sqrt{n}} \right) + \left(1 - \frac{1}{n} \right) \mathrm{Unif}_{[-\frac{1}{n}, \frac{1}{n}]}, \qquad n \in \mathbb{N}.$$

Determine:

(i) the mean and variance of X_n;

(ii) the CHF of X_n and deduce that $X_n \xrightarrow{d} 0$;

(iii) if X_n also converges in L^2.

Solution

(i) We have

$$E[X_n] = 0, \quad \text{var}(X_n) = \int_{\mathbb{R}} x^2 \mu_n(dx)$$

$$= 1 + \left(1 - \frac{1}{n}\right) \frac{n}{2} \int_{-\frac{1}{n}}^{\frac{1}{n}} x^2 dx = 1 + \frac{1}{3n^2}\left(1 - \frac{1}{n}\right).$$

(ii) Recalling the expression of the uniform CHF we have

$$\varphi_{X_n}(\eta) = \frac{1}{2n}\left(e^{i\eta\sqrt{n}} + e^{-i\eta\sqrt{n}}\right) + \left(1 - \frac{1}{n}\right) \frac{e^{i\frac{\eta}{n}} - e^{-i\frac{\eta}{n}}}{i\eta\frac{2}{n}} \xrightarrow[n\to\infty]{} 1.$$

Now the constant function 1 is the CHF of the Dirac delta centered at zero, which proves the thesis by Lévy's continuity theorem.

(iii) There is no convergence in $L^2(\Omega, P)$ because, as seen in point (i),

$$\|X_n\|^2_{L^2(\Omega, P)} = \text{var}(X_n) \xrightarrow[n\to\infty]{} 1.$$

Appendix A
Dynkin's Theorems

Let Ω be a generic non-empty set. As anticipated in Section 1.4.1, it is difficult to give an explicit representation of the σ-algebra $\sigma(\mathscr{A})$ generated by a family $\mathscr{A}$ of subsets of Ω. The results of this section, of a rather technical nature, allow us to prove that if a certain property holds for the elements of a family $\mathscr{A}$ then it also holds for all the elements of $\sigma(\mathscr{A})$.

Definition A.0.1 (MSect. Onotone Family of Sets) A family $\mathscr{M}$ of subsets of Ω is a monotone family if it has the following properties:

  (i) $\Omega \in \mathscr{M}$;
  (ii) if $A, B \in \mathscr{M}$ and $A \subseteq B$, then $B \setminus A \in \mathscr{M}$;
  (iii) if $(A_n)_{n \in \mathbb{N}}$ is an *increasing* sequence of elements of $\mathscr{M}$, then $\bigcup_{n \in \mathbb{N}} A_n \in \mathscr{M}$.

Every σ-algebra is a monotone family, while the converse is not necessarily true since property (iii) of "closure with respect to countable union" holds only for increasing sequences, i.e., such that $A_n \subseteq A_{n+1}$ for every $n \in \mathbb{N}$. However, we have the following result.

Lemma A.0.2 *If a monotone family $\mathscr{M}$ is $\cap$-closed[1] then it is a σ-algebra.*

Proof If $\mathscr{M}$ is monotone, it verifies the first two properties of the definition of a σ-algebra: it remains only to prove (ii-b) of Definition 1.1.1, i.e., that the countable union of elements of $\mathscr{M}$ belongs to $\mathscr{M}$. First, given $A, B \in \mathscr{M}$, since $A \cup B =$

---

[1] That is, such that $A \cap B \in \mathscr{M}$ for every $A, B \in \mathscr{M}$.

© The Editor(s) (if applicable) and The Author(s), under exclusive license
to Springer Nature Switzerland AG 2024
A. Pascucci, *Probability Theory I*, La Matematica per il 3+2 165,
https://doi.org/10.1007/978-3-031-63190-0

$(A^c \cap B^c)^c$, the hypothesis of closure with respect to intersection implies that $A \cup B \in \mathcal{M}$. Now, given a sequence $(A_n)_{n \in \mathbb{N}}$ of elements of $\mathcal{M}$, we define the sequence

$$\bar{A}_n := \bigcup_{k=1}^{n} A_k, \qquad n \in \mathbb{N},$$

which is increasing and such that $\bar{A}_n \in \mathcal{M}$ as just demonstrated. Then we conclude that

$$\bigcup_{n \in \mathbb{N}} A_n = \bigcup_{n \in \mathbb{N}} \bar{A}_n \in \mathcal{M}$$

by (iii) of Definition A.0.1. $\qquad\qquad\qquad\qquad\qquad\qquad\qquad\qquad\qquad\qquad\square$

We observe that the intersection of monotone families is a monotone family. Given a family $\mathcal{A}$ of subsets of Ω, we denote by $\mathcal{M}(\mathcal{A})$ the intersection of all monotone families that contain $\mathcal{A}$: we say that $\mathcal{M}(\mathcal{A})$ is the *monotone family generated by* $\mathcal{A}$, i.e., the smallest monotone family that contains $\mathcal{A}$.

Theorem A.0.3 (First Dynkin's Theorem) *[!] Let $\mathcal{A}$ be a family of subsets of Ω. If $\mathcal{A}$ is $\cap$-closed then $\mathcal{M}(\mathcal{A}) = \sigma(\mathcal{A})$.*

Proof $\sigma(\mathcal{A})$ is monotone and therefore $\sigma(\mathcal{A}) \supseteq \mathcal{M}(\mathcal{A})$. Conversely, if we prove that $\mathcal{M}(\mathcal{A})$ is $\cap$-closed then from Lemma A.0.2 it follows that $\mathcal{M}(\mathcal{A})$ is a σ-algebra and therefore $\sigma(\mathcal{A}) \subseteq \mathcal{M}(\mathcal{A})$.

Let us then prove that $\mathcal{M}(\mathcal{A})$ is $\cap$-closed. We set

$$\mathcal{M}_1 = \{A \in \mathcal{M}(\mathcal{A}) \mid A \cap I \in \mathcal{M}(\mathcal{A}), \ \forall I \in \mathcal{A}\},$$

and prove that $\mathcal{M}_1$ is a monotone family: since $\mathcal{A} \subseteq \mathcal{M}_1$, it follows $\mathcal{M}(\mathcal{A}) \subseteq \mathcal{M}_1$ and therefore $\mathcal{M}(\mathcal{A}) = \mathcal{M}_1$. We have:

(i) $\Omega \in \mathcal{M}_1$;

(ii) for every $A, B \in \mathcal{M}_1$ with $A \subseteq B$, we have

$$(B \setminus A) \cap I = (B \cap I) \setminus (A \cap I) \in \mathcal{M}(\mathcal{A}), \qquad I \in \mathcal{A},$$

and therefore $B \setminus A \in \mathcal{M}_1$;

(iii) let (A_n) be an increasing sequence in $\mathcal{M}_1$ and denote by A the union of the A_n. Then we have

$$A \cap I = \bigcup_{n \geq 1} (A_n \cap I) \in \mathcal{M}(\mathcal{A}), \qquad I \in \mathcal{A},$$

and therefore $A \in \mathcal{M}_1$.

This proves that $\mathscr{M}(\mathscr{A}) = \mathscr{M}_1$. Now let

$$\mathscr{M}_2 = \{A \in \mathscr{M}(\mathscr{A}) \mid A \cap I \in \mathscr{M}(\mathscr{A}), \, \forall I \in \mathscr{M}(\mathscr{A})\}.$$

We have already proved that $\mathscr{A} \subseteq \mathscr{M}_2$. Moreover, in a similar way we can prove that $\mathscr{M}_2$ is a monotone family: it follows that $\mathscr{M}(\mathscr{A}) \subseteq \mathscr{M}_2$ and therefore $\mathscr{M}(\mathscr{A}) = \mathscr{M}_2$, i.e., $\mathscr{M}(\mathscr{A})$ is $\cap$-closed. □

The following result is a direct consequence of Theorem A.0.3.

Corollary A.0.4 *Let $\mathscr{M}$ be a monotone family. If $\mathscr{M}$ contains a $\cap$-closed family $\mathscr{A}$, then it also contains $\sigma(\mathscr{A})$.*

As a second corollary, we prove the uniqueness part of Carathéodory's Theorem 1.4.29 (see Remark A.0.6).

Corollary A.0.5 [!] *Let μ, ν be finite measures on $(\Omega, \sigma(\mathscr{A}))$ where $\mathscr{A}$ is a $\cap$-closed family and $\Omega \in \mathscr{A}$. If $\mu(A) = \nu(A)$ for every $A \in \mathscr{A}$ then $\mu = \nu$.*

Proof Let

$$\mathscr{M} = \{A \in \sigma(\mathscr{A}) \mid \mu(A) = \nu(A)\}.$$

We verify that $\mathscr{M}$ is a monotone family: from Dynkin's first theorem it will follow that $\mathscr{M} \supseteq \mathscr{M}(\mathscr{A}) = \sigma(\mathscr{A})$, which proves the thesis.

Of the three conditions in Definition A.0.1, (i) is true by hypothesis. As for (ii), if $A, B \in \mathscr{M}$ with $A \subseteq B$ then we have

$$\mu(B \setminus A) = \mu(B) - \mu(A) = \nu(B) - \nu(A) = \nu(B \setminus A)$$

and therefore $(B \setminus A) \in \mathscr{M}$. Finally, if $(A_n)_{n \in \mathbb{N}}$ is an increasing sequence in $\mathscr{M}$ and $A = \bigcup_{n \in \mathbb{N}} A_n$, then by the continuity from below of the measures (cf. Proposition 1.1.32) we have

$$\mu(A) = \lim_{n \to \infty} \mu(A_n) = \lim_{n \to \infty} \nu(A_n) = \nu(A)$$

from which $A \in \mathscr{M}$ and this concludes the proof. □

Remark A.0.6 The uniqueness part of Carathéodory's Theorem 1.4.29 follows easily from Corollary A.0.5: the claim is that if μ, ν are σ-finite measures on an algebra $\mathscr{A}$ and coincide on $\mathscr{A}$ then they also coincide on $\sigma(\mathscr{A})$.

By assumption, there exists a sequence $(A_n)_{n \in \mathbb{N}}$ in $\mathscr{A}$ such that $\mu(A_n) = \nu(A_n) < \infty$ and $\Omega = \bigcup_{n \in \mathbb{N}} A_n$. Fixed $n \in \mathbb{N}$, since $\mathscr{A}$ is $\cap$-closed, using Corollary A.0.5 it is easy to prove that

$$\mu(A \cap A_n) = \nu(A \cap A_n), \qquad \forall A \in \sigma(\mathscr{A}).$$

Taking the limit in n, the thesis follows from the continuity from below of the measures.

Definition A.0.7 (Monotone Family of Functions) A family $\mathscr{H}$ of *bounded* functions, defined from a set Ω to real values, is *monotone* if it has the following properties:

  (i) $\mathscr{H}$ is a real vector space;
 (ii) the constant function 1 belongs to $\mathscr{H}$;
(iii) if $(X_n)_{n\in\mathbb{N}}$ is a sequence of non-negative functions in $\mathscr{H}$ such that $X_n \nearrow X$ and X is bounded, then $X \in \mathscr{H}$.

Theorem A.0.8 (Second Dynkin's Theorem) *[!] Let $\mathscr{A}$ be a $\cap$-closed family of subsets of Ω. If $\mathscr{H}$ is a monotone family that contains the indicator functions of elements of $\mathscr{A}$, then $\mathscr{H}$ also contains all bounded and $\sigma(\mathscr{A})$-measurable functions.*

Proof Let

$$\mathscr{M} = \{H \subseteq \Omega \mid \mathbb{1}_H \in \mathscr{H}\}.$$

By assumption, $\mathscr{A} \subseteq \mathscr{M}$ and, using the fact that $\mathscr{H}$ is a monotone family, it is easy to prove that $\mathscr{M}$ is a monotone family of sets. Then $\mathscr{M} \supseteq \mathscr{M}(\mathscr{A}) = \sigma(\mathscr{A})$, where the equality is a consequence of Dynkin's first theorem. Therefore, $\mathscr{H}$ contains the indicator functions of elements of $\sigma(\mathscr{A})$.

Given $X \in m\sigma(\mathscr{A})$, non-negative and bounded, by Lemma 2.2.3 there exists a sequence $(X_n)_{n\in\mathbb{N}}$ of simple $\sigma(\mathscr{A})$-measurable and non-negative functions such that $X_n \nearrow X$. Each X_n is a linear combination of indicator functions of elements of $\sigma(\mathscr{A})$ and therefore belongs to $\mathscr{H}$, being $\mathscr{H}$ a vector space: by property iii) of $\mathscr{H}$, we have that $X \in \mathscr{H}$. Finally, to prove that every $\sigma(\mathscr{A})$-measurable and bounded function belongs to $\mathscr{H}$, it is sufficient to decompose it into the sum of its positive and negative parts. $\square$

Appendix B
Absolute Continuity

B.1 Radon-Nikodym Theorem

In this section we deepen the concept of *absolute continuity between measures* of which we had considered a particular case (absolute continuity with respect to the Lebesgue measure) in Sect. 1.4.5. As the main result, we prove that the existence of a density is a necessary and sufficient condition for absolute continuity: this is the content of the classic Radon-Nikodym theorem.

Definition B.1.1 Let μ, ν be σ-finite measures on $(\Omega, \mathscr{F})$. We say that ν is μ-absolutely continuous on $\mathscr{F}$, and we write $\nu \ll \mu$, if every μ-negligible set of $\mathscr{F}$ is also ν-negligible. When it is important to specify the σ-algebra considered, we also write

$$\nu \ll_{\mathscr{F}} \mu.$$

Obviously, if $\mathscr{F}_1 \subseteq \mathscr{F}_2$ are σ-algebras, then $\nu \ll_{\mathscr{F}_2} \mu$ implies $\nu \ll_{\mathscr{F}_1} \mu$ but the converse is not true in general.

Example B.1.2 Definition 1.4.18 of absolute continuity is a particular case of the previous definition: in fact, if μ is an absolutely continuous distribution, then $\mu(H) = 0$ for every $H \in \mathscr{B}$ such that $\mathrm{Leb}(H) = 0$ or, in other words,

$$\mu \ll_{\mathscr{B}} \mathrm{Leb}$$

that is, μ is absolutely continuous with respect to the Lebesgue measure.

A. Pascucci, *Probability Theory I*, La Matematica per il 3+2 165,
https://doi.org/10.1007/978-3-031-63190-0

Theorem B.1.3 (Radon-Nikodym Theorem) *[!] If μ, ν are σ-finite measures on $(\Omega, \mathscr{F})$ and $\nu \ll \mu$, then there exists $g \in m\mathscr{F}^+$ such that*

$$\nu(A) = \int_A g d\mu, \qquad A \in \mathscr{F}. \tag{2.1.1}$$

Moreover, if $\tilde{g} \in m\mathscr{F}^+$ satisfies (2.1.1), then $g = \tilde{g}$ almost everywhere with respect to μ. We say that g is a density (or Radon-Nikodym derivative) of ν with respect to μ and we write

$$d\nu = g d\mu \quad or \quad g = \frac{d\nu}{d\mu} \quad or \quad g = \frac{d\nu}{d\mu}|_{\mathscr{F}}.$$

Remark B.1.4 Let μ, ν be measures as in the previous statement, defined on $(\Omega, \mathscr{F})$, and $f \in m\mathscr{F}^+$: approximating f with an increasing sequence of simple non-negative functions as in Lemma 2.2.3, thanks to Beppo Levi's theorem we have

$$\int_\Omega f d\nu = \lim_{n\to\infty} \int_\Omega f_n d\nu =$$

(by (2.1.1) and indicating with $\frac{d\nu}{d\mu}$ the Radon-Nikodym derivative of ν with respect to μ)

$$= \lim_{n\to\infty} \int_\Omega f_n \frac{d\nu}{d\mu} d\mu =$$

(reapplying Beppo Levi's theorem)

$$= \int_\Omega f \frac{d\nu}{d\mu} d\mu.$$

Thus, the following formula for the change of measure of integration holds

$$\int_\Omega f d\nu = \int_\Omega f \frac{d\nu}{d\mu} d\mu$$

for every $f \in m\mathscr{F}^+$.

***Proof of Theorem B.1.3* [Uniqueness]** If $g, \tilde{g} \in m\mathscr{F}^+$ satisfy (2.1.1), then we have

$$\int_A (g - \tilde{g}) d\mu = 0, \qquad A \in \mathscr{F}. \tag{B.2}$$

In particular, setting $A = \{g - \tilde{g} > 0\} \in \mathscr{F}$, we get $\mu(A) = 0$ that is $g \le \tilde{g}$ μ-a.e. because otherwise we would have

$$\int_A (g - \tilde{g})d\mu > 0$$

which contradicts (B.2). Similarly, we prove that $g \ge \tilde{g}$ μ-a.e.

[Existence] First, assume that μ, ν are finite. We give a proof based on the Riesz representation theorem[1] for linear and continuous functionals on a Hilbert space. Consider the linear operator

$$L(f) := \int_\Omega f d\mu$$

defined on the Hilbert space $L^2(\Omega, \mathscr{F}, \mu + \nu)$ equipped with the usual inner product

$$\langle f, g \rangle = \int_\Omega fg\, d(\mu + \nu).$$

The operator L is bounded and therefore continuous: in fact, applying the triangle inequality and then the Hölder inequality, we have

$$|L(f)| \le \int_\Omega |f|d\mu \le \int_\Omega |f|d(\mu + \nu) \le \|f\|_{L^2}\sqrt{(\mu + \nu)(\Omega)}.$$

Then, by the Riesz representation theorem, there exists $\varphi \in L^2(\Omega, \mathscr{F}, \mu + \nu)$ such that

$$\int_\Omega f d\mu = \int_\Omega f\varphi\, d(\mu + \nu), \qquad f \in L^2(\Omega, \mathscr{F}, \mu + \nu). \tag{B.3}$$

Let us prove that $0 < \varphi < 1$ μ-almost everywhere: for this purpose, let $A_0 = \{\varphi < 0\}$, $A_1 = \{\varphi > 1\}$ and $f_i = \mathbb{1}_{A_i} \in L^2(\Omega, \mathscr{F}, \mu + \nu)$, for $i = 0, 1$. If it were $\mu(A_i) > 0$, from (B.3) we would have

$$\mu(A_0) = \int_\Omega f_0 d\mu = \int_{A_0} \varphi\, d(\mu + \nu) \le \int_{A_0} \varphi\, d\mu < 0,$$

---

[1]
Theorem B.1.5 (Riesz Representation Theorem) *If L is a linear and continuous operator on a Hilbert space $(\mathbb{H}, \langle \cdot, \cdot \rangle)$, then there exists and is unique $y \in \mathbb{H}$ such that*

$$L(x) = \langle x, y \rangle, \qquad x \in \mathbb{H}.$$

For the proof of Theorem B.1.5, and more generally for a simple but complete introduction to Hilbert spaces, see Chapter 4 in [40].

$$\mu(A_1) = \int_\Omega f_1 d\mu = \int_{A_1} \varphi d(\mu + \nu) \geq \int_{A_1} \varphi d\mu > \mu(A_1),$$

which is absurd.

Now, (B.3) is equivalent to

$$\int_\Omega f\varphi d\nu = \int_\Omega f(1 - \varphi) d\mu, \qquad f \in L^2(\Omega, \mathscr{F}, \mu + \nu),$$

and by Lemma 2.2.3 and Beppo Levi's theorem (which applies since $0 < \varphi < 1$ μ-almost everywhere and therefore also ν-almost everywhere), this equality extends to every $f \in m\mathscr{F}^+$. In particular, for $f = \frac{\mathbb{1}_A}{\varphi}$ we obtain

$$\nu(A) = \int_A \frac{1 - \varphi}{\varphi} d\mu, \qquad A \in \mathscr{F}.$$

This proves the thesis with $g = \frac{1-\varphi}{\varphi} \in m\mathscr{F}^+$.

Now consider the general case where μ, ν are σ-finite. Then there exists an increasing sequence $(A_n)_{n \in \mathbb{N}}$ in $\mathscr{F}$, which covers Ω and such that $(\mu+\nu)(A_n) < \infty$ for each $n \in \mathbb{N}$. Consider the finite measures

$$\mu_n(A) := \mu(A \cap A_n), \quad \nu_n(A) := \nu(A \cap A_n), \qquad A \in \mathscr{F}, \; n \in \mathbb{N}.$$

It is easy to see that $\nu_n \ll \mu_n$ and therefore there exists $g_n \in m\mathscr{F}^+$ such that $\nu_n = g_n d\mu_n$. Moreover, as in the proof of uniqueness, we see that $g_n = g_m$ on A_n for $n \leq m$. Then consider $g \in m\mathscr{F}^+$ defined by $g = g_n$ on A_n. For each $A \in \mathscr{F}$ we have

$$\nu(A \cap A_n) = \nu_n(A) = \int_A g_n d\mu_n = \int_{A \cap A_n} f d\mu$$

and the thesis follows by taking the limit as $n \to +\infty$. $\qquad\qquad\square$

B.2 Representation of Open Sets in $\mathbb{R}$

Lemma B.2.1 *Every open subset A of $\mathbb{R}$ can be written as a countable union of disjoint open intervals:*

$$A = \biguplus_{n \geq 1}]a_n, b_n[. \tag{B.4}$$

Proof Let A be an open set in $\mathbb{R}$. For any $x \in A$, we define

$$a_x = \inf\{a \in \mathbb{R} \mid \text{there exists } b \text{ such that } x \in]a_x, b[\subseteq A\} \quad \text{and}$$

$$b_x = \sup\{b \in \mathbb{R} \mid]a_x, b[\subseteq A\}.$$

Then it is clear that $x \in I_x :=]a_x, b_x[\subseteq A$. On the other hand, if $x, y \in A$ and $x \neq y$ then either $I_x \cap I_y = \emptyset$ or $I_x \equiv I_y$. In fact, if for the sake of contradiction $I_x \cap I_y \neq \emptyset$ and $I_x \neq I_y$, then $I := I_x \cup I_y$ would be an open interval, included in A and such that $x \in I_x \subset I$: this would contradict the definition of a_x and b_x.

We have therefore proved that A can be written as a union of disjoint open intervals: each of them contains a different rational number and therefore it is a countable union.   □

Remark B.2.2 [!] As a consequence of Lemma B.2.1, we have that if μ is a distribution on $\mathbb{R}$ and A is an open set, then by (B.4) we have

$$\mu(A) = \sum_{n \geq 1} \mu(]a_n, b_n[).$$

Combining this result with Corollary 1.4.10, we conclude that two distributions μ_1 and μ_2 on $\mathbb{R}$ are equal if and only if $\mu_1(I) = \mu_2(I)$ for every open interval I.

Lemma B.2.1 does not extend to the multi-dimensional case (or, even worse, to the case of a generic metric space). It would seem natural to replace the intervals of $\mathbb{R}$ with disks. However, by doing so the result becomes false even in dimension one (at least if we assume that the radius of the disks must be finite): just consider, for example, $A =]0, +\infty[$. Similarly, a disjoint union of open disks in $\mathbb{R}^2$ is a connected set if and only if it consists of a single disk: therefore there is no hope of representing a generic connected open set in $\mathbb{R}^2$ as a countable union of disjoint open disks.

In the proof of Lemma B.2.1 we used the density of rational numbers in $\mathbb{R}$: given the subtlety of the arguments, one must be careful with what seems intuitive, as shown by the following

Example B.2.3 Let $(x_n)_{n \in \mathbb{N}}$ be an enumeration of the points of $H :=]0, 1[\cap \mathbb{Q} \in \mathscr{B}$. Fixed $\varepsilon \in]0, 1[$, let $(r_n)_{n \in \mathbb{N}}$ be a sequence of positive real numbers such that the series

$$\sum_{n \geq 1} r_n < \frac{\varepsilon}{2}.$$

We define

$$A := \bigcup_{n \geq 1}]x_n - r_n, x_n + r_n[\cap]0, 1[.$$

Then A is open, $H \subseteq A$ and by sub-additivity (cf. Proposition 1.1.22-(ii))

$$\text{Leb}(A) \leq \sum_{n \geq 1} \text{Leb}(]x_n - r_n, x_n + r_n[) < \varepsilon.$$

It also follows that A is strictly included in $]0, 1[$ (because it has Lebesgue measure less than 1) even though it is open and dense in $]0, 1[$.

B.3   Differentiability of Integral Functions

The starting point of the results in this section is the classic Lebesgue theorem on the differentiability of monotone functions.

Theorem B.3.1 (Lebesgue) *[!!] Every monotone increasing function*

$$F : [a, b] \longrightarrow \mathbb{R}$$

is a.e. differentiable and

$$\int_a^b F'(x)dx \leq F(b) - F(a). \tag{B.5}$$

The inequality in (B.5) can be strict (think of piecewise constant functions): the Cantor-Vitali function of Example 1.4.36 is monotone, *continuous* and satisfies (B.5) with a strict inequality.

The standard proof of Theorem B.3.1 is based on Vitali's covering theorem and can be found in [3], Theorem 14.18. Another more direct proof, but under the additional assumption of continuity, is due to Riesz (cf. Chapter 1.3 in [39]).

Proposition B.3.2 *If* $\gamma \in L^1([a, b])$ *and*

$$\int_a^x \gamma(t)dt = 0 \quad \text{for every } x \in [a, b], \tag{B.6}$$

then $\gamma = 0$ *a.e.*

Proof By assumption, we have

$$\int_{x_0}^x \gamma(t)dt = \int_a^x \gamma(t)dt - \int_a^{x_0} \gamma(t)dt = 0 \qquad a \leq x_0 < x \leq b.$$

Moreover, by Lemma B.2.1 every open set $A \subseteq [a, b]$ can be written in the form (B.4) and therefore

$$\int_A \gamma(t)dt = \sum_{n=1}^{\infty} \int_{a_n}^{b_n} \gamma(t)dt = 0. \tag{B.7}$$

Now let $H \in \mathscr{B}$, with $H \subseteq [a, b]$: by Proposition 1.4.9 on the regularity of Borel measures, for every $n \in \mathbb{N}$ there exists an open set A_n such that $H \subseteq A_n$ and $\mathrm{Leb}(A_n \setminus H) \leq \frac{1}{n}$. Then we have

$$\int_H \gamma(t)dt = \int_{A_n} \gamma(t)dt - \int_{A_n \setminus H} \gamma(t)dt =$$

(by (B.7))

$$= -\int_{A_n \setminus H} \gamma(t)dt \xrightarrow[n \to +\infty]{} 0$$

by the dominated convergence theorem. Hence $\int_H \gamma(t)dt = 0$ for every $H \in \mathscr{B}$.

Then, for every $n \in \mathbb{N}$, let $H_n = \{x \in [a, b] \mid \gamma(x) \geq \frac{1}{n}\} \in \mathscr{B}$: we have

$$0 = \int_{H_n} \gamma(t)dt \geq \frac{\mathrm{Leb}(H_n)}{n}$$

from which $\mathrm{Leb}(H_n) = 0$ and therefore also

$$\{x \in [a, b] \mid \gamma(x) > 0\} = \bigcup_{n=1}^{\infty} H_n$$

has Lebesgue measure zero, i.e., $\gamma \leq 0$ almost everywhere. Similarly, we prove that $\gamma \geq 0$ almost everywhere and this concludes the proof. □

Proposition B.3.3 *If*

$$F(x) = F(a) + \int_a^x \gamma(t)dt, \qquad x \in [a, b],$$

with $\gamma \in L^1([a, b])$, then $F' = \gamma$ almost everywhere.

Proof Without loss of generality, we can assume $\gamma \geq 0$ almost everywhere (and therefore F is monotone increasing). First, we observe that F is continuous since[2]

$$F(x + h) - F(x) = \int_x^{x+h} \gamma(t)dt \xrightarrow[h \to 0]{} 0$$

by the dominated convergence theorem.

First, we assume that $\gamma \in L^\infty$: then we have

$$\left| \frac{F(x + h) - F(x)}{h} \right| = \left| \frac{1}{h} \int_x^{x+h} \gamma(t)dt \right| \leq \|\gamma\|_\infty$$

and on the other hand, by Lebesgue's Theorem B.3.1, since F is monotone increasing, we have that there exists

$$\lim_{h \to 0} \frac{F(x + h) - F(x)}{h} = F'(x) \quad \text{almost everywhere.}$$

Hence, again by the dominated convergence theorem, for $a < x_0 < x < b$ we have

$$\int_{x_0}^x F'(t)dt = \lim_{h \to 0} \int_{x_0}^x \frac{F(t + h) - F(t)}{h} dt$$

$$= \lim_{h \to 0} \frac{1}{h} \left(\int_x^{x+h} F(t)dt - \int_{x_0}^{x_0+h} F(t)dt \right)$$

(since F is continuous)

$$= F(x) - F(x_0).$$

It follows that

$$\int_a^x \left(F'(t) - \gamma(t) \right) dt = 0, \qquad x \in [a, b]$$

and therefore, by Proposition B.3.2, $F' = \gamma$ almost everywhere.

---

[2] If $h < 0$ we define

$$\int_x^{x+h} \gamma(t)dt = -\int_{x+h}^x \gamma(t)dt.$$

Now consider the case where $\gamma \in L^1([a, b])$. For $n \in \mathbb{N}$, consider the sequence

$$\gamma_n(t) = \begin{cases} \gamma(t) & \text{if } 0 \le \gamma(t) \le n, \\ 0 & \text{if } \gamma(t) > n. \end{cases}$$

Then we have $F = F_n + G_n$ where

$$F_n(x) = \int_a^x \gamma_n(t)dt, \qquad G_n(x) = \int_a^x (\gamma(t) - \gamma_n(t))\, dt.$$

On one hand, G_n is an increasing function (and therefore a.e. differentiable with $G_n' \ge 0$) since $\gamma - \gamma_n \ge 0$ and on the other hand, as just proven, there exists $F_n' = \gamma_n$ a.e. Thus, we have

$$F' = \gamma_n + G' \ge \gamma_n \quad \text{a.e.}$$

and, taking the limit as $n \to \infty$, $F' \ge \gamma$ a.e. Then we have

$$\int_a^b F'(t)dt \ge \int_a^b \gamma(t)dt = F(b) - F(a).$$

But the opposite inequality comes from Lebesgue's Theorem B.3.1 (see (B.5)) and therefore

$$\int_a^b F'(t)dt = F(b) - F(a).$$

Then we still have

$$\int_a^b \left(F'(t) - \gamma(t)\right)dt = 0$$

and, since $F' \ge \gamma$ a.e., we conclude that $F' = \gamma$ a.e.   □

B.4 Absolutely Continuous Functions

Definition B.4.1 (Absolutely Continuous Function) We say that

$$F : [a, b] \longrightarrow \mathbb{R}$$

is *absolutely continuous*, and we write $F \in AC([a, b])$, if, for every $\varepsilon > 0$ there exists $\delta > 0$ such that

$$\sum_{n=1}^{N} |F(b_n) - F(a_n)| < \varepsilon \tag{B.8}$$

for every choice of a finite number of disjoint intervals $[a_n, b_n] \subseteq [a, b]$ such that

$$\sum_{n=1}^{N} (b_n - a_n) < \delta.$$

Exercise B.4.2 Prove that if $F \in AC([a, b])$ then, for every $\varepsilon > 0$ there exists $\delta > 0$ such that

$$\sum_{n=1}^{\infty} |F(b_n) - F(a_n)| < \varepsilon$$

for every sequence of disjoint intervals $[a_n, b_n] \subseteq [a, b]$ such that

$$\sum_{n=1}^{\infty} (b_n - a_n) < \delta.$$

The importance of absolutely continuous functions lies in the fact that *they are the functions for which the fundamental theorem of calculus holds*. The main result of this section is the following

Theorem B.4.3 *[!] A function F is absolutely continuous on* $[a, b]$ *if and only if F is a.e. differentiable with* $F' \in L^1([a, b])$ *and*

$$F(x) = F(a) + \int_a^x F'(t)dt, \qquad x \in [a, b]. \tag{B.9}$$

We establish some preliminary results to the proof of Theorem B.4.3.

Definition B.4.4 (Function of Bounded Variation) We say that

$$F : [a, b] \longrightarrow \mathbb{R}$$

has *bounded variation*, and we write $F \in BV([a, b])$, if

$$\bigvee_a^b (F) := \sup_{\sigma \in \mathscr{P}_{[a,b]}} \sum_{k=1}^{q} |F(t_k) - F(t_{k-1})| < \infty$$

where $\mathscr{P}_{[a,b]}$ denotes the set of partitions σ of the interval $[a, b]$, that is the choices of a finite number of points $\sigma = \{t_0, t_1, \ldots, t_q\}$ such that

$$a = t_0 < t_1 < \cdots < t_q = b.$$

A presentation of the main results on functions of bounded variation can be found in [26]. Here we only recall that for every $F \in BV([a, b])$ we have

$$\bigvee_a^b (F) = \bigvee_a^c (F) + \bigvee_c^b (F), \qquad c \in]a, b[, \tag{B.10}$$

and moreover F can be written as the difference of *monotone increasing* functions in the following way: for $x \in [a, b]$

$$F(x) = u(x) - v(x), \qquad u(x) := \bigvee_a^x (F), \quad v(x) := u(x) - F(x). \tag{B.11}$$

Lemma B.4.5 *If $F \in AC([a, b])$ then $F \in BV([a, b])$ and in the decomposition (B.11), the functions u, v are monotone increasing and absolutely continuous.*

Proof Since $F \in AC([a, b])$, there exists $\delta > 0$ such that

$$\sum_{n=1}^N |F(b_n) - F(a_n)| < 1$$

for every choice of a finite number of disjoint intervals $[a_n, b_n] \subseteq [a, b]$ such that

$$\sum_{n=1}^N (b_n - a_n) < \delta.$$

This implies that $F \in BV$ on every sub-interval of $[a, b]$ of length less than or equal to δ. Then the fact that $F \in BV([a, b])$ follows from (B.10), dividing $[a, b]$ into a finite number of intervals of length less than or equal to δ.

Now we prove that $u \in AC([a, b])$ (and therefore also $v \in AC([a, b])$). By assumption, $F \in AC([a, b])$ and therefore given $\varepsilon > 0$ there exists $\delta > 0$ as in Definition B.4.1. Let $[a_n, b_n] \subseteq [a, b], n = 1, \ldots, N$, be disjoint intervals such that

$$\sum_{n=1}^N (b_n - a_n) < \delta.$$

We have

$$\sum_{n=1}^{N}(u(b_n) - u(a_n)) = \sum_{n=1}^{N}\bigvee_{a_n}^{b_n}(F) = \sum_{n=1}^{N}\sup_{\sigma \in \mathscr{P}_{[a_n,b_n]}}\sum_{k=1}^{q_n}\left|F(t_{n,k}) - F(t_{n,k-1})\right| < \varepsilon$$

since, based on (B.8), we have

$$\sum_{n=1}^{N}\sum_{k=1}^{q_n}\left|F(t_{n,k}) - F(t_{n,k-1})\right| < \varepsilon$$

for every partition $(t_{n,0}, \dots, t_{n,q_n}) \in \mathscr{P}_{[a_n,b_n]}$.   □

Proof of Theorem B.4.3 If F admits a representation of the type

$$F(x) = F(a) + \int_{a}^{x}\gamma(t)dt, \qquad x \in [a, b],$$

with $\gamma \in L^1([a, b])$ then clearly F is absolutely continuous by the Lebesgue dominated convergence theorem. Moreover, $F' = \gamma$ a.e. by Proposition B.3.3.

Conversely, if $F \in AC([a, b])$, by Lemma B.4.5 it is not restrictive to also assume that F is monotone increasing. Then we can consider the measure μ_F defined as in Theorem 1.4.33-(i):

$$\mu_F(]x, y]) = F(y) - F(x), \qquad a \leq x < y \leq b.$$

We have to prove that μ_F is absolutely continuous with respect to the Lebesgue measure, i.e., $\mu_F \ll \text{Leb}$. Consider $B \in \mathscr{B}$ such that $\text{Leb}(B) = 0$: by definition of the Lebesgue measure,[3] for every $\delta > 0$ there exists a sequence $(]a_n, b_n])_{n \in \mathbb{N}}$ of disjoint intervals such that

$$A \supseteq B, \qquad \text{Leb}(A) < \delta, \qquad A := \bigcup_{n=1}^{\infty}]a_n, b_n]). \tag{B.12}$$

Consequently, for every $\varepsilon > 0$ there exist $\delta > 0$ and A as in (B.12) for which we have

$$\mu_F(B) \leq \mu_F(A \cap [a, b]) \leq \varepsilon,$$

---

[3] We recall that (cf. (1.5.5))

$$\text{Leb}(B) = \inf\{\text{Leb}(A) \mid B \subseteq A \in \mathscr{U}\}$$

where $\mathscr{U}$ denotes the family of countable unions of disjoint intervals of the form $]a, b]$.

where the first inequality is due to the monotonicity of μ_F and the second comes from the fact that $F \in AC([a, b])$ and $\text{Leb}(A) < \delta$ (cf. Exercise B.4.2). Given the arbitrariness of ε, we conclude that $\mu_F(B) = 0$

By the Radon-Nikodym Theorem B.1.3, there exists $\gamma \in L^1([a, b])$ such that

$$F(x) - F(a) = \mu_F(]a, x]) = \int_a^x \gamma(t)dt, \qquad x \in [a, b],$$

and thanks to Proposition B.3.3 we conclude that $F' = \gamma$ a.e.   □

Appendix C
Uniform Integrability

We introduce a useful tool for analyzing sequences of random variables: Vitali's theorem. It extends the Lebesgue dominated convergence theorem. In this section, $X = (X_t)_{t \in I}$ is a family of random variables on the space $(\Omega, \mathscr{F}, P)$ with values in $\mathbb{R}^d$, with I any set of indices: we refer to X as a *stochastic process*.

Definition C.0.1 (Uniform Integrability) A stochastic process $(X_t)_{t \in I}$ on the space $(\Omega, \mathscr{F}, P)$ is *uniformly integrable* if

$$\lim_{R \to \infty} \sup_{t \in I} E\left[|X_t| \mathbb{1}_{(|X_t| \geq R)}\right] = 0,$$

or, in other words, if for every $\varepsilon > 0$ there exists $R > 0$ such that $E\left[|X_t| \mathbb{1}_{(|X_t| \geq R)}\right] < \varepsilon$ for every $t \in I$.

Theorem C.0.2 (Vitali's Convergence Theorem) *If* $X_n \xrightarrow{a.s.} X$ *and* $(X_n)_{n \in \mathbb{N}}$ *is uniformly integrable then* $X_n \xrightarrow{L^1} X$.

Proof We prove the thesis in the case $X = 0$. Fixed $\varepsilon > 0$, there exists $R > 0$ such that $E\left[|X_n| \mathbb{1}_{(|X_n| \geq R)}\right] < \frac{\varepsilon}{2}$ for every $n \in \mathbb{N}$; moreover, by the dominated convergence theorem, there exists $\bar{n}$, which depends on ε and R, such that $E\left[|X_n| \mathbb{1}_{(|X_n| < R)}\right] < \frac{\varepsilon}{2}$ for every $n \geq \bar{n}$. In conclusion,

$$E\left[|X_n|\right] = E\left[|X_n| \mathbb{1}_{(|X_n| \geq R)}\right] + E\left[|X_n| \mathbb{1}_{(|X_n| < R)}\right] < \varepsilon$$

for every $n \geq \bar{n}$.

We will see shortly, in Corollary C.0.5, that the sum of uniformly integrable processes is uniformly integrable. Therefore, to conclude it suffices to consider the process $Y_n = X_n - X$ which is uniformly integrable and such that $Y_n \xrightarrow{a.s.} 0$. $\quad\square$

We give a characterization of uniform integrability.

A. Pascucci, *Probability Theory I*, La Matematica per il 3+2 165, https://doi.org/10.1007/978-3-031-63190-0

Definition C.0.3 (Uniform Absolute Continuity) A process $(X_t)_{t \in I}$ on the space $(\Omega, \mathscr{F}, P)$ is *uniformly absolutely continuous* if for every $\varepsilon > 0$ there exists $\delta > 0$ such that $E\left[|X_t|\mathbb{1}_A\right] < \varepsilon$ for every $t \in I$ and $A \in \mathscr{F}$ such that $P(A) < \delta$.

Proposition C.0.4 *The following properties are equivalent:*

(i) *the process $(X_t)_{t \in I}$ is uniformly integrable;*
(ii) *the process $(X_t)_{t \in I}$ is uniformly absolutely continuous and* $\sup_{t \in I} E\left[|X_t|\right] < \infty$.

Proof If $(X_t)_{t \in I}$ is uniformly integrable, there exists $R > 0$ such that

$$\sup_{t \in I} E\left[|X_t|\mathbb{1}_{(|X_t| \geq R)}\right] \leq 1.$$

Then we have

$$E\left[|X_t|\right] \leq 1 + E\left[|X_t|\mathbb{1}_{(|X_t| \leq R)}\right] \leq 1 + R.$$

Similarly, given $\varepsilon > 0$ there exists R such that $E\left[|X_t|\mathbb{1}_{(|X_t| \geq R)}\right] < \frac{\varepsilon}{2}$ for every $t \in I$: then for every $A \in \mathscr{F}$ such that $P(A) < \frac{\varepsilon}{2R}$, we have

$$E\left[|X_t|\mathbb{1}_A\right] = E\left[|X_t|\mathbb{1}_{A \cap (|X_t| \geq R)}\right] + E\left[|X_t|\mathbb{1}_{A \cap (|X_t| < R)}\right] < \frac{\varepsilon}{2} + RP(A) < \varepsilon.$$

Conversely, by hypothesis, given $\varepsilon > 0$ there exists $\delta > 0$ such that $E\left[|X_t|\mathbb{1}_A\right] < \varepsilon$ for every $t \in I$ and $A \in \mathscr{F}$ such that $P(A) < \delta$. By Markov's inequality, there exists R such that

$$P(|X_t| \geq R) \leq \frac{1}{R} \sup_{t \in I} E\left[|X_t|\right] < \delta$$

and consequently

$$E\left[|X_t|\mathbb{1}_{(|X_t| \geq R)}\right] < \varepsilon$$

for every $t \in I$.   □

Corollary C.0.5 *If $(X_t)_{t \in I}$ and $(Y_t)_{t \in I}$ are uniformly integrable, then $(X_t + Y_t)_{t \in I}$ is uniformly integrable.*

Proof With the characterization provided in Proposition C.0.4, verifying the thesis is straightforward.   □

We now give some examples.

Proposition C.0.6 *If there exists $Y \in L^1(\Omega, P)$ such that $|X_t| \leq Y$ for every $t \in I$, then $(X_t)_{t \in I}$ is uniformly integrable.*

Proof By the absolute continuity of the expectation (Corollary 2.2.13), for every $\varepsilon > 0$ there exists $\delta > 0$ such that $E[|Y|\mathbb{1}_A] < \varepsilon$ for every $A \in \mathcal{F}$ such that $P(A) < \delta$. Now, by Markov's inequality, we have

$$P(|X_t| \geq R) \leq \frac{E[|X_t|]}{R} \leq \frac{E[|Y|]}{R} < \delta, \qquad \text{if } R > \frac{E[|Y|]}{\delta}.$$

Then

$$E\left[|X_t|\mathbb{1}_{(|X_t| \geq R|)}\right] \leq E\left[|Y|\mathbb{1}_{(|X_t| \geq R|)}\right] < \varepsilon.$$

$\square$

From Proposition C.0.6 we deduce that:

- a process consisting of a single r.v. $X \in L^1(\Omega, P)$ is uniformly integrable;
- Lebesgue's dominated convergence theorem is a corollary of Vitali's theorem.

Proposition C.0.7 *Let* $X \in L^1(\Omega, \mathcal{F}, P)$ *and* $(\mathcal{F}_t)_{t \in I}$ *be a family of sub-σ-algebras of* $\mathcal{F}$. *The process defined by* $X_t = E[X \mid \mathcal{F}_t]$ *is uniformly integrable.*

Proof The proof is similar to that of Lemma C.0.6. Fix $\varepsilon > 0$, let $\delta > 0$ be such that $E[|X|\mathbb{1}_A] < \varepsilon$ for every $A \in \mathcal{F}$ such that $P(A) < \delta$. By combining the Markov and Jensen inequalities, we obtain

$$P(|X_t| \geq R) \leq \frac{E[|X_t|]}{R} \leq \frac{E[|X|]}{R} < \delta, \qquad \text{if } R > \frac{E[|X|]}{\delta}.$$

Again, by Jensen's inequality, we have

$$E\left[|X_t|\mathbb{1}_{(|X_t| \geq R)}\right] \leq E\left[E[|X| \mid \mathcal{F}_t]\mathbb{1}_{(|X_t| \geq R)}\right] =$$

(by the properties of conditional expectation, since $\mathbb{1}_{(|X_t| \geq R)} \in b\mathcal{F}_t$)

$$= E\left[|X|\mathbb{1}_{(|X_t| \geq R)}\right] < \varepsilon.$$

$\square$

Remark C.0.8 [!] Proposition C.0.7 finds frequent application in the analysis of the convergence of specific stochastic processes known as *martingales*. The typical situation is when there is a sequence $(X_n)_{n \in \mathbb{N}}$ that converges *pointwise*; if X_n takes the form $X_n = E[X \mid \mathcal{F}_n]$ for a certain $X \in L^1(\Omega, P)$ and a family $(\mathcal{F}_n)_{n \in \mathbb{N}}$ of sub-σ-algebras of $\mathcal{F}$, then Proposition C.0.7 implies that $(X_n)_{n \in \mathbb{N}}$ is uniformly integrable. Then, Vitali's convergence theorem ensures that $(X_n)_{n \in \mathbb{N}}$ also converges in the L^1-norm.

Proposition C.0.9 *If there exists an increasing function*

$$\varphi : \mathbb{R}_{\geq 0} \longrightarrow \mathbb{R}_{\geq 0}$$

such that $\lim\limits_{r \to +\infty} \frac{\varphi(r)}{r} = +\infty$ *and* $\sup\limits_{t \in I} E\left[\varphi(|X_t|)\right] < \infty$ *then* $(X_t)_{t \in I}$ *is uniformly integrable.*

Proof For every $\varepsilon > 0$ there exists $r_\varepsilon > 0$ such that $\frac{\varphi(r)}{r} > \frac{1}{\varepsilon}$ for every $r \geq r_\varepsilon$. Then, for $R > r_\varepsilon$ we have

$$E\left[|X_t| \mathbb{1}_{(|X_t| \geq R)}\right] = E\left[\frac{|X_t|}{\varphi(|X_t|)} \varphi(|X_t|) \mathbb{1}_{(|X_t| \geq R)}\right] \leq \varepsilon \sup_{t \in I} E\left[\varphi(|X_t|)\right]$$

which proves the thesis due to the arbitrariness of ε. $\qquad\qquad\qquad\qquad\square$

Remark C.0.10 Let $p > 1$. According to Proposition C.0.9 with $\varphi(r) = r^p$, if $(X_t)_{t \in I}$ is bounded in norm $L^p(\Omega, P)$, meaning $\sup\limits_{t \in I} E\left[|X_t|^p\right] < \infty$, then it is uniformly integrable.

Tables for the Main Distributions

Name	Symbol	Distribution function $\tilde{\mu}(k)$	Expectation	Variance	Characteristic function	Properties: see page
Dirac Delta	δ_{x_0}	$\mathbb{1}_{\{x_0\}}(k)$	x_0	0	$e^{ix_0\eta}$	60, 68, 132
Bernoulli	Be_p	$\begin{cases} p & \text{if } k=1 \\ 1-p & \text{if } k=0 \end{cases}$	p	$p(1-p)$	$1+p\left(e^{i\eta}-1\right)$	62, 106, 132, 197
Uniform	$Unif_n$	$\frac{1}{n}\mathbb{1}_{I_n}(k)$	$\frac{n+1}{2}$	$\frac{n^2-1}{12}$	$\frac{e^{i\eta}\left(e^{in\eta}-1\right)}{n\left(e^{i\eta}-1\right)}$	62
Binomial	$Bin_{n,p}$	$\binom{n}{k}p^k(1-p)^{n-k},\, 0\le k\le n$	np	$np(1-p)$	$\left(1+p\left(e^{i\eta}-1\right)\right)^n$	26, 62, 107
Poisson	$Poisson_\lambda$	$\frac{e^{-\lambda}\lambda^k}{k!},\, k\in\mathbb{N}_0$	λ	λ	$e^{\lambda\left(e^{i\eta}-1\right)}$	62, 108, 132, 138, 198
Geometric	$Geom_p$	$p(1-p)^{k-1},\, k\in\mathbb{N}$	$\frac{1}{p}$	$\frac{1-p}{p^2}$	$\frac{p}{e^{-i\eta}-1+p}$	110, 113
Hypergeometric	$Hyper_{n,b,N}$	$\frac{\binom{b}{k}\binom{N-b}{n-k}}{\binom{N}{n}},\, 0\le k\le n\wedge b$	$\frac{bn}{N}$	$\frac{bn(N-b)(N-n)}{N^2(N-1)}$		27, 112
Uniform on $[a,b]$	$Unif_{[a,b]}$	$\frac{1}{b-a}\mathbb{1}_{[a,b]}(x)$	$\frac{a+b}{2}$	$\frac{(b-a)^2}{12}$	$\frac{e^{ib\eta}-e^{ia\eta}}{i\eta(b-a)}$	66, 77, 170
Exponential	Exp_λ	$\lambda e^{-\lambda x}\mathbb{1}_{\mathbb{R}_{\ge0}}$	$\frac{1}{\lambda}$	$\frac{1}{\lambda^2}$	$\frac{\lambda}{\lambda-i\eta}$	66, 69, 116, 113, 140, 201
Normal real	$\mathcal{N}_{\mu,\sigma^2}$	$\frac{1}{\sqrt{2\pi\sigma^2}}e^{-\frac{1}{2}\left(\frac{x-\mu}{\sigma}\right)^2}$	μ	σ^2	$e^{i\mu\eta-\frac{\sigma^2\eta^2}{2}}$	66, 69, 114, 118, 139, 199
Gamma	$Gamma_{\alpha,\lambda}$	$\frac{\lambda^\alpha e^{-\lambda x}}{\Gamma(\alpha)x^{1-\alpha}}\mathbb{1}_{\mathbb{R}>0}(x)$	$\frac{\alpha}{\lambda}$	$\frac{\alpha}{\lambda^2}$	$\left(\frac{\lambda}{\lambda-i\eta}\right)^\alpha$	116, 117, 139, 118
Chi-square with n degrees	$\chi^2(n)=Gamma_{\frac{n}{2},\frac{1}{2}}$	$\frac{1}{2^{\frac{n}{2}}\Gamma\left(\frac{n}{2}\right)}\frac{e^{-\frac{x}{2}}}{x^{1-\frac{n}{2}}}\mathbb{1}_{\mathbb{R}>0}(x)$	n	$2n$	$(1-2i\eta)^{-\frac{n}{2}}$	200, 118

References

1. P. Baldi, *Introduzione alla probabilità con elementi di statistica - Seconda edizione* (McGraw-Hill, New York, 2012)
2. R.F. Bass, *Probabilistic Techniques in Analysis*. Probability and its Applications (New York) (Springer, New York, 1995)
3. R.F. Bass, *Real Analysis for Graduate Students* (2013). http://bass.math.uconn.edu/real.html
4. H. Bauer, *Probability Theory*. De Gruyter Studies in Mathematics, vol. 23 (Walter de Gruyter, Berlin, 1996). Translated from the fourth (1991) German edition by Robert B. Burckel and revised by the author
5. F. Biagini, M. Campanino, *Elements of Probability and Statistics*. Unitext, vol. 98 (Springer, Cham, 2016). An introduction to probability with de Finetti's approach and to Bayesian statistics, Translated from the 2006 Italian original, La Matematica per il 3+2
6. P. Billingsley, *Probability and Measure*. Wiley Series in Probability and Mathematical Statistics, 3rd edn. (Wiley, New York, 1995). A Wiley-Interscience Publication
7. P. Billingsley, *Convergence of Probability Measures*. Wiley Series in Probability and Statistics: Probability and Statistics, 2nd edn. (Wiley, New York, 1999). A Wiley-Interscience Publication
8. F. Caravenna, P. Dai Pra, *Probabilità - Un'introduzione Attraverso Modelli e Applicazioni* (Springer, Berlin, 2013)
9. D. Costantini, *Introduzione Alla Probabilità*. Testi e manuali della scienza contemporanea. Serie di logica matematica (Bollati Boringhieri, Turin, 1977)
10. J. Dieudonné, Sur le théorème de Lebesgue-Nikodym. III. Ann. Univ. Grenoble. Sect. Sci. Math. Phys. **23**, 25–53 (1948)
11. J.L. Doob, *Stochastic Processes* (Wiley, New York, 1953)
12. R. Durrett, *Probability: Theory and Examples*. Cambridge Series in Statistical and Probabilistic Mathematics, vol. 49 (Cambridge University Press, Cambridge, 2019). https://services.math.duke.edu/~rtd/PTE/pte.html
13. V. D'Urso, F. Giusberti, *Esperimenti di Psicologia - Seconda Edizione* (Zanichelli, Bologna, 2000)
14. A.M. Faden, The existence of regular conditional probabilities: necessary and sufficient conditions. Ann. Probab. **13**, 288–298 (1985)
15. W. Feller, *An Introduction to Probability Theory and its Applications*, vol. II, 2nd edn. (Wiley, New York, 1971)
16. H. Fischer, *A History of the Central Limit Theorem*. Sources and Studies in the History of Mathematics and Physical Sciences (Springer, New York, 2011). From classical to modern probability theory

© The Editor(s) (if applicable) and The Author(s), under exclusive license to Springer Nature Switzerland AG 2024
A. Pascucci, *Probability Theory I*, La Matematica per il 3+2 165,
https://doi.org/10.1007/978-3-031-63190-0

17. P. Glasserman, *Monte Carlo Methods in Financial Engineering*. Applications of Mathematics, vol. 53 (Springer, New York, 2004). Stochastic Modelling and Applied Probability
18. P. Glasserman, B. Yu, Number of paths versus number of basis functions in American option pricing. Ann. Appl. Probab. **14**, 2090–2119 (2004)
19. I. Goodfellow, Y. Bengio, A. Courville, *Deep Learning* (MIT Press,Cambridge, 2016). http://www.deeplearningbook.org
20. P.R. Halmos, *Measure Theory* (D. Van Nostrand Company, New York, 1950)
21. J. Jacod, P. Protter, *Probability Essentials*. Universitext (Springer, Berlin, 2000)
22. J. Jacod, A.N. Shiryaev, *Limit Theorems for Stochastic Processes*. Grundlehren der Mathematischen Wissenschaften [Fundamental Principles of Mathematical Sciences], vol. 288, 2nd edn. (Springer, Berlin, 2003)
23. O. Kallenberg, *Foundations of Modern Probability*. Probability and its Applications, 2nd edn. (Springer, New York, 2002)
24. A. Klenke, *Probability Theory*. Universitext, 2nd edn. (Springer, London, 2014). A comprehensive course
25. E. Lanconelli, *Lezioni di Analisi Matematica 1* (Pitagora Editrice, Bologna, 1994)
26. E. Lanconelli, *Lezioni di Analisi Matematica 2* (Pitagora Editrice, Bologna, 1995)
27. E. Lanconelli, *Lezioni di Analisi Matematica 2 - Seconda parte* (Pitagora Editrice, Bologna, 1997)
28. G. Letta, *Probabilità elementare. Compendio di teorie. Problemi risolti* (Zanichelli, Bologna, 1993)
29. P.A. Meyer, Stochastic processes from 1950 to the present. J. Électron. Hist. Probab. Stat. **5**, 42 (2009). Translated from the French [MR1796860] by Jeanine Sedjro
30. D. Mumford, The dawning of the age of stochasticity. Atti Accad. Naz. Lincei Cl. Sci. Fis. Mat. Natur. Rend. Lincei. Mat. Appl. **11**, 107–125 (2000). Mathematics towards the third millennium (Rome, 1999)
31. J. Neveu, *Mathematical Foundations of the Calculus of Probability*. Translated by Amiel Feinstein (Holden-Day, San Francisco, 1965)
32. B. Oksendal, *Stochastic Differential Equations*. Universitext, 5th edn. (Springer, Berlin, 1998). An introduction with applications
33. A. Pascucci, *PDE and Martingale Methods in Option Pricing*. Bocconi & Springer Series, vol. 2 (Springer, Milan, 2011)
34. A. Pascucci, *Probability Theory.Volume 2 - Stochastic Calculus*. Unitext (Springer, Milan, 2024)
35. J.A. Paulos, *A Mathematician Reads the Newspaper* (Basic Books, New York, 2013). Paperback edition of the 1995 original with a new preface
36. N. Pintacuda, *Probabilità* (Zanichelli, Bologna, 1995)
37. G. Pólya, Über den zentralen Grenzwertsatz der Wahrscheinlichkeitsrechnung und das Momentenproblem. Math. Z. **8**, 171–181 (1920)
38. C.E. Rasmussen, C.K.I. Williams, *Gaussian Processes for Machine Learning* (MIT Press, Cambridge, 2006). http://www.gaussianprocess.org/gpml/
39. F. Riesz, B. Sz.-Nagy, *Functional Analysis* (Frederick Ungar, New York, 1955). Translated by Leo F. Boron
40. W. Rudin, *Real and Complex Analysis*, 3rd edn. (McGraw-Hill, New York, 1987)
41. D. Salsburg, *The Lady Tasting Tea: How Statistics Revolutionized Science in the Twentieth Century* (Henry Holt and Company, New York, 2002)
42. A.N. Shiryaev, *Probability. 1*. Graduate Texts in Mathematics, vol. 95, 3rd edn. (Springer, New York, 2016). Translated from the fourth (2007) Russian edition by R. P. Boas and D. M. Chibisov
43. Y.G. Sinai, *Probability Theory*. Springer Textbook (Springer, Berlin, 1992). An introductory course, Translated from the Russian and with a preface by D. Haughton
44. D.W. Stroock, S.R.S. Varadhan, *Multidimensional Diffusion Processes*. Classics in Mathematics (Springer, Berlin, 2006). Reprint of the 1997 edition

45. G. Vitali, *Sul problema della misura dei gruppi di punti di una retta*, Bologna, Tipografia Gamberini e Parmeggiani (1905)
46. D. Williams, *Probability with Martingales*. Cambridge Mathematical Textbooks (Cambridge University Press, Cambridge, 1991)
47. A. Zolotukhin, S. Nagaev, V. Chebotarev, On a bound of the absolute constant in the Berry-Esseen inequality for i.i.d. Bernoulli random variables. Mod. Stoch. Theory Appl. **5**, 385–410 (2018)

Index